Fractals in the Earth Sciences

Fractals in the Earth Sciences

Edited by

Christopher C. Barton
U.S. Geological Survey
Denver, Colorado

and

Paul R. La Pointe
Golder Associates Inc.
Redmond, Washington

Plenum Press • New York and London

Library of Congress Cataloging-in-Publication Data

Fractals in the earth sciences / edited by Christopher C. Barton and
  Paul R. La Pointe.
       p.   cm.
    Includes bibliographical references and index.
    ISBN 0-306-44865-3
      1. Fractals.   2. Geology--Mathematics.   I. Barton, Christopher
  Cramer. II. La Pointe, P. R.
  QE33.2.F73F73 1995
  551'.01'51474--dc20   94-45969
                                                                      CIP

Geology
QE
33.2
.F73
F73
1995

Front Cover: Without an object of known size, such as a rock hammer or lens cap, it is often difficult or impossible to determine whether a photograph portrays a geological feature at the scale of centimeters, meters, or tens of meters.

Back Cover: Most geologic features have heterogeneous properties that are scale invariant, i.e., have no characteristic length scale, a fundamental characteristic of fractals. Length of hammer is 0.74 meters.

(Photographs by Christopher C. Barton and Sarah F. Tebbens, U.S. Geological Survey and University of South Florida, St. Petersburg, Florida)

ISBN 0-306-44865-3

© 1995 Plenum Press, New York
A Division of Plenum Publishing Corporation
233 Spring Street, New York, N. Y. 10013

10 9 8 7 6 5 4 3 2 1

All rights reserved

No part of this book may be reproduced, stored in a retrieval system, or transmitted in any form or by any means, electronic, mechanical, photocopying, microfilming, recording, or otherwise, without written permission from the Publisher

Printed in the United States of America

Dedication

This volume is dedicated to Benoit B. Mandelbrot,
the "father of fractal geometry," on his
70th birthday

Benoit B. Mandelbrot was born in Warsaw, Poland, November 20, 1924 and is best known as the author of the books *Les Objets Fractals*, 1975, 1984, and 1989 (translated into Italian, Spanish, Portuguese, and Basque) and *The Fractal Geometry of Nature*, 1982 (translated into Japanese and German). He is Robinson Professor of Mathematical Sciences at Yale University and IBM Fellow emeritus at the IBM Thomas J. Watson Research Center, Fellow of the American Academy of Arts and Sciences, Foreign Associate of the U.S. National Academy of Sciences, Member of the European Academy of Arts, Sciences, and Humanities. Dr. Mandelbrot is a graduate of the Paris Ecole Polytechnique; has a M.S. and Ae.E. in Aeronautics from the California Institute of Technology; and a Docteur d'Etat ès Sciences Mathématiques from the University of Paris. He has been awarded honorary degrees of Doctor of Science from Syracuse University, Laurentian University, Boston University, State University of New York, University of Guelph, Universität Bremen, Union College, University of Buenos Aires; and an honorary degree, Doctor of Humane Letters, from Pace University. He has received the Distinguished Service Award from the California Institute of Technology, and the Humboldt Preis from A. von Humboldt, Stiftung. His awards include the 1985 *Barnard Medal* granted jointly by the U.S. National Academy of Sciences and Columbia University, the 1986 *Franklin Medal* by the Franklin Institute of Philadelphia, the 1988 *Charles Proteus Steinmetz Medal* by the Institute of Electrical and Electronics Engineers, the 1988 *Science for Art Prize* by the Foundation Moët-Hennessy-Louis Vuitton, the 1989 *Harvey Prize for Science and Technology* by the Technion-Israel Institute of Technology in Haifa, the 1991 Nevada Prize conferred by the Desert Research Institute, the 1993 Wolf Prize in Physics, and the 1994 Honda Prize.

Before joining IBM, Dr. Mandelbrot's positions were with the CNRS in Paris, Philips Electronics, the Massachusetts Institute of Technology, the Institute for Advanced Study at Princeton, the University of Geneva, the University of Lille, and Ecole Polytechnique, Paris. On leave from IBM, he has been an Institute Lecturer at Massachusetts Institute of Technology and a Visiting Professor of Economics, of Mathematics, of Applied Mathematics and of the Practice of Mathematics at Harvard University; of Engineering at Yale University; of Physiology at the Albert Einstein College of Medicine; and of Mathematics at Université de Paris-Sud.

Mandelbrot has long been interested in earth science starting with a paper that concerned Hurst's law, published in *Comptes Rendus* in 1965. It was followed by "How long is the coast of Britain? Statistical self-similarity and fractional dimensions," in *Science*, 1967. He published four papers with J. R. Wallis in *Water Resources Research* in 1968 and 1969; one of these papers is revisited in his contribution to this volume. His pioneering paper on multifractals, published in the *Journal of Fluid Mechanics* in 1974 was motivated in part by the spatial distribution of rare minerals.

Contributors

J. Angelier • Laboratoire de Tectonique Quantitative, Université P. & M., Paris, France

P. Bak • Department of Physics, Brookhaven National Laboratory, Upton, New York 11973

C. C. Barton • US Geological Survey, MS 940, Denver, Colorado 80225

S. R. Brown • Geomechanics Division, Sandia National Laboratories, Albuquerque, New Mexico 87185

K. Chen • Department of Physics, Brookhaven National Laboratory, Upton, New York 11973

A. D. Fowler • Ottawa–Carleton Geoscience Centre, University of Ottawa, Ottawa, Canada K1N 6N5

J. Huang • Department of Geological Sciences, Cornell University, Ithaca, New York 14853; *present address*: Exxon Production Research, Houston, Texas 77252

A. Malinverno • Lamont–Doherty Earth Observatory, Columbia University, Palisades, New York, 10964; *present address*: Schlumberger–Doll Research, Ridgefield, Connecticut 06877

B. B. Mandelbrot • Yale University, New Haven, Connecticut 06520 and IBM T. J. Watson Research Center, Yorktown Heights, New York 10598

W. L. Power • Department of Geological Sciences, Brown University, Providence, Rhode Island 78712; *present address*: CSIRO Exploration and Mining, Nedlands, Western Australia, Australia 6009

S. A. Pruess • Department of Mathematics, Colorado School of Mines, Golden Colorado 80401

C. G. Sammis • Department of Geological Sciences, University of Southern California, Los Angeles, California 90089

J. A. Saunders • Department of Geology, Auburn University, Auburn, Alabama 36849

P. A. Schoenly • Department of Geology, Auburn University, Auburn, Alabama 36849

C. H. Scholz • Lamont–Doherty Earth Observatory and Department of Geological Sciences, Columbia University, Palisades, New York 10964

S. J. Steacy • Department of Geological Sciences, University of Southern California, Los Angeles, California 90089

C. Sunwoo • Korean Institute of Energy and Resources, Daejeon, Korea

D. L. Turcotte • Department of Geological Sciences, Cornell University, Ithaca, New York 14853

T. E. Tullis • Department of Geological Sciences, Brown University, Providence, Rhode Island 78712

T. Villemin • Laboratoire de Géologie Structuraleet Appliqueé, Université de Savoie, Chambery, France

J. R. Wallis • IBM T. J. Watson Research Center, Yorktown Heights, New York 10598

Foreword

Fractals have changed the way we understand and study nature. This change has been brought about mainly by the work of B. B. Mandelbrot and his book *The Fractal Geometry of Nature*. Now here is a book that collects articles treating fractals in the earth sciences. The themes chosen span, as is appropriate for a discourse on fractals, many orders of magnitude; including earthquakes, ocean floor topography, fractures, faults, mineral crystallinity, gold and silver deposition. There are also chapters on dynamical processes that are fractal, such as rivers, earthquakes, and a paper on self-organized criticality. Many of the chapters discuss how to estimate fractal dimensions, Hurst exponents, and other scaling exponents.

This book, in a way, represents a snapshot of a field in which fractals has brought inspiration and a fresh look at familiar subjects. New ideas and attempts to quantify the world we see around us are found throughout. Many of these ideas will grow and inspire further work, others will be superseded by new observations and insights, most probably with future contributions by the authors of these chapters.

It is wonderful to witness that the idea of fractal geometry has lead to such a high rate of discovery in the earth sciences. I believe that this book will also inspire others to take part in the scientific process of revealing the unseen, clarifying what is incomplete, discarding views that turn out not to stand up to the test of experiments and observations and thereby extending our understanding of the earth.

Jens Feder

Department of Physics, University of Oslo

Preface

It is not surprising that many earth scientists are drawn to fractal geometry. As the contributions in this volume show, fractal geometry is the natural mathematical language to describe much of what geologists observe. During their earliest field training, geologists are taught to place an object of known size, such as a rock hammer or lens cap, in photographs of outcrops in order to convey a sense of scale. Why? Without such an object, it is not always easy to determine whether the photograph portrays a feature at the scale of centimeters, meters, or tens of meters. Likewise, the magnitude of geological events and the time scale in which they occur underpins geologists's understanding of great floods, earthquakes, and mass extinctions. The geological record as we know it suggests that there are abundant small events, occurring close to one another in time or space, with fewer large events occurring in the same temporal or spatial region. Such relations are indicative of power law scaling, which in turn suggests that fractal geometry might prove useful for quantifying geological patterns.

This book contains contributions by many of the pioneers of fractal geometry, not just those who first applied fractal geometry to describe geological features and processes. Although the articles may be read in any order, they have been organized to follow the path by which many scientists evolve in their understanding and investigation of a new branch of knowledge. The first papers provide a broad overview of fractals, demonstrating the ubiquity of fractal geometry in geology. After becoming convinced that fractal geometry does seem to be important, a geologist would typically tend to try to measure the fractal properties of a geological feature in their own field, whether it be seismology, geomorphology, or structural geology. In order to do this, the geologist must learn how to measure fractal properties, and to be aware of all of the different types of fractal methods that are available and their pitfalls. The middle part of the volume contains several contributions that discuss fractal methodology. The volume ends with the natural next stage: application, and a deeper understanding of geological processes.

The first two contributions provide both a widespread overview of the temporal and spatial fractal character of geological features and processes, and the historical context of fractal geometry and its use in the earth sciences. Turcotte and Huang outline many of the key spatial and temporal geological features and processes that have fractal characteristics. They examine such features as rock breakage, ore and petroleum concentrations in the earth's crust, seismicity and tectonics, fracture intensity, volcanic eruptions, surface topography, and drainage basin morphology.

Mandelbrot and Wallis reprise a seminal article first appearing over twenty years ago that gave rise to the fractal description of self-affine systems which characterize much of the earth sciences. Subsequent experience has shown that many of the most provocative and useful applications of fractal geometry in the earth sciences concern self-affine objects or processes. At a time when geologists knew little of fractal geometry, Mandelbrot and Wallis were laying the groundwork for the revolution to come. This updated and revised version of a key paper in the science of fractal geometry will be of interest to many in the earth sciences.

Although the practice and pitfalls of studying fractal patterns are described throughout this book, those interested in applying fractal geometry to their own data sets must delve into the pragmatic and theoretical considerations surrounding the estimation of the fractal dimension. Pruess provides a lucid exposition on the calculation of the fractal box dimension, a dimension used by many geologists to measure the fractal properties of self-similar phenomena.

Brown complements Pruess' work with a clear and thorough description of the correct approach for calculating the fractal dimension of self-affine patterns. His exposition identifies the problems with using self-similar methods, such as the divider or box methods, for studying self-affine objects. These pitfalls were not fully appreciated in some of the early published work applying fractal geometry to geology, and are important for any new practitioner of fractals to understand.

Fractal geometry describes temporal or spatial properties of geological patterns. It does not describe the mechanism that produces the fractal scaling, but it nonetheless helps to sort out possible mechanisms or explanations. Some dynamical systems produce fractal processes; others do not. Power and Tullis begin a series of chapters that examine the relation between the fractal characteristics and the geological processes that give rise to them. In so doing, these papers touch upon most of the fundamental issues in estimating the fractal dimension and applying fractal geometry to geological problems. For example, Power and Tullis's study shows the usefulness of spectral methods for studying earth science systems and the fact that the roughness of fracture surfaces scales over six orders of magnitude. Moreover, their work shows why geologists have observed a relation between width and displacement in fracture zones.

Malinverno also connects the fractal description of a roughness profile with a possible geological mechanism. He shows that self-affine fractals are good models of sea-floor topography. But the estimation of a self-affine system, such as bathymetry, has pitfalls which must be recognized and avoided, as the author explains. Malinverno's work addresses how sea-floor topography might have evolved, given that the resulting profiles are fractal.

Scholz also considers the fractal nature of geological surfaces as an important object lesson—a single fractal dimension may not characterize a geological feature over all scale ranges of practical interest. Fractal theory often treats its object as if the same fractal scaling exponent extends over all scales. In fact, gradual and abrupt transitions are common and important. Scholz shows how the transitions are very important for interpreting earthquake data.

Barton uses fractal geometry to confront a problem that nearly all geologists face in one form or another; the quantification of heterogeneous data. Every geologist who has studied fracture patterns in outcrop has observed that the pattern in one outcrop is not identical to the pattern in any other exposure. Although they may show statistical similarities, they are not exactly the same. The question arises as to when these differences are sufficient to suggest that the difference is due to different mechanisms, rather than the expected variation at some scale. Moreover, if two fracture network patterns are different yet are in the same rock unit and have been subjected to the same geological history, how can their similarities be quantified for practical studies, such as the construction of a radioactive waste repository? Barton uses fractal geometry to characterize the outcrop scale heterogeneity of fracture network patterns over nearly 10 orders of magnitude in length scale, and to examine the relevance of acquiring fracture data from drill holes.

Linear differential equations are often used to describe the deformation of steel, rubber or other industrial products. The earth did not emerge from a factory, no do linear equations describe much of how it evolves. The earth's rheology is distinctly non-linear. Sammis and Steacy investigate the failure of the earth's crust, focusing on the fractal geometry of particles formed in fault gouges. These authors show that the geometry of communition is different depending upon the mechanism of fragmentation. The existence of fractal distributions of particles in fault gouges suggests that communition takes place according to specific processes, and not others that have been proposed.

Villemin, Angelier and Sunwoo also study the geometry of faults using fractals, though for a very different purpose. Their study of the Lorraine coal basin in France shows that the relation between fault geometry and displacement is fractal. More importantly, this observation makes it possible to calculate the total rock mass deformation from observations of a limited size range of faults. Fractal geometry makes it possible to extend a geological model for faulting and rock mass behavior to different scales of observation, and in so doing, solve an important problem in structural geology.

Fractal shapes and patterns abound. Bak and Chen show that there are dynamical reasons why this should be so. Taking earthquakes as an example of a dynamic geological system with abundant, well-documented fractal characteristics, the authors demonstrate how the earth's crust is in a self-organized critical state. Bak and Chen show that there is good reason to expect that self-organized critical systems should abound in nature, and that it is not surprising that fractal geometry is an important mathematical language in the earth sciences.

The final two chapters in this book illustrate how petrologists and ore geologists have exploited fractal geometry. Fowler focuses on textures in igneous rock. He shows that igneous textures, characterized by jagged outlines, fail to be adequately quantified by conventional textural descriptions. Fractal geometry proves to be useful, and provides insight into how crystalline textures develop.

Earth science is a practical enterprise, concerned with finding water, identifying earthquake risk, assessing the stability of foundations, mitigating pollution, and finding useful minerals. The pragmatism of earth scientists' work makes them both skeptical and open to new ideas and tools. Concepts are winnowed through application to real problems; the chaff quickly disappears. Saunders and Schoenly use fractal geometry to understand the formation of gold-silver deposits. Exploration for and exploitation of bonanza epithermal gold deposits improves as the understanding of how such deposits form improves. Funded by a major mining company, these authors show how the fractal structure of such deposits suggests how they could have formed.

The chapters in this volume do not exhaust all of the many ways in which fractal geometry benefits the earth sciences. Yet the contributions are comprehensive, illustrating a broad range of important fractal techniques, pitfalls, theoretical considerations and practical applications that should resonate with pragmatic field geologists and researchers alike. It is our hope that this book will help earth scientists not yet skilled in fractal geometry to rapidly assimilate these tools and to look at important geological problems in new ways.

Christopher C. Barton
Paul R. La Pointe

Contents

1. **Fractal Distributions in Geology, Scale Invariance, and Deterministic Chaos**

   D. L. Turcotte and J. Huang

1.1.	Introduction	1
1.2.	Definition of a Fractal Distribution	2
1.3.	Fragmentation	3
1.4.	Fractal Porosity	6
1.5.	Ore and Petroleum Reserves	7
1.6.	Seismicity and Tectonics	11
1.7.	Perimeter Relations	14
1.8.	Probability	18
1.9.	Cluster Analysis	19
1.10.	Self-Affine Fractals	20
1.11.	Chaos	32
1.12.	Conclusions	37
	References	38

2. **Some Long-Run Properties of Geophysical Records**

   B. B. Mandelbrot and J. R. Wallis

Presentation		41
2.1.	Introduction	42
2.2.	R/S and Pox Diagrams	46
2.3.	Pitfalls in the Graphic Estimation of H	47
2.4.	Determining H for a Specific Hydrological Project	48
	References	61

3. **Some Remarks on the Numerical Estimation of Fractal Dimension**

   S. A. Pruess

3.1.	Introduction	65
3.2.	Mathematical Background	66

3.3.	Some Sample Results	69
3.4.	Conclusions	74
	References	75

4. Measuring the Dimension of Self-Affine Fractals: Example of Rough Surfaces

S. R. Brown

4.1.	Introduction	77
4.2.	Fractures in Rock	78
4.3.	Scaling, Fractals, and Crossover Length	79
4.4.	Estimating the Fractal Dimension	81
4.5.	Conclusions	86
	References	87

5. Review of the Fractal Character of Natural Fault Surfaces with Implications for Friction and the Evolution of Fault Zones

W. L. Power and T. E. Tullis

5.1.	Introduction	89
5.2.	Techniques for Observing and Characterizing Fractal Surfaces	91
5.3.	Observations of Natural Rock Surfaces	94
5.4.	Discussion and Implications	95
5.5.	Conclusions	103
	References	103

6. Fractals and Ocean Floor Topography: A Review and a Model

A. Malinverno

6.1.	Introduction	107
6.2.	Elementary Theory and Examples	109
6.3.	Using Fractals to Parameterize Topography	114
6.4.	Fractals and Relief-Forming Processes	120
6.5.	Conclusions and Speculations	126
	References	128

7. Fractal Transitions on Geological Surfaces

C. H. Scholz

7.1.	Introduction	131
7.2.	Observations	132
7.3.	Discussion	139
	References	140

12. Mineral Crystallinity in Igneous Rocks: Fractal Method

A. D. Fowler

12.1.	Introduction	237
12.2.	Fractals	238
12.3.	Experimental Techniques	239
12.4.	Crystallinity	241
12.5.	Fractal Growth Simulation	245
12.6.	Summary	248
	References	248

13. Fractal Structure of Electrum Dendrites in Bonanza Epithermal Au-Ag Deposits

J. A. Saunders and P. A. Schoenly

13.1.	Introduction	251
13.2.	Geologic Setting	252
13.3.	Electrum Textures	253
13.4.	Fractal Dimensions of Natural Dendrites	255
13.5.	Computer Simulations	257
13.6.	Conclusions	260
	References	260

Index 263

8. Fractal Analysis of Scaling and Spatial Clustering of Fractures

C. C. Barton

8.1.	Introduction	141
8.2.	Sampling Spatial Clustering and Scaling of Fractures in Rock	144
8.3.	Fractal Measure of Spatial and Scaling Properties	145
8.4.	Methods of Measuring the Fractal Dimension of Fracture Networks	145
8.5.	Previous Fractal Studies of the Scaling and Spatial Distribution of Fractures	148
8.6.	One-Dimensional Sampling and Analysis of Fracture Networks	150
8.7.	Development Pattern of Fracture Networks	162
8.8.	Discussion	171
8.9.	Conclusions	176
	References	177

9. Fractal Fragmentation in Crustal Shear Zones

C. G. Sammis and S. J. Steacy

9.1.	Introduction	179
9.2.	Discussion	202
	References	203

10. Fractal Distribution of Fault Length and Offsets: Implications of Brittle Deformation Evaluation—The Lorraine Coal Basin

T. Villemin, J. Angelier, and C. Sunwoo

10.1.	Introduction	205
10.2.	Generalized Geology	207
10.3.	Data Collection	209
10.4.	Data Distribution and Relationships and Their Implications	209
10.5.	Fault Deformation and Self-Similarity	218
10.6.	Conclusions	223
	Appendix	232
	References	225

11. Fractal Dynamics of Earthquakes

P. Bak and K. Chen

11.1.	Introduction	227
11.2.	Models and Simulations	229
11.3.	Discussion and Conclusions	233
	References	235

1

Fractal Distributions in Geology, Scale Invariance, and Deterministic Chaos

D. L. Turcotte and J. Huang

1.1. INTRODUCTION

The scale invariance of geological phenomena is one of the first concepts taught to a student of geology. It is pointed out that an object with a scale, i.e., a coin, a rock hammer, a person, must be included whenever a photograph of a geological feature is taken. Without the scale, it is often impossible to determine whether the photograph covers 10 cm or 10 km. A specific example is folded, layered sedimentary rocks that occur over this range of scales. Another example is an aerial photograph of a rocky coastline. Without an object with a characteristic dimension, such as a tree or house, the elevation of the photograph cannot be determined. In this context, Mandelbrot (1967) introduced the concept of fractals. Because of scale invariance, the length of a coastline increases as the length of the measuring rod decreases according to a power law; the power determines the fractal dimension of the coastline.

Fractal concepts can be applied to geological problems in a variety of ways. One is the frequency-size distribution. Under a variety of circumstances, the frequency-size distributions of fragments, faults, mineral deposits, oil fields, and earthquakes have been shown to be fractal. Fractal concepts can also be applied to continuous distributions; an example is topography. Mandelbrot (1982) used fractal concepts to generate synthetic landscapes that look remarkably similar to actual landscapes. The fractal dimension is a measure of the roughness of the features. The earth's topography is a composite of many competing influences. Topography is created by tectonic processes, including faulting, folding, and

---

D. L. Turcotte and J. Huang • Department of Geological Sciences, Cornell University, Ithaca, New York; *present address of J.H.*: Exxon Production Research, Houston, Texas 77252.

Fractals in the Earth Sciences, edited by Christopher C. Barton and Paul R. La Pointe. Plenum Press, New York, 1995.

flexure; topography is modified and destroyed by erosion and sedimentation. Some aspects of topography are deterministic; the flexure of the elastic lithosphere is an example. However much of the Earth's topography is complex and chaotic; yet there is order in the chaos as shown by the applicability of fractal statistics. Many of the geological applications of fractals are discussed by Turcotte (1989a).

A fractal distribution is the only statistical distribution that is scale invariant. An important question however is how fractal distributions are related to the governing physical laws. Historically geophysics has been the study of the classical linear equations of physics, La Place's equation for gravity and magnetics, the wave equation for propagating seismic waves, and the heat equation for thermal problems. These linear equations cannot give scale invariance and fractal statistics; nonlinear equations are required. Lorenz (1963) derived a set of total, nonlinear differential equations that approximate thermal convection in a fluid heated from below. This set of equations was the first shown to exhibit chaotic behavior. Infinitesimal variations in initial conditions led to order one differences in the solutions obtained. Stewart and Turcotte (1989) showed that equations governing mantle convection yield chaotic solutions. Random reversals of the Earth's magnetic field are characteristic of chaotic behavior. In fact a parameterized set of dynamo equations proposed by Rikitake (1958) has been shown to exhibit deterministic chaos and spontaneous reversals (Cook and Roberts, 1970). Another example of chaotic behavior is the logistic map studied by May (1976). This type of recursion relation is applicable to population dynamics. Slider-block models obey similar mathematics, and these are a simple analog for the stick-slip behavior of a fault. Huang and Turcotte (1990b) have shown that an asymmetric pair of slider blocks behave chaotically. Chaotic solutions to deterministic equations obey fractal statistics in a variety of ways.

Another class of models that yield fractal statistics involves the cellular automata concept. Bak and Tang (1989) gave a cellular automata model for seismicity; Chase (1992) gave a cellular automata model for erosion.

1.2. DEFINITION OF A FRACTAL DISTRIBUTION

For the objectives of Chapter 1, a fractal set is defined by the following relation (Pfeifer and Obert, 1989):

$$N_i = \frac{C}{r_i^D} \quad (1)$$

where N_i is the number of objects with a characteristic linear dimension r_i, D is the fractal dimension, and C is a constant of proportionality. Relation (1) is applicable to deterministic fractal sets, an example is given in Fig. 1.1. The triadic Koch island is constructed by placing a triangle with length ⅓ in the center of each side. For the three examples illustrated, $r_1 = ⅓, N_1 = 3; r_2 = ⅑, N_2 = 12; r_3 = 1/27, n_3 = 48$, where N_i is the number of sides. For this example Eq. (1) is satisfied by $D = \ln 4/\ln 3 = 1.26186$. The implication of a fractal dimension between 1 and 2 is that the perimeter of the region is no longer a continuous differentiable line.

Fractal concepts can also be applied to a statistical distribution of objects. If the number of objects N with a characteristic linear dimension greater than r satisfies the relation

1. FRACTAL DISTRIBUTIONS IN GEOLOGY

FIGURE 1.1. Triadic Koch island. A triangle with length ⅓ is placed in the center of each side, $P_{i+1}/P_i = 4/3$, $r_{i+1}/r_i = 1/3$, $D = \ln 4/\ln 3$.

$$N = \frac{C}{r^D} \qquad (2)$$

then a fractal distribution is defined. Geological applications are generally statistical, since they are generally not ordered, even though scale-invariant.

The essential feature of the fractal distribution is its scale invariance. No characteristic length scale enters the Eq. (1) or (2). As mathematical representations, Eq. (1) or (2) could be valid over an infinite range; however for any physical application, there are upper and lower limits on the applicability of the fractal distribution. If scale invariance extends over a sufficient range of length scales, then the fractal distribution provides a useful description of the applicable statistics. The fractal dimension D provides a measure of the relative importance of large versus small objects.

Other definitions of fractal distributions can be derived from Eqs. (1) and (2), however in a number of applications, the basic definitions can be applied directly. For example, the Korcak empirical relation for the number of islands with an area greater than a specified value fits Eq. (2) with $D = 1.30$ (Mandelbrot, 1975). We turn next to examples of direct applicability in fragmentation, seismology, and tectonics.

1.3. FRAGMENTATION

A material can be fragmented in many ways (Grady and Kipp, 1987; Clark, 1987). Rocks can be fragmented by weathering; the resulting distribution of fragment sizes is likely to be related to the distribution of joints and other preexisting planes of weakness in the rock. Fragments can be produced by explosives; again preexisting planes of weakness may determine the distribution of fragment sizes. Fragments can also be produced by impacts; the rocks in active tectonic zones are fragmented in the upper crust. A variety of statistical relationships are used to correlate data on the size distribution of fragments; two of the most widely used distributions are the log normal and the power law. Turcotte (1986a) showed that the power law distribution is equivalent in a variety of forms to the fractal distribution given in Eq. (2).

Several examples of power law fragmentation are given in Fig. 1.2. A classic study of the size-frequency distribution for broken coal was carried out by Bennett (1936). The size-frequency distribution for chimney rubble above the Piledriver nuclear explosion was given by Schoutens (1979). This was a 61-kiloton event at a depth of 457 m in granite. The

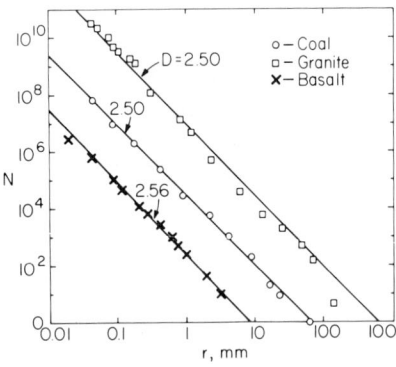

FIGURE 1.2. The number of fragments N with mean radius greater than r for broken coal (Bennett, 1936), broken granite from a 61-kiloton underground nuclear detonation (Schoutens, 1979), and impact ejecta due to a 2.6 km/s polycarbonate projectile impacting on basalt (Fujiwara and others, 1977). The best fit fractal distribution from Eq. (2) is shown for each data set.

size-frequency distribution for fragments resulting from the high-velocity impact of a projectile on basalt was given by Fujiwara and others (1977). In each example, the fractal dimension for the distribution is near $D = 2.5$. Other examples of power law distributions of fragments are given in Table 1.1.

We see that a great variety of fragmentation processes can be interpreted in terms of a fractal dimension. Values of the fractal dimension vary considerably, but most lie in the $2 < D < 3$ range. If the fractal dimension is less than 3, the volume integral diverges for large particles ($V \to \infty$); that is, the volume (mass) of small particles is negligible. If the fractal dimension is greater than 2, the area integral diverges for small particles ($V \to 0$); that is, the area of small particles dominates. Volume (mass) is conserved on fragmentation while area is not; however the creation of area requires energy. Thus it is reasonable to hypothesize that the fractal dimension increases with an increase in the energy density

TABLE 1.1. Fractal Dimensions for a Variety of Fragmented Objects

Object	Reference	Fractal Dimension D
Projectile fragmentation of gabbro with lead	Lange and others (1984)	1.44
Projectile fragmentation of gabbro with steel	Lange and others (1984)	1.71
Artificially crushed quartz	Hartmann (1969)	1.89
Disaggregated gneiss	Hartmann (1969)	2.13
Disaggregated granite	Hartmann (1969)	2.22
FLAT TOP I (chemical explosion, 0.2 kiloton)	Schoutens (1979)	2.42
PILEDRIVER (nuclear explosion, 62 kiloton)	Schoutens (1979)	2.50
Broken coal	Bennett (1936)	2.50
Projectile fragmentation of quartzite	Curran and others (1977)	2.55
Projectile fragmentation of basalt	Fujiwara (1977)	2.56
Fault gauge	Sammis and Biegel (1989)	2.60
Sandy clays	Hartmann (1969)	2.61
Terrace sands and gravels	Hartmann (1969)	2.82
Glacial till	Hartmann (1969)	2.88
Ash and pumice	Hartmann (1969)	3.54

1. FRACTAL DISTRIBUTIONS IN GEOLOGY

available for fragmentation. In all cases, we expect upper and lower limits on the validity of the fractal or power law relation for fragmentation. The upper limit is generally controlled by the size of the object or region being fragmented. The lower limit is likely to be controlled by the scale of the heterogeneities responsible for fragmentation, i.e., grain size.

A simple model, as shown in Fig. 1.3, illustrates how fragmentation can result in a fractal distribution. At each scale, two diagonally opposed blocks are retained. In this configuration, no two blocks of equal size are in direct contact with one another. This is the comminution model for fragmentation proposed by Sammis and others (1987). It is based on the hypothesis that the direct contact between two fragments of near equal size during the fragmentation process result in the breakup of one of the fragments. It is unlikely that small fragments break large fragments or a large fragment breaks small fragments.

In Fig. 1.3 $N_1 = 2$ for $r_1 = h/2$, $N_2 = 12$ for $r_2 = h/4$, and $N_3 = 72$ for $r_3 = h/8$. From Eq. (1) we find that $D = \ln 6/\ln 2 = 2.5850$. This comminution model is discrete in that fragment sizes are in powers of 2. This is clearly an approximation to a continuum distribution of fragment sizes, but it should still be a reasonable model. This model was originally developed for a fault gouge. Sammis and Biegel (1989) have shown that a fault gouge obtained from the Lopez fault zone, San Gabriel Mountains, California has a fractal dimension $D = 2.6 \pm 0.11$ on scales from 0.5μ–10mm. A synthetic fault gouge produced in the laboratory has also been shown to have a fractal dimension $D = 2.6$ (Biegel and others, 1986). In both cases, results are in excellent agreement with the comminution model that gives $D = 2.585$. Fig. 1.2 and Table 1.1 show that many observed distributions of fragments have fractal distributions near $D = 2.5$. This is evidence that scale invariant comminution may be widely applicable to rock fragmentation. The comminution model may also be applicable to tectonic fragmentation on larger scales.

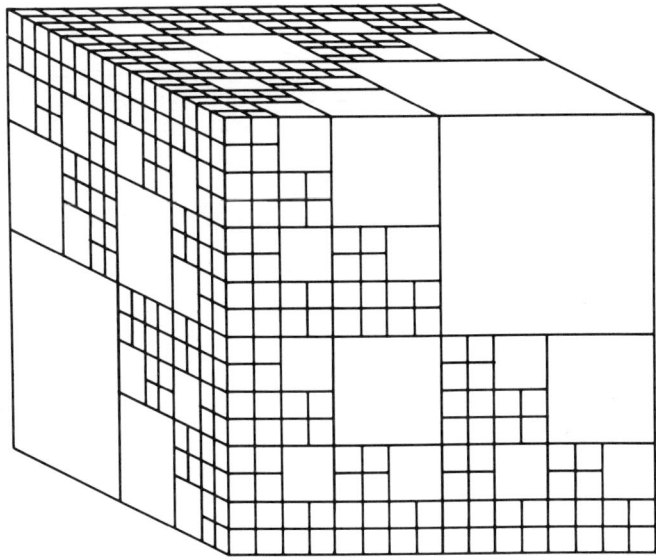

FIGURE 1.3. Illustration of a fractal model for fragmentation. Two diagonally opposite cubes are retained at each scale. With $r_1 = h/2$, $N_1 = 2$ and $r_2 = h/4$, $N_2 = 12$, we have $D = \ln 6/\ln 2 = 2.5850$.

1.4. FRACTAL POROSITY

A number of authors (Katz and Thompson, 1985; Krohn and Thompson, 1986; Daccord and Lenormand, 1987; Krohn, 1988a, b) have suggested that sandstones have a fractal distribution of porosity. In many cases, the distribution of fractures in rocks may also have a fractal distribution. The Menger sponge in Fig. 1.4 is a simple model for fractal porosity. A solid cube of unit dimensions has square passages with $r_1 = \frac{1}{3}$ cut through the center of the six sides. Twenty solid cubes with dimension $r_1 = \frac{1}{3}$ remain of the original 27, so that $N_1 = 20$. From Eq. (1) we find $D = \ln 20/\ln 3 = 2.7268$. This construction can be repeated in a scale-invariant manner, creating small and smaller channels.

We consider the porosity of the Menger sponge at various scales. We assume that at the highest order n, each cube has no porosity, so that its density is ρ_0 and its size is r_n. At the next larger scale, $r_{n-1} = 3r_n$ and the density is $\rho_{n-1} = 20\rho_0/27$. Continuing the construction, the size of the cube is $r_{n-2} = 9r_n$ and the density if $\rho_{n-2} = 400\rho_0/729$. The density dependence of the Menger sponge is illustrated in Fig. 1.5. The density of the nth order Menger sponge is

$$\frac{\rho}{\rho_0} = \left(\frac{r_n}{r}\right)^{3-\ln 20/\ln 3} \tag{3}$$

But the fractal dimension of the Menger sponge is $D = \ln 20/\ln 3$. Thus for a fractal distribution of porosity, the density of a rock scales with its size according to

$$\frac{\rho}{\rho_0} = \left(\frac{r_0}{r}\right)^{3-D} \tag{4}$$

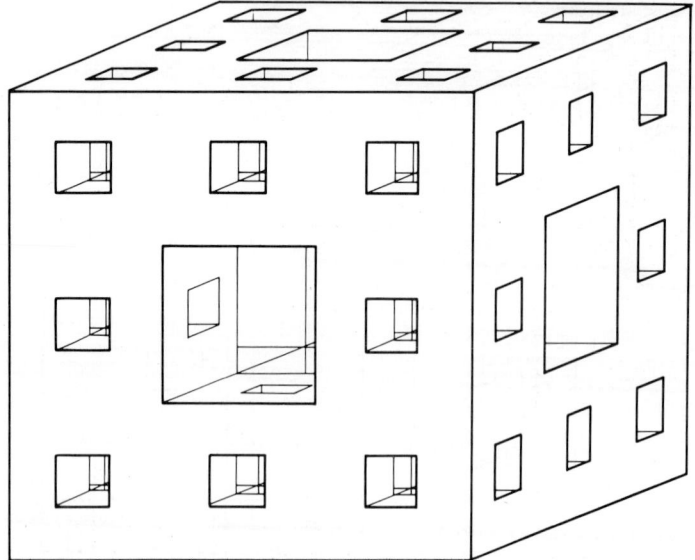

FIGURE 1.4. Illustration of the Menger sponge. A solid cube of unit dimensions has square passages with $r_1 = \frac{1}{3}$ cut through the center of the six sides. Twenty solid cubes remain with $D = \ln 20/\ln 3$. The construction is repeated in scale-invariant manner.

1. FRACTAL DISTRIBUTIONS IN GEOLOGY

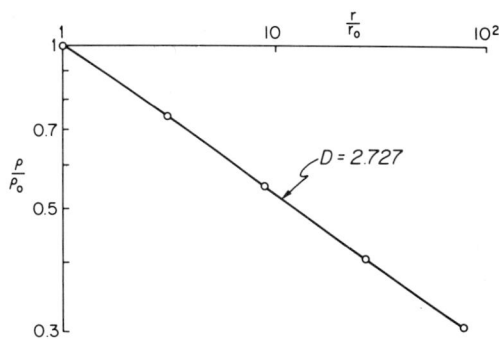

FIGURE 1.5. Density dependence ρ/ρ_0 of a Menger sponge as a function of the size of the sponge r/r_0. A fractal decrease in the density is found with $D = 2.727$ from Eq. (4).

where r is the linear dimension of the sample considered. The density of a fractal solid decreases systematically with an increase in the size of the sample considered.

A number of studies of the densities of soil aggregates as a function of size have been carried out. These studies show a systematic decrease in density as the size of the aggregate increases. Such studies involve a sieve analysis, which is carried out on a soil to obtain the mean density of each aggregate size. Results for a sandy loam given by Chepil (1950) are shown in Fig. 1.6. Although there is considerable scatter, results agree reasonably well with fractal soil porosity of dimension $D = 2.869$.

Volumetric fractals, such as the Menger sponge, can be used to develop models for porous flow through fractal distributions of permeability. This is particularly important in studies of dispersion (Adler, 1985a–d, 1986; Nolte and other, 1989), Curl (1986) has determined the fractal dimension of caves and applied the Menger sponge as a model. Laverty (1987) considered fractals in karst.

1.5. ORE AND PETROLEUM RESERVES

Statistical relations for the tonnage of ore with a grade above a specified value have provided a basis for estimating ore reserves. Available data have generally been correlated using either log-normal or power law relations. This work has been reviewed by Harris (1984). Turcotte (1986b) showed that power law relations were equivalent to the fractal

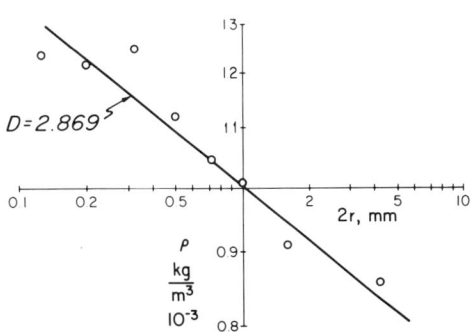

FIGURE 1.6. Density of soil aggregates as a function of their size (Chepil, 1950). The solid line is from Eq. (4) with $D = 2.869$.

relation in Eq. (2). If M is the tonnage of ore with an average Grade $\bar{C}$, the distribution is a fractal if

$$\bar{C} \sim M^{-D/3} \tag{5}$$

It is interesting to compare Eq. (5) with available data.

The cumulative tonnage of mercury mined in the United States prior to a specified date was divided by the cumulative tonnage of ore by Cargill and others (1980) to give the cumulative average (weight) grade. Their results, given in Fig. 1.7, are for 5-year intervals. An excellent correlation with Eq. (5) is found for $D = 2.01$. A similar correlation for copper as given by Cargill and other (1981) is illustrated in Fig. 1.8 and compared with Eq. (5) by taking $D = 1.16$. Data obtained prior to 1920 fall systematically low compared with the fractal correlation; however this deviation can be attributed to a more efficient mineral extraction process that was introduced at that time.

The simple model illustrated in Fig. 1.9 gives a fractal distribution of ore reserves. In this model, an original mass of rock M_0 is divided into two equal-sized cells, each with a mass $M_1 = M_0/2$. The original mass of the rock has a mean mineral concentration C_0; this is the ratio of the mass of mineral to mass of rock. It is hypothesized that the mineral is concentrated in one of the first-order cells such that its concentration in the enriched cell is

$$C_{11} = \phi_2 C_0 \tag{6}$$

The first subscript on C refers to the order of cell and the second to the enriched or depleted cell. The enrichment factor ϕ_2 is greater than 1, since C_{11} must be greater than C_0. A simple mass balance shows that the concentration in the depleted zero-order cell is

$$C_{12} = (2 - \phi_2)C_0 \tag{7}$$

The process is repeated for the enriched cell in a scale-invariant manner. The enriched first-order cell is divided into two second-order cells of equal mass $M_2 = M_1/2 = M_0/4$. The mineral is again concentrated by the same enrichment factor ϕ_2. The enriched second-order cell in the enriched first-order cell has a concentration

$$C_{21} = \phi_2 C_{11} = \phi_2^2 C_0 \tag{8}$$

The depleted second-order cell in the enriched first-order cell has a concentration

$$C_{22} = (2 - \phi_2)C_{11} = \phi_2(2 - \phi_2)C_0 \tag{9}$$

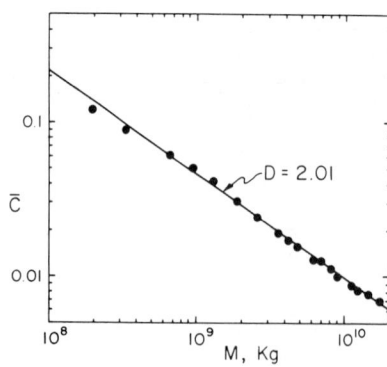

FIGURE 1.7. Dependence of ore grade on tonnage for mercury production in the United States (Cargill and others, 1980).

1. FRACTAL DISTRIBUTIONS IN GEOLOGY

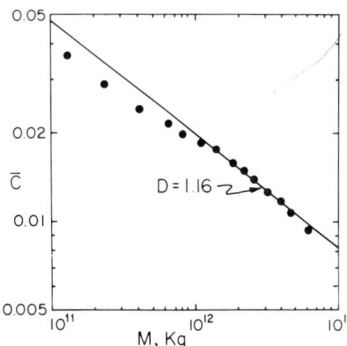

FIGURE 1.8. Dependence of ore grade on tonnage for copper production in the United States (Cargill and others, 1981).

These results can be generalized to the nth-order cell with the result

$$C_{nl} = \phi_2^n C_0 \tag{10}$$

where C_{nl} is the mean grade associated with the mass

$$M_n = \frac{1}{2^n} M_0 \tag{11}$$

Taking the natural logarithms of Eqs. (10) and (11) gives

$$\ln\left(\frac{C_{nl}}{C_0}\right) = n \ln \phi_2 \tag{12}$$

$$\ln\left(\frac{M_n}{M_0}\right) = -n \ln 2 \tag{13}$$

ϕ_2	$(2 - \phi_2)$

ϕ_2^2	$\phi_2(2 - \phi_2)$	$(2 - \phi_2)$

ϕ_2^3	$\phi_2^2(2 - \phi_2)$	$\phi_2(2 - \phi_2)$	$(2 - \phi_2)$

ϕ_2^4	$\phi_2^3(2-\phi_2)$	$\phi_2^2(2 - \phi_2)$	$\phi_2(2 - \phi_2)$	$(2 - \phi_2)$

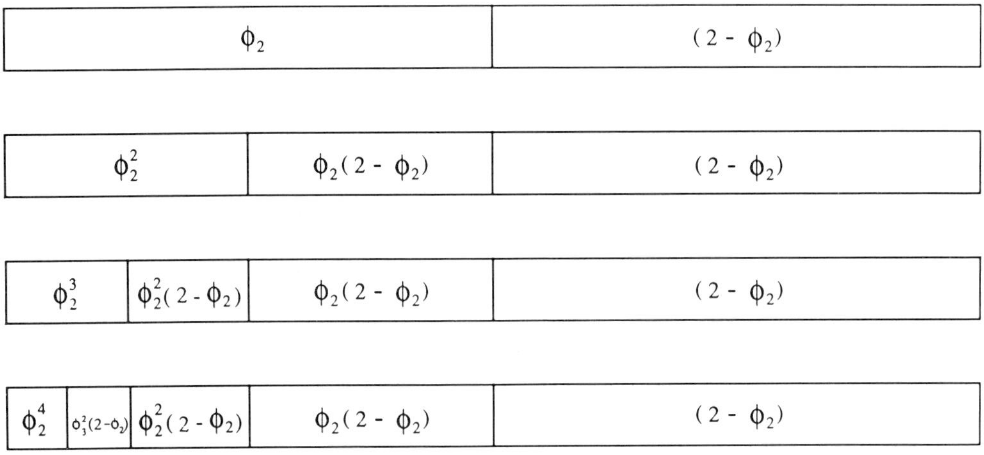

FIGURE 1.9. Model for ore concentration that gives a fractal dependence of grade on tonnage. The enrichment factor ϕ_2 is applied at all scales.

And eliminating n between Eqs. (12) and (13) gives

$$\ln\left(\frac{C_{n1}}{C_0}\right) = -\left(\frac{\ln \phi_2}{\ln 2}\right) \ln\left(\frac{M_n}{M_0}\right) \tag{14}$$

or

$$\frac{C_{n1}}{C_0} = \left(\frac{M_n}{M_0}\right)^{\ln \phi_2/\ln 2} = \left(\frac{r_n}{r_0}\right)^{3 \ln \phi_2/\ln 2} \tag{15}$$

since the density is assumed constant and $M \sim r^3$, where r is the linear dimension of the ore sample considered. Comparison with Eq. (2) shows that this is a power law or fractal distribution with

$$D = 3\frac{\ln \phi_2}{\ln 2} \tag{16}$$

To be fractal the distribution must be scale-invariant, and scale invariance is clearly illustrated in Fig. 1.9. The concentration of ore could be started at any order, and the same result would be obtained. The left-half at order 2 looks like order 1, the left-half at order 3 looks like order 2, etc.

As in previous examples of naturally occurring fractal distributions, there are limits to the applicability of Eq. (5). The lower limit on the ore grade is clearly the regional background grade C_0 that has been concentrated. However there is also an upper limit—the grade $\bar{C}$ cannot exceed unity, which would be the pure mineral.

There is also evidence that the frequency-size distribution of oil fields obeys fractal statistics. Drew and others (1982) used the relation $N_{i-1} = 1.67N_i$ to estimate the number of fields of order i, N_i, in the western Gulf of Mexico. Since the volume of oil in a field of order i is a factor of two greater than the volume of oil in a field or order $i-1$, their relation is equivalent to a fractal distribution with $D = 2.22$. The number-size statistics for oil fields in the United States (excluding Alaska) as compiled by Ivanhoe (1976) are given in Fig. 1.10. A reasonably good correlation with the fractal relation in Eq. (2) is obtained by taking $D = 4.6$. The large difference between these two values for the fractal dimension can be attributed to differences in regional geology, but it may also be due to the data, since it is often difficult to determine when adjacent fields are truly separate; thus data on reserves are

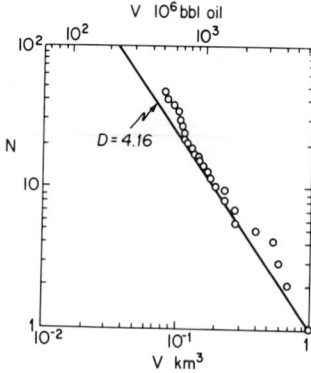

FIGURE 1.10. Dependence of the number of oil fields in the United States with volume greater than V as a function of V. The equivalent number of barrels is also given. Circles indicate data from Ivanhoe (1976). The solid line shows the correlation with Eq. (2), taking $D = 4.16$.

1. FRACTAL DISTRIBUTIONS IN GEOLOGY

often poorly constrained. Nevertheless the applicability of fractal statistics to petroleum reserves can have important implications. Reserve estimates for petroleum have been obtained by using power law (fractal) statistics and log-normal statistics. Accepting power law statistics leads to considerably higher estimates for available reserves.

The high value of the fractal dimension given in Fig. 1.10 is also indicative of another characteristic of power laws and fractal dimensions. There is no reason for the power in a power law associated with a statistical distribution to be limited to a particular range. Whether a power law that does not fall within a particular range should be called a fractal distribution is a matter of choice. Scale invariance implies a power law distribution if the scale invariance is related to a geometrical relation such as the fragmentation model given in Fig. 1.3, where we see from Eq. (4) that $0 < D < 3$. However the concentration relation given in Eq. (16) does not imply limits on the values of D, so that in principle D can take on any value for a power law dependence of ore grade on tonnage or in other cases.

1.6. SEISMICITY AND TECTONICS

Under many circumstances the number of earthquakes occurring in a specified length of time $\dot{N}$ with a magnitude greater than m satisfies the empirical relation (Gutenberg and Richter, 1954).

$$\log \dot{N} = -bm + \log \dot{a} \tag{17}$$

where $\dot{a}$ and b are constants. This relationship has been found to be applicable both regionally and worldwide. Aki (1981) showed that Eq. (17) is equivalent to the definition of a fractal distribution.

The moment of an earthquake is defined by

$$M = \mu \delta A \tag{18}$$

where μ is the shear modulus, A the area of the fault break, and δ the mean displacement on the fault break. The moment of the earthquake can be related to its magnitude by

$$\log M = cm + d \tag{19}$$

where c and d are constants. Kanamori and Anderson (1975) established a theoretical basis for taking $c = 1.5$. These authors also showed that it is a good approximation to take

$$M = \alpha r^3 \tag{20}$$

were $r = A^{1/2}$ is the linear dimension of the fault break. Combining Eqs. (17), (19), and (20) and taking $c = 1.5$ gives

$$\log \dot{N} = -2b \log r + \log \dot{\beta} \tag{21}$$

with

$$\log \dot{\beta} = \frac{bd}{1.5} + \log \dot{a} - \frac{b}{1.5} \log \alpha \tag{22}$$

and Eq. (21) can be rewritten as

$$\dot{N} = \dot{B} r^{-2b} \tag{23}$$

A comparison with the definition of a fractal given in Eq. (2) shows that

$$D = 2b \qquad (24)$$

Thus the fractal dimension of regional or worldwide seismic activity is simply twice the b-value. The empirical frequency-magnitude relation given in Eq. (17) is entirely equivalent to a fractal distribution.

As a specific application, we consider regional seismicity in southern California. The frequency-magnitude distribution of seismicity in southern California as summarized by Main and Burton (1986) is given in Fig. 1.11. Based on data from 1932 to 1972, the number of earthquakes per year $\dot{N}$ is given as a function of magnitude m in (open squares). In the magnitude range of $4.25 < m < 6.5$, data are in excellent agreement with Eq. (17), taking $b = 0.89$ ($D = 1.78$) and $\dot{a} = 2 \times 10^5$ yr^{-1}. In terms of the linear dimension of the fault break, this magnitude range corresponds to a size range of $0.9 < r < 12$ km.

Also included in Fig. 1.11 is the $\dot{N}$ associated with great earthquakes on the southern section of the San Andreas fault as given by Sieh (1978), based on dated seismic events in trenches; the solid dot corresponds to $m = 8.05$ and $\dot{N} = 0.006$ yr^{-1} (a repeat time of 163 years). An extrapolation of the fractal relation for regional seismicity appears to make a reasonable prediction of great earthquakes on this section of the San Andreas fault. Since this extrapolation is based on 40 years of data from 1932–1972, a relatively small fraction of the mean interval of 163 years, it suggests that $\dot{a}$ for this region may not have a strong dependence on time during the earthquake cycle. This conclusion has a number of important implications. If a great earthquake substantially relieved the regional stress, then we would expect the regional seismicity systematically to increase as stress increased before the next great earthquake. An alternative hypothesis is that an active tectonic zone is continuously in a critical state and the fractal frequency-magnitude statistics are evidence for this critical behavior. In the critical state, the background seismicity (small earthquakes not associated with aftershocks), has little time dependence. Accepting this hypothesis allows regional background seismicity to be used to assess seismic hazards. The regional

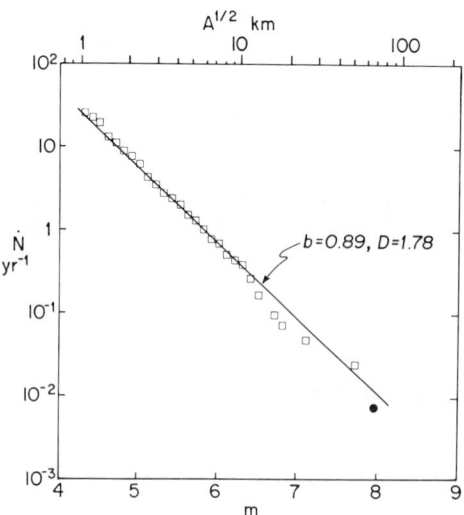

FIGURE 1.11. Number of earthquakes occurring per year $\dot{N}$ with a surface wave magnitude greater than m as a function of m. Open squares indicate data for Southern California from 1932–1972 (Main and Burton, 1986). The solid circle indicates the expected rate of occurrence of great earthquakes in Southern California (Sieh, 1978).

1. FRACTAL DISTRIBUTIONS IN GEOLOGY

frequency-magnitude statistics can be extrapolated to estimate recurrence times for larger magnitude earthquakes.

The empirical use of distributed seismicity has been described by several authors (Smith, 1976; Molnar, 1979; Anderson and Lucco, 1983) and used either to determine regional strain (Anderson, 1986; Youngs and Coppersmith, 1985) or compare levels of seismicity with known strain rates (Hyndman and Weichert, 1983; Singh and others, 1983). This approach in a fractal context has been discussed by Turcotte (1989b).

As an example of this approach, let us consider seismicity in the eastern United States. Since the eastern United States is a plate interior, the concept of rigid plates precludes seismicity in the region. However plates act as stress guides, so they can transmit high stresses associated with forces applied at plate boundaries. Since plates have weakness zones, they deform under stresses and earthquakes result. Chinnery (1979) compiled frequency-magnitude statistics for three areas in the eastern United States, his results are given Fig. 1.12. The Mississippi Valley region includes the site of the great New Madrid, Missouri earthquakes of 1811–12. This was a sequence of three earthquakes felt throughout the eastern United States that disrupted the flow of the Mississippi River. Because these earthquakes occurred before the invention of the seismograph and no well-defined fault break has been found, it is difficult to estimate the magnitude of these earthquakes. The southern Appalachian region includes the site of the 1886 Galveston earthquake. This earthquake, which did considerable damage, is estimated to have had a magnitude of 7.5. A sizeable earthquake occurred near Boston in 1755; this is in the New England region, but little information is available on its magnitude. Since data in Fig. 1.12 are based on felt intensities, they are not as reliable as data in Fig. 1.11, which are based on instrumental observations. Nevertheless, comparing Figure 1.12 with Fig. 1.11 indicates that the probability of an earthquake occurring in the eastern United States is about 1/100 the probability that the same earthquake will occur in southern California. Taking the New Madrid, Missouri, earthquakes to have an equivalent magnitude $m = 8$, we can estimate that the recurrence time from data in Fig. 1.12 is about 5000 years.

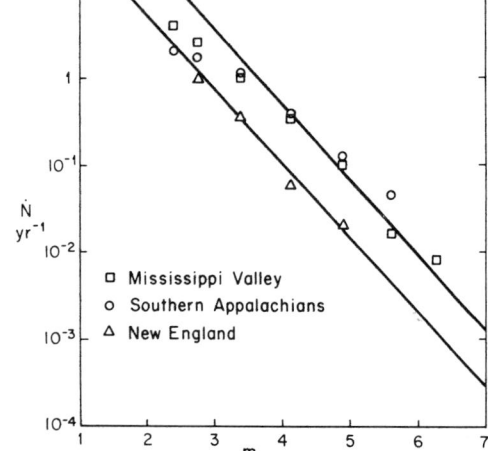

FIGURE 1.12. Number of earthquakes occurring per year $\dot{N}$ in the eastern United States with magnitudes greater than m as a function of m. Open squares represent the Mississippi region; open circles represent the southern Appalachian region; open triangles represent the New England region. Solid lines are from Eq. (17) with $b = 0.85$.

Tectonic models to explain the fractal distribution of seismicity have been proposed by King (1983, 1986), Turcotte (1986c), King and others (1988), and Hirata (1989a). Fractal models of faults that give well-defined *b*-values have been proposed by Huang and Turcotte (1988) and Hirata (1989b). Hirata and others (1987) have found a fractal distribution of microfractures in laboratory experiments on unfractured granite.

It is generally difficult to quantify the frequency-size distribution of faults because the surface exposure is generally limited. Many faults are not recognized until earthquakes occur on them. A systematic study of the statistics of exposed joints and fractures has been given by Barton and Hsieh (1989). Basement rock near Yucca Mountain, Nevada, was cleared and the fracture distribution mapped. The results of their exposures are given in Fig. 1.13. This exposure was located in the densely welded orange brick unit of the Topopah Spring member of the Miocene Paintbrush Tuff. These authors used a box-counting algorithm to determine the fractal dimension of the exposure. The result for the pavement illustrated in Fig. 1.13 is given in Fig. 1.14; the mean fractal dimension is $D = 1.7$.

Two-dimensional exposures can be used to infer the three-dimensional distribution of fracture (fault) sizes. We hypothesize that the model for fragmentation in Fig. 1.3 is also applicable to tectonic fragmentation. For a two-dimensional exposure of this model, we have $N_1 = 1$ for $r_1 = h/2$, $N_2 = 3$ for $r_2 = h/4$, and $N_3 = 9$ for $r_3 = h/8$. From Eq. (1) we find that $D = \ln 3/\ln 2 = 1.5850$. This could have been obtained by subtracting unity from the fractal dimension given in Fig. 1.3. The fragmentation model is in quite good agreement with observed results given in Fig. 1.14.

The fractal distribution of earthquake magnitudes is a universal feature of distributed seismicity on a variety of scales. There is extensive evidence that this result implies a fractal distribution of fault sizes. Although major plate boundaries are characterized by a dominant fault, such as the San Andreas, the evolution of plates is not consistent with displacements on a single fault (Dewey, 1975); thus regional deformation must occur. The hypothesis that this deformation follows the comminution model provides a fractal distribution with a fractal dimension consistent with observations.

We turn next to volcanic eruptions. It is considerably more difficult to quantify a volcanic eruption than to quantify an earthquake. Nevertheless McClelland and others (1989) have published frequency-volume statistics for volcanic eruptions. Their results for eruptions during the period from 1975–1985 and for historic (last 200 years) eruptions are given in Fig. 1.15. The numbers of eruptions with a volume of tephra greater than a specified value are given as a function of the volume. A reasonably good correlation is obtained with the fractal relation in Eg. (2), taking $D = 2.14$. It also appears that volcanic eruptions are scale-invariant over a significant range of sizes.

1.7. PERIMETER RELATIONS

Although we suggest that the basic definition of a fractal is the number-size relation given in Eq. (1), the original definition of a fractal given by Mandelbrot (1967) is the length of a trail or perimeter as a function of step length. If the length of the step is r_i and N_i is the number of steps required to obtain the length of the trail or perimeter P_i, we have

$$P_i = N_i r_i \tag{25}$$

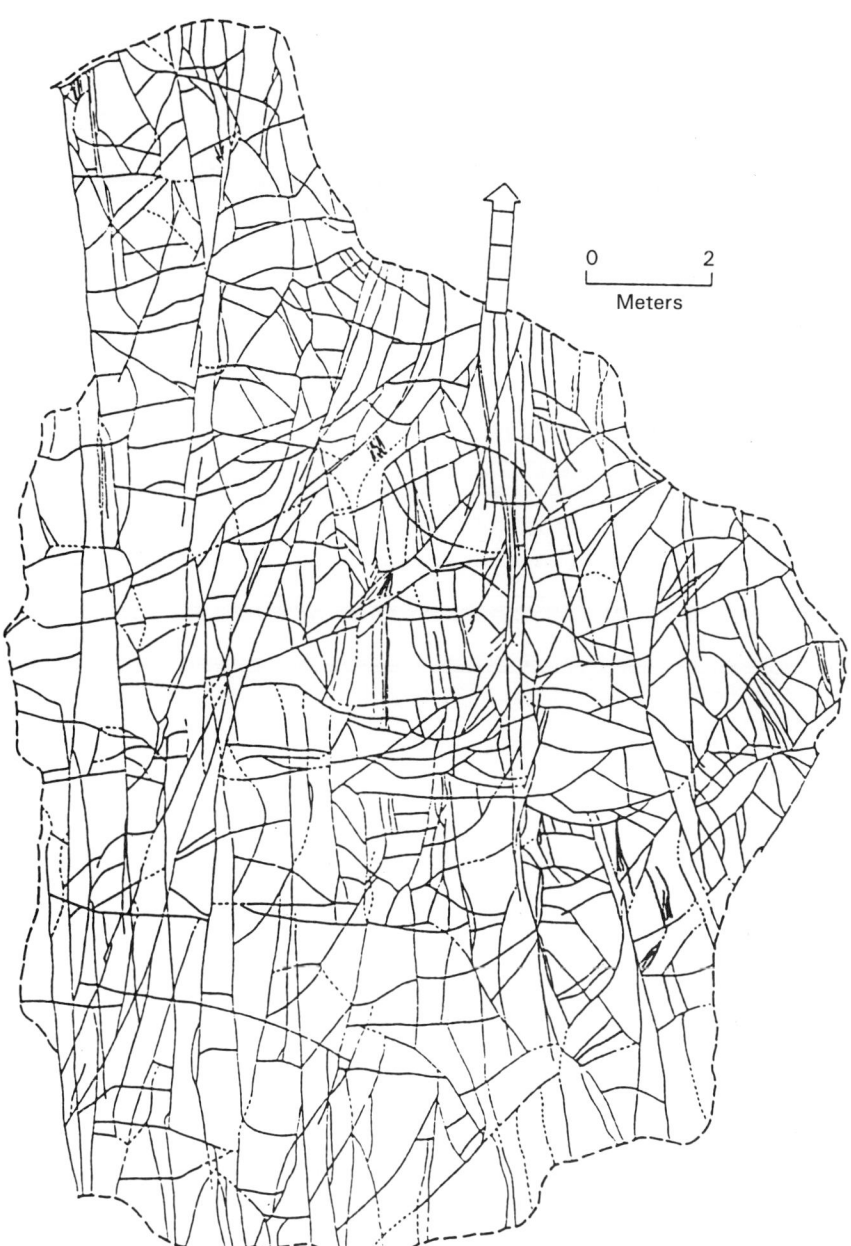

FIGURE 1.13. Fracture trace map of pavement 1000, Yucca Mountain, Nevada (Barton and Hsieh, 1989). This exposure was located in the densely welded orange brick unit of the Topopah Spring Member of the Miocene Paintbrush Tuft.

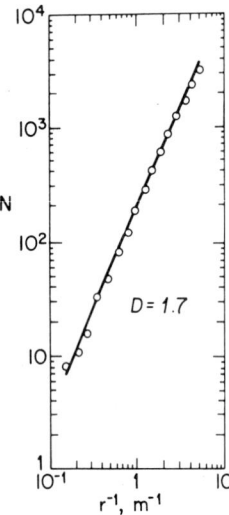

FIGURE 1.14. Statistics using the box-counting algorithm on the exposed fracture network at Yucca Mountain, Nevada (Barton and Hsieh, 1989) given in Fig. 1.13. A correlation with Eq. (2) is used to obtain the fractal dimension $D = 1.7$.

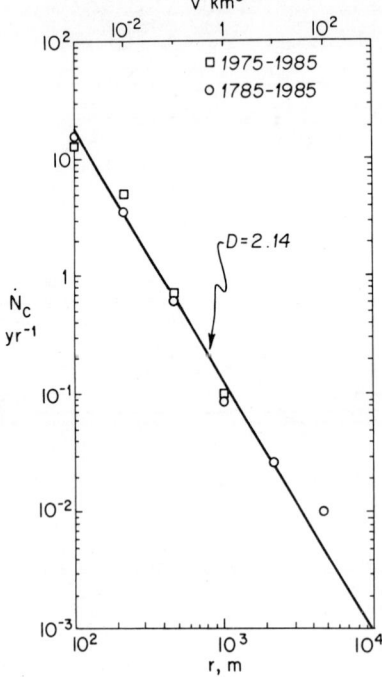

FIGURE 1.15. Number of volcanic eruptions per year N_c with tephra volume greater than V as a function of V for the period 1975–1985 (*open squares*) and for the last 200 years (*open circles*) (McClelland and others, 1989). The solid line is the correlation with Eq. (2), taking $D = 2.14$.

1. FRACTAL DISTRIBUTIONS IN GEOLOGY

Substituting Eq. (1) gives

$$P_i = Cr_i^{1-D} \qquad (26)$$

This relation is applicable to the triadic Koch island in Fig. 1.1. For this example, $r_1 = \frac{1}{3}$, $P_1 = 1; r_2 = \frac{1}{9}, P_2 = \frac{4}{3}; r_3 = \frac{1}{27}, P_3 = \frac{16}{9}$. These values are consistent with Eq. (26), taking $D = \ln 4/\ln 3$. Mandelbrot (1967) showed that the west coast of Britain satisfied Eq. (26), with $D = 1.25$.

Fractal dimensions for topography can be easily obtained from topographic maps. Several examples are given in Fig. 1.16. The length along specified contours P is obtained by using dividers of different lengths r. Results generally satisfy Eq. (26), and a fractal dimension can be obtained. However it should be emphasized that not all regions yield well-defined fractal dimensions (Goodchild, 1980); young volcanic islands are an example. An important question is whether there are systematic variations in the fractal dimension; for example, does it depend systematically on the age of the geological province. Figure 1.7 shows that quite diverse geological provinces have very similar fractal dimensions. Since most fractal dimensions are near $D = 1.2$, it does not appear possible to use fractal dimensions obtained by this method to characterize geological terrains.

Scholz and Aviles (1986) and Aviles and others (1987) used the length-of-trail technique to determine the fractal dimension of the San Andreas fault. They found values ranging from $D = 1.0008–1.0191$. A similar analysis was carried out by Okubo and Aki (1987).

Many relationships in geomorphology have a fractal nature. An example is the power law relationship between the drainage area of a basin and the length of the principal river. Data for a number of drainage basins in the northeastern United States (Hack, 1957) are given in Fig. 1.17. Snow (1989) considered the fractal sinuosity of stream channels.

FIGURE 1.16. Fractal dimensions for specified topographic contours in several mountain belts. (a) 3000-foot contour of the Cobblestone Mountain Quadrangle Transverse ranges, California ($D = 1.21$); (b) 5400-foot contour of the Tatooh Buttes quadrangle, Cascade Mountains, Washington ($D = 1.21$); (c) 10,000-foot contour of the Byers Peak quadrangle, Rocky Mountains, Colorado ($D = 1.15$); (d) 1000-foot contour of the Silver Bay quadrangle, Adirondack Mountains, New York ($D = 1.19$).

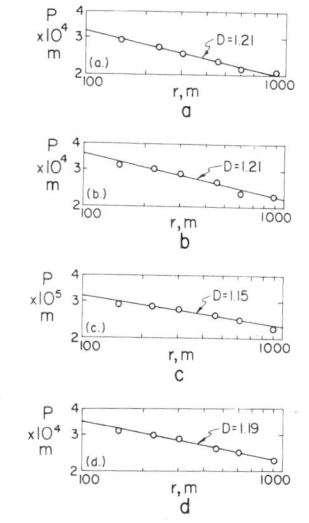

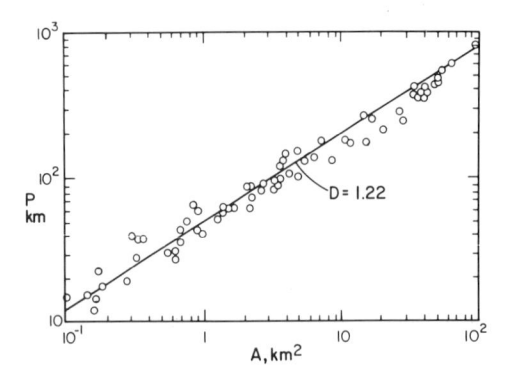

FIGURE 1.17. Dependence of the length of the principal river P on the area of the drainage basin A for several drainage basins in the northeastern United States (Hack, 1957).

1.8. PROBABILITY

Next we relate fractal distributions to probability. A line has a Euclidian dimension of 1, while a point has a zero Euclidian dimension. The Cantor dust illustrated in Fig. 1.18a has a fractal dimension between 0 and 1. The solid line of unit length is divided into three parts ($R = \frac{1}{3}$, and the center third is removed ($N = 2$), from Eq. (1), $D = \ln 2/\ln 3 = 0.6309$. The process is repeated, and total line length decreases; this is referred to as curdling.

The probability that a step of length r includes a solid line is $\Pr = 1$ when $r = 1$; $\Pr = \frac{2}{3}$ when $r = \frac{1}{3}$, and $\Pr = \frac{4}{9}$ when $r = \frac{1}{9}$. This is generalized to

$$\Pr_i = N_i r_i \tag{27}$$

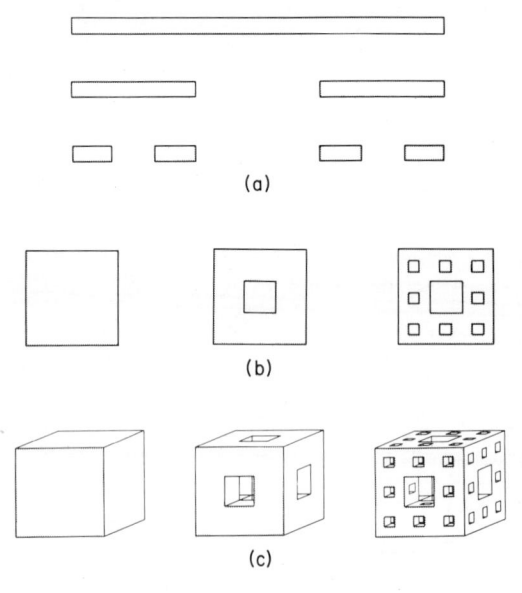

FIGURE 1.18. (a) Cantor dust. The middle one-third of each solid line is removed at each stage, $N_{i+1}/N_i = 2$, $r_{i+1}/r_i = \frac{1}{3}$, $D = \ln 2/\ln 3$. (b) Sierpinski carpet. The middle one-third of each solid square is removed at each stage $N_{i+1}/N_i = 8$, $r_{i+1}/r_i = \frac{1}{3}$, $D = \ln 8/\ln 3$. (c) Mengor sponge. Each solid cube has square passages cut through the center third of each face at each stage, $N_{i+1}/N_i = 20$, $r_{i+1}/r_i = \frac{1}{3}$, $D = \ln 20/\ln 3$.

1. FRACTAL DISTRIBUTIONS IN GEOLOGY

and substituting Eq. (1) gives

$$\Pr_i = r_i^{1-D} \tag{28}$$

Although the Cantor dust is a determinant set, the probability relation in Eq. (28) can be applied to statistical clustering.

The two-dimensional analog of the Cantor dust is the Sierpinski carpet illustrated in Fig. 1.18b. The solid square of unit dimension has a square with dimensions $r = 1/3$ removed from its center. Eight solid squares with dimensions $r = 1/3$ remain, so that $N = 8$. Thus from Eq. (1), $D = \ln 8/\ln 3 = 1.8928$. The process is repeated. The fractal dimension lies between 1 (the Euclidian dimension of a line) and 2 (the Euclidian dimension of an area). The probability that a square with dimensions r includes solids is $\Pr = 1$ when $r = 1$; $\Pr = 8/9$ when $r = 1/3$, and $\Pr = 68/81$ when $r = 1/9$. This is generalized to

$$\Pr_i = N_i r_i^2 \tag{29}$$

and substituting Eq. (1) gives

$$\Pr_i = r_i^{2-D} \tag{30}$$

This result can be applied to clustering in an area.

The three-dimensional analog of the Cantor dust and Sierpinski carpet is the Menger sponge illustrated in Fig. 1.18c. This construction was previously discussed as a model for porosity. The probability that a cube with dimensions r includes solid is $\Pr = 1$ when $r = 1$; $\Pr = 20/27$ when $r = 1/3$, and $\Pr = 400/729$ when $r = 1/9$. This is generalized to

$$\Pr_i = N_i r_i^3 \tag{31}$$

and substituting Eq. (1) gives

$$\Pr_i = r_i^{3-D} \tag{32}$$

The generalization of Eqs. (28), (30), and (32) is

$$\Pr_i = r_i^{d-D} \tag{33}$$

where d is the Euclidian dimension of the problem being considered. Probability methods introduced in this section are used to obtain the fractal dimension of stationary, self-similar time series.

1.9. CLUSTER ANALYSIS

The Cantor dust, Sierpinski carpet, and Menger sponge form the basis for cluster analyses in one, two, and three dimensions. As a specific example of fractal cluster analysis in one dimension, we consider the temporal distribution of N earthquakes occurring in a time interval t_0. The number of earthquakes is determined by the area of the region considered, and a magnitude cut off is made to ensure that all earthquakes in the region are detected by the regional network. The natural period for the data set is t_0/N.

To carry out a fractal cluster analysis, we divide the interval t_0 into a series of smaller intervals.

$$t = \frac{t_0}{n}, \quad n = 1, 3, 4$$

Our measure of clustering is the fraction Pr of intervals in which an earthquake occurs as a function of t. If Pr has a power law dependence on t, a fractal dimension for the clustering is defined.

This approach to fractal clustering of seismicity was used by Sadovskiy and others (1985) and by Smalley and others (1987). The latter considered the temporal variation of seismicity in several regions near Efate Island in the New Hebrides Island arc for the period from 1978–1984; one of their examples is given in Fig. 1.19. During the period, 49 events occurred in the region considered that satisfied magnitude criteria. Time intervals between 8 minutes $< t <$ 524,288 minutes were considered. The fraction of intervals with earthquakes Pr as a function of the interval length t is shown in Fig. 1.19a as open circles. The solid line shows the correlation of the fractal relation in Eq. (28) with $D = 0.255$; the dashed line is the result for uniformly spaced events. Results of a simulation for a random distribution of the 49 events in the time interval studied is given in Fig. 1.19b. The random simulation (Poisson distribution) is significantly different than the earthquake data, and it is close to the uniform distribution.

1.10. SELF-AFFINE FRACTALS

Up to this point, we have considered self-similar fractals; we now turn to self-affine fractals (Mandelbrot, 1985). Topography is an example of both fractals: In the two horizontal directions, topography is self-similar, but in the vertical direction, it is self-affine. The ruler method can be applied to a topographic map, but it cannot be meaningfully

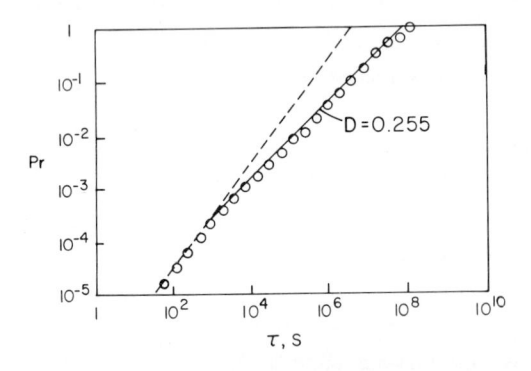

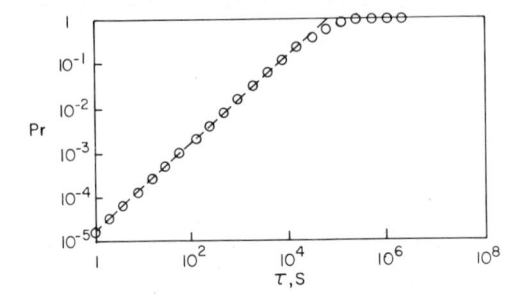

FIGURE 1.19. Fractal cluster analysis of 49 earthquakes that occurred near Efate Island, New Hebrides, from 1978–1984 (Smalley and others, 1987). (a) Open circles represent the fraction of intervals Pr of length t with earthquakes as a function of t. The solid line represents the correlation with Eq. (28), taking $D = 0.255$. (b) Open circles represent a random simulation of 49 events assuming the Poisson process.

1. FRACTAL DISTRIBUTIONS IN GEOLOGY

applied to a vertical cross section. Brown (1987) showed that this can be done only if vertical coordinates are scaled upward relative to the horizontal.

Spectral methods are generally applied to self-affine fractals. Consider a single-valued function of a variable—say, $x(t)$—that is random but has a specified spectrum. We consider the specific case in which increments $[x(t + T) - x(t)]$ satisfy the probability condition

$$\Pr\left[\frac{x(t + T) - x(t)}{T^H} < x'\right] = F(x') \tag{34}$$

where H is a constant. In many examples, $F(x')$ is a Gaussian distribution. If this is the case and $0 < H < 1$, then the random signal is known as fractional Brownian motion. For $H = \frac{1}{2}$, we have Brownian motion.

Although the signal is continuous, it is not differentiable; thus it may be appropriate to consider the noise to be fractal in character. For the time series, the fractal dimension is between 1 and 2, so that the appropriate Euclidian dimension is 2. With $d = 2$, Eq. (33) can be written as

$$\frac{\Pr}{T^{2-D}} = \text{constant} \tag{35}$$

where the appropriate scale for the time series is T. Comparing Eqs. (34) and (35), we define

$$H = 2 - D \tag{36}$$

This is the basic definition of the fractal dimension for fractional Brownian motion.

The variance of the increment is defined by

$$V(T) = \frac{1}{T}\int_0^T [x(t) - \bar{x}]^2 \, dt \tag{37}$$

where T is the time over which the profile is specified and $\bar{x}$ is the mean value of x over this time. A necessary condition for the time series to be a fractal is that the variance $V(T)$ has a power law dependence on T (Voss, 1985a, b, 1988).

$$V(T) \sim T^{2H} \tag{38}$$

The standard deviation is related to the length of the profile T by

$$\sigma(T) = [V(T)]^{1/2} \sim T^H \tag{39}$$

An alternative derivation of the relevant fractal dimension is obtained by introducing a reference "box" of width T and height σ. If the fractal were self-similar, the box would be square; however, for self-affine fractals, arbitrary rectangular boxes must be used. Consider a set of nth-order smaller boxes with width $T_n = T/n$ and height $x_n = \sigma/n$, where n is an integer. The number of nth-order boxes N_n required to cover width T and height $\sigma_n = \sigma(T/n)$ is

$$N_n = \frac{T\sigma_n}{T_n x_n} = n^2 \frac{\sigma_n}{\sigma} \tag{40}$$

Using Eq. (39), we find

$$\frac{\sigma_n}{\sigma} = \frac{\sigma(T/n)}{\sigma(T)} = \frac{1}{n^H} \tag{41}$$

and combining Eqs. (40) and (41) gives

$$N_n = n^{2-H} = \left(\frac{T}{T_n}\right)^{2-H} \tag{42}$$

Comparing Eq. (42) with the definition of the fractal dimension in Eq. (1), again yields the definition given in Eq. (36).

Random functions in time are often characterized by their spectral energy densities $S(f)$. If we define $\tilde{x}(f,t)$ as the Fourier transform of $x(t)$ for $0 < t < T$,

$$\tilde{x}(f,t) = \frac{1}{T}\int_0^T x(t)e^{2\pi i f t}\, dt \tag{43}$$

then

$$S(f) \sim T\tilde{x}^2(f,t) \tag{44}$$

as $T \to \infty$. For Gaussian white noise, $S(f) = $ constant. For a self-affine fractal, the spectral energy has the following power law dependence on frequency:

$$S(f) \sim f^{-\beta} \tag{45}$$

The relationship between β and the fractal dimension is obtained from Eqs. (36), (38), (39), and (45), with the following result:

$$S \sim f^{-\beta} \sim T^{\beta} \sim TV \sim T^{2H+1} \sim T^{5-2D} \sim f^{5-2D} \tag{46}$$

or

$$\beta = 5 - 2D \tag{47}$$

For a true fractal, $1 < D < 2$ and $1 < \beta < 3$; for Brownian motion, $\beta = 2$ and $D = 1.5$. Examples of synthetically generated fractional Brownian motion are given in Fig. 1.20.

The method used to generate these random time series is:

1. The time interval is divided into a large number of increments. Each increment is given a random number based on a Gaussian probability distribution. This is then a Gaussian white noise sequence.
2. The Fourier transform of the sequence is taken.
3. The resulting Fourier coefficients are filtered by multiplying by $f^{-\beta/2}$.
4. An inverse Fourier transform is taken using the filtered coefficients.
5. To remove edge effects (periodicities), only the central portion of the time series is retained.

Figure 1.20d illustrates Brownian motion. Brown and Scholtz (1985) have applied spectral methods to fault surfaces.

An important question is whether climate obeys fractal statistics (Nicolis and Nicolis, 1984). Fluigeman and Snow (1989) showed that oxygen isotope records in sea floor cores obey fractal statistics. Plotnick (1986) argued that the distribution of stratigraphic hiatuses is fractal. Hewett (1986) showed that porosity logs in oil fields obey fractal statistics and he has used this fact to develop an inversion technique for obtaining a three-dimensional permeability structure from well logs.

It is common practice to expand data sets on the surface of the earth in terms of spherical harmonics; examples include topography and the geoid. Using topography as an example, the appropriate expansion for the radius r of the Earth is

1. FRACTAL DISTRIBUTIONS IN GEOLOGY

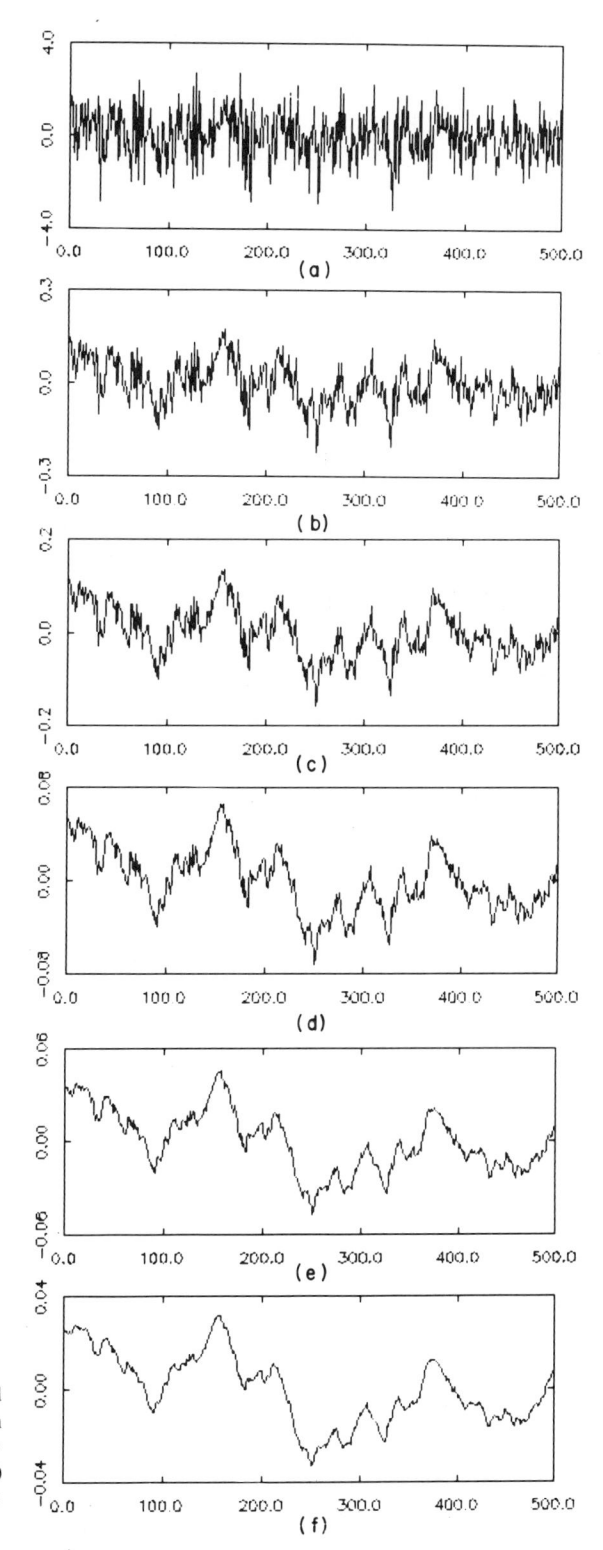

FIGURE 1.20. Synthetically generated fractional Brownian noise. (a) Gaussian white noise sequence; (b) $\beta = 1.2$ ($D = 1.9$); (c) $\beta = 1.6$ ($D = 1.7$); (d) $\beta = 2.0$ ($D = 1.5$), brown noise; (e) $\beta = 2.4$ ($D = 1.3$); (f) $\beta = 2.8$ ($D = 1.1$).

$$r(\theta,\phi) = a_0\left[1 + \sum_{l=1}^{\infty}\sum_{m=0}^{1}(C_{lm}\cos m\phi + S_{lm}\sin m\phi)P_{lm}(\sin\theta)\right] \quad (48)$$

where a_0 is a reference radius, θ and ϕ are polar coordinates, C_{lm} and S_{lm} are coefficients, and P_{lm} are associated Legendre functions fully normalized so that

$$\frac{1}{4\pi}\int_0^{2\pi}\int_{-\pi/2}^{\pi/2} P_{lm}(\sin\theta)\left(\begin{array}{c}\cos^2 m\phi\\ \sin^2 m\phi\end{array}\right)d(\sin\theta)\,d\phi = 1 \quad (49)$$

The variance of the spectra for order l is defined by

$$V_1 = a_0^2 \sum_{m=0}^{1}(C_{lm}^2 + S_{lm}^2) \quad (50)$$

The power spectral density is defined by

$$s(k) = \frac{1}{k_0}V(k) = \lambda_0 V(k) \quad (51)$$

where k_0 is the wave number and $\lambda_0 = 1/k_0$ is the wave length over which data is included in the expansion. With $\lambda_0 = 2\pi a_0$, we have

$$S_1 = 2\pi a_0^3 \sum_{m=0}^{1}(C_{lm}^2 + S_{lm}^2) \quad (52)$$

$$k_1 = \frac{1}{2\pi a_0} \quad (53)$$

A fractal dependence can be defined if S_1 has a power law relation to k_1.

The power spectral density of the Earth's topography is given in Fig. 1.21. This is based on spherical harmonic expansion of the Earth's topography to order 180 by Rapp (1989). Except for low-degree harmonics, an excellent correlation with Eq. (45) is obtained with $b = 2$ ($D = 1.5$). The spectral dependence of topography corresponds to Brown noise. This correlation has been previously noted by Bell (1975, 1979), Berkson and Matthews (1983), Barenblatt and others (1984), Fox and Hayes (1985), Gilbert and Malinverno (1988), Fox (1989), Gilbert (1989), Malinverno (1989), and Mareschal (1989). For Brown noise, the amplitude is proportional to the wavelength; in this sense, topography is truly self-similar.

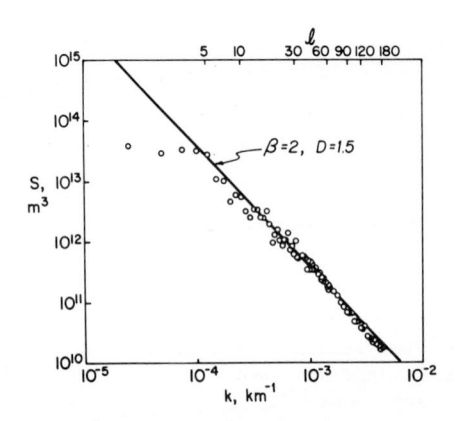

FIGURE 1.21. Power spectral density of the Earth's topography as a function of wave number. Circles represent data compilation by Rapp (1989). The solid line is Eq. (45) with $\beta = 2$ ($D = 1.5$).

1. FRACTAL DISTRIBUTIONS IN GEOLOGY

The Fourier spectral approach to fractal analysis for one-dimensional profiles discussed in the preceding section can be extended to two-dimensional image analysis. Consider an $N \times N$ grid of equally spaced data points in a square with linear size L so that the grid spacing is L/N. The N^2 data points are denoted by h_{nm}, where (n,m) specify the position in the x and y directions, respectively. Figure 1.22a illustrates a case with $N = 8$.

The first step is to carry out a two-dimensional discrete Fourier transform on the N^2 set of h_{nm} data points. An $N \times N$ array of complex coefficients H_{st} is obtained by the usual definition

$$H_{st} = \sum_{n=0}^{N-1} \sum_{m=0}^{N-1} h_{nm} \exp\left[\frac{2\pi i}{N}(sn + tm)\right] \tag{54}$$

	1 8	2 8	3 8	4 8	5 8	6 8	7 8	8 8
	1 7	2 7	3 7	4 7	5 7	6 7	7 7	8 7
	1 6	2 6	3 6	4 6	5 6	6 6	7 6	8 6
m	1 5	2 5	3 5	4 5	5 5	6 5	7 5	8 5
	1 4	2 4	3 4	4 4	5 4	6 4	7 4	8 4
	1 3	2 3	3 3	4 3	5 3	6 3	7 3	8 3
	1 2	2 2	3 2	4 2	5 2	6 2	7 2	8 2
	1 1	2 1	3 1	4 1	5 1	6 1	7 1	8 1

n

(a) The 64 nm coefficients for an 8 × 8 sub-set of raw data.

	8	8	8	8	8				
	7	7	7	7	8	8			
	6	6	6	7	7	7	8		
	5	5	5	5	6	7	7	8	
t	4	4	4	5	5	6	7	8	8
	3	3	3	4	5	5	7	7	8
	2	2	2	3	4	5	6	7	8
	1	1	2	3	4	5	6	7	8
	0	1	2	3	4	5	6	7	8

s

(b) Equivalent radial coefficients r for various coefficients s and t in spatial frequency space.

FIGURE 1.22. Illustrations of subscript arrangement in two-dimensional spectral analysis. (a) The 64-nm coefficients for an 8 × 8 subset of raw data. (b) Equivalent radial coefficients r for various coefficients s and t in spatial-frequency space.

where s denotes the transform in the x direction ($s = 0, 1, 2, \ldots, N - 1$) and t denotes the transform in the y direction ($t = 0, 1, 2, \ldots, N - 1$). Then each transform coefficient H_{st} is assigned an equivalent radial number using the relation

$$r = (s^2 + t^2)^{1/2} \tag{55}$$

The mean spectral energy density S_j for each radial wave number k_j is given by

$$S_j = \frac{L}{N_j} \sum_j^{N_j} |H_{st}|^2 \tag{56}$$

where N_j is the number of coefficients satisfying the condition $j < r < j + 1$ and the summation is carried out over the coefficients H_{st} in this range. Coefficients assigned to each interval for the example given in Fig. 1.22a are illustrated in Fig. 1.22b.

The dependence of the mean spectral energy density on the radial wave number for a fractal distribution is

$$S_j \sim k_j^{-\beta - 1} \tag{57}$$

instead of Eq. (45). The fractal dimension D is obtained from the relation

$$D = \frac{7 - \beta}{2} \tag{58}$$

This is the generationalization of Eq. (47) to two-dimensional spectral analysis.

Synthetic fractal images can be generated by the following technique, which is used to generate synthetic fractional Brownian motion:

1. Each pixel is given a random number based on a Gaussian probability distribution.

$$\Pr(x)\, dx = \frac{1}{\sqrt{2\pi}} e^{-x^2/2}\, dx \tag{59}$$

2. Using Eq. (54), a set of two-dimensional Fourier coefficients H_{st} are obtained from the N^2 set of random numbers.
3. A fractal dimension D is specified, and the corresponding value for β is obtained from Eq. (58). A new set of complex coefficients are obtained from the relation

$$H_{st}^* = \frac{H_{st}}{k_{st}^{(\beta+1)/2}} \tag{60}$$

4. An inverse two-dimensional Fourier transform is carried out to generate a new image. Three examples of synthetic images are given in Fig. 1.23. In Fig. 1.23a, the original random data are given without fractal filtering; this is white noise, so that $\beta = 0$. In Fig. 1.23b, the result is given for $\beta = 2$ ($D = 2.5$), and in Fig. 1.23c, for $\beta = 2.8$ ($D = 2.1$).

As an example, we consider the digitized topography of the state of Oregon illustrated in Fig. 1.24a (Huang and Turcotte, 1990a). Combining defense mapping agency (DMA) $1° \times 1°$ data with topographical maps, the U.S. Geological Survey (USGS) (Flagstaff) has produced digitized topography on a grid scale of about seven points per kilometer. We previously used this technique to study the topography of Arizona (Huang and Turcotte, 1989).

1. FRACTAL DISTRIBUTIONS IN GEOLOGY

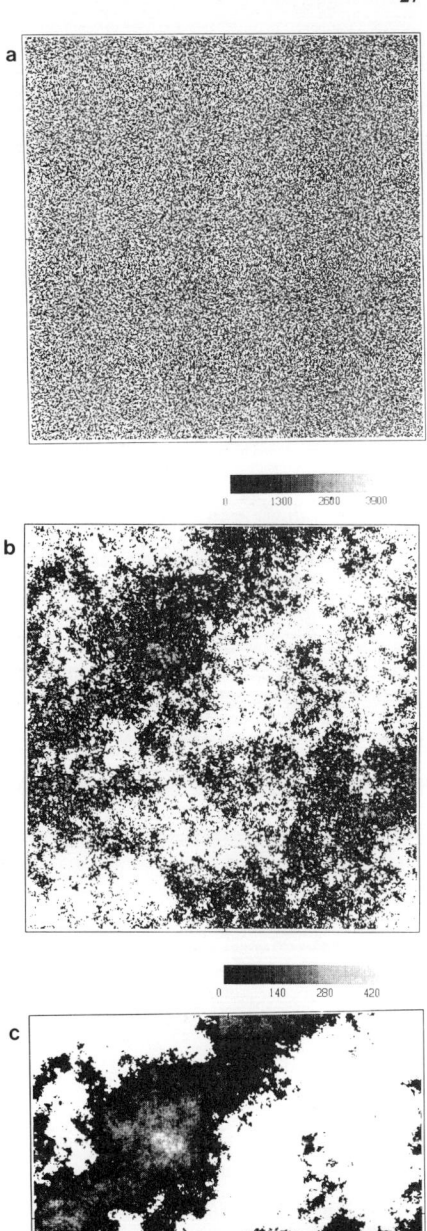

FIGURE 1.23. Synthetic fractal image on a 512 × 512 grid. (a) White noise without fractal filtering; (b) filtered with $\beta = 1$ and $D = 3$; (c) filtered with $\beta = 1.8$ and $D = 2.6$.

FIGURE 1.24. (a) Color-coded digitized topography for Oregon. Data resolution is about 7 point/km. (b) Color-coded maps of fractal dimension and (c) roughness amplitude for Oregon. There is relatively little variation in the fractal dimension; however the roughness amplitude is very sensitive to texture changes.

To examine whether data are fractal, one-dimensional spectral analyses are carried out on three areas with different tectonic settings. For each of the three regions, eight randomly chosen one-dimensional profiles with 512 data points are spectrally analyzed using the classical periodogram approach. Log-log plots of spectral energy are given in Fig. 1.25 for (a) the Willamette lowland, which is dominated by sedimentary processes; (b) the Wallowa Mountains, associated with a major tectonic uplift; and (c) the Klamath Falls area, a basin and range regime. In each case, Eq. (45) is a good approximation; the mean fractal dimensions are 1.47, 1.48 and 1.50, respectively.

The fact that fractal statistics are a good approximation for topography allows us to make fractal maps of a region of diverse tectonics. Using the digitized topography of Oregon, plots of spectral density versus wave number are made for subregions. From these plots, a fractal dimension (slope) and unit wave number amplitude are obtained for each subregion. The amplitude is a measure of roughness. Basically we are carrying out a texture analysis using the fractal statistics as a basis.

1. FRACTAL DISTRIBUTIONS IN GEOLOGY

FIGURE 1.24. (*Continued*)

In this study, fractal dimensions and roughness amplitudes are obtained using subregions of 32 × 32 data points. Thus fractal dimensions and roughness amplitudes are obtained for each 4.5 × 4.5 km subregion in the state. We also carried out studies using square subregions with 16 × 16 and 64 × 64 data points. The 32 × 32 set was chosen because it generally gives well-defined fractal spectra; for smaller regions, errors in fractal dimension and roughness become substantially larger. For larger regions, the spatial resolution of the map is degraded.

The following technique is used to obtain a fractal dimension and roughness amplitude for each subregion:

1. The 32 × 32 set of digitized elevations forms a set of coefficients as illustrated in Fig. 13, with $N = 32$.
2. The mean and linear trends for each subset of data are removed.
3. A two-dimensional discrete Fourier transform is carried out, and an $N \times N$ array of complex Fourier coefficients H_{st} is obtained using Eq. (54).
4. Each coefficient H_{st} is assigned an equivalent radial wave number r using Eq. (55). The mean spectral energy density S_j is obtained for each radial integer wave number k_j using Eq. (56).

FIGURE 1.24. (*Continued*)

5. The mean slope of a log-log plot of S_j versus k_j obtained by a least squares regression yields a fractal dimension D using Eq. (58). The intercept at $k_j = 1$ cycle/km yields a roughness amplitude.

Maps of fractal dimension and roughness amplitude are given in Figs. 1.24b and 1.24c. There is relatively little variation in the fractal dimension around the mean value $D = 2.586$, although the range is from about $2.40 < D < 2.90$. The variation in the roughness amplitude in Fig. 1.24c is much more impressive. The sedimentary Willamette lowland shows low overall roughness, while the erosional system associated with the nearby mountain ranges and the Wallowa Mountains in the northeast stand out as regions of high roughness. Roughness contrasts in the southern basin and range region are also quite impressive. Fractal analysis gives a qualitative measure of roughness.

Since landscapes evolve primarily through erosional processes, we conclude that erosional processes generate scale-invariant, fractal topography. Very little progress has been made in understanding the underlying physical processes leading to erosional topography. Culling (1960) proposed that erosional and depositional processes satisfy the heat equation. This approach is applicable to the erosion of fault scarps and the deposition of

1. FRACTAL DISTRIBUTIONS IN GEOLOGY

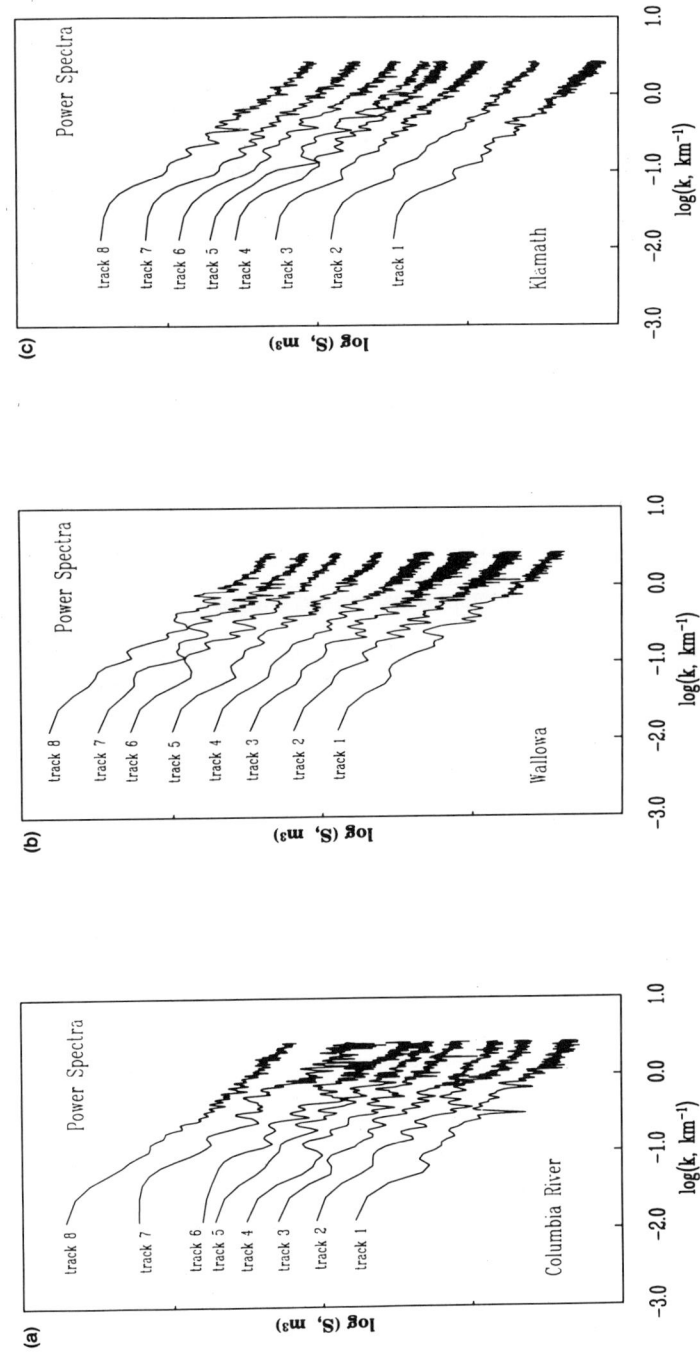

FIGURE 1.25. Plots of one-dimensional spectral energy density versus wave number selected from three regions with different tectonic settings in Oregon: (a) Willamette lowland, (b) Wallowa Mountains, (c) Klamath Falls.

alluvial fans; however it cannot be universally applicable to the generation of erosional landforms because it does not generate self-similar, fractal topography. In fact no linear equation generates fractal topography; the applicable theory must be nonlinear. Culling (1986, 1988), Culling and Datko (1987), Chase (1992), and Newman and Turcotte (1990) have considered how fractal topography can be generated.

A related question is whether storms, and in particular floods, have a fractal distribution. If floods have a fractal distribution, the ratio of the greatest mean flood in a thousand years to the greatest mean flood in a hundred years is equal to the ratio of the hundred-year flood to the ten-year flood, etc. If a large fraction of erosion takes place in the very largest floods, it is reasonable to argue that the fractal structure of topography is evidence of a fractal distribution of erosion in time, i.e. great floods.

1.11. CHAOS

The applicability of fractal statistics to a variety of geological problems has been demonstrated. One direct implication is scale invariance, but this has little to do with the underlying physics that yields fractal distributions. It is now recognized that either sets of differential equations or recurrence relations exhibiting deterministic chaos also yield fractal statistics (Devaney, 1988). In the last ten years, many problems have been shown to have chaotic solutions (Thompson and Stewart, 1986). In some cases, these solutions are applicable to problems of geological interest.

The simplest demonstration of deterministic chaos is the logistic map (May, 1976), this is the simple recursion relation.

$$x_{n+1} = ax_n(1 - x_n) \qquad 0 < a < 4 \tag{61}$$

For $0 < a < 1$, Eq. (61) gives $x = 0$ at the limit $n \to \infty$; this behavior is illustrated for $a = 0.8$ in the iterative map in Fig. 1.26a. For $1 < a < 3$, Eq. (61) gives $x_\infty = 1 - 1/a$; this behavior is illustrated for $a = 2.5$, $x_\infty = 0.6$ in Fig. 1.26b. At $a = 3$, a flip bifurcation occurs, and for $3 < a < 3.544090$, a two-root ($n = 2$) limit cycle is found; this behavior for $a = 3.1$, $x_\infty = 0.558, 0.765$ is illustrated in Fig. 1.26c. Further bifurcations occur, and for $3.569946 < a < 4$, windows of chaotic behavior and multiroot cycles appear. Fully chaotic behavior for $a = 3.9$ is illustrated in Fig. 1.26d, 1000 cycles are given. The systematic bifurcations of the solution strongly resemble the fractal Cantor set. The logistic map can be associated with the diversity of species.

A simple analog model for earthquakes is a spring, slider-block model. Huang and Turcotte (1990b) showed that the two-block model in Fig. 1.27 gives deterministic chaos. The two-slider blocks are coupled to each other and the constant velocity driver by springs. The springs extend until the pulling force exceeds the frictional static resisting force F_S. Once sliding begins, the motion of the block is resisted by the dynamic friction F_D. A necessary and sufficient condition for stick-slip behavior is that static friction exceeds dynamic friction $F_S < F_D$. Static conditions for the onset of sliding are

$$k_1 y_1 + k_c(y_1 - y_2) = F_{S1} \tag{62}$$

$$k_2 y_2 + k_c(y_2 - y_1) = F_{S2} \tag{63}$$

1. FRACTAL DISTRIBUTIONS IN GEOLOGY

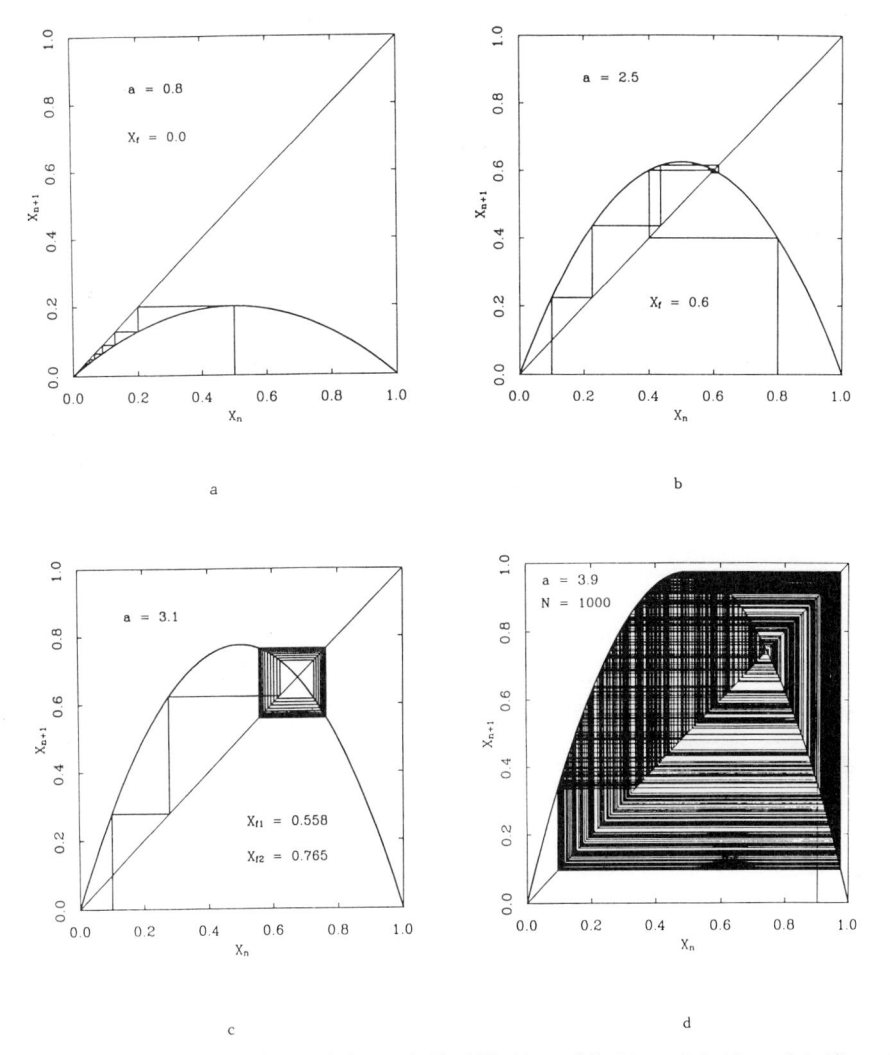

FIGURE 1.26. Iterative maps of the logistic map in Eq. (61); (a) $s = 0.8$, (b) $a = 2.5$, (c) $a = 3.1$, (d) $a = 3.9$.

Once sliding begins, the applicable equations of motion are

$$m_1 \frac{d^2 y_1}{dt^2} + k_1 y_1 + k_c(y_1 - y_2) = F_{D1} \tag{64}$$

$$m_2 \frac{d^2 y_2}{dt^2} + k_2 y_2 + k_c(y_2 - y_1) = F_{D2} \tag{65}$$

To simplify the model further, we assume that $m_1 = m_2 = m$, $k_1 = k_2 = k$, and $F_{S1}/F_{D1} = F_{S2}/F_{D2} = \phi$.

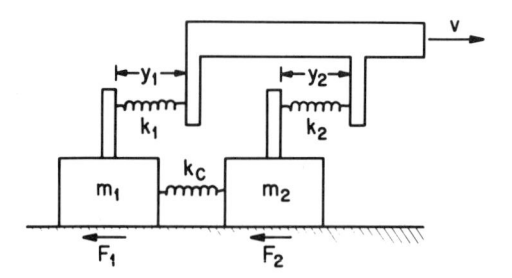

FIGURE 1.27. Illustration of the two-block model. The constant velocity driver extends the springs until a block commences to slide. In some cases, one block sliding induces the second block to slide.

It is convenient to introduce the nondimensional variables

$$\tau = t\left(\frac{k}{m}\right)^{1/2} \quad Y = \frac{ky}{F_{S1}} \quad \alpha = \frac{k_c}{k} \quad \beta = \frac{F_{S2}}{F_{S1}} \tag{66}$$

In terms of nondimensional parameters, sliding conditions in Eqs. (62) and (63) become

$$Y_1 + \alpha(Y_1 - Y_2) = 1 \tag{67}$$
$$Y_2 + \alpha(Y_2 - Y_1) = \beta \tag{68}$$

And the equations of motion in Eqs. (64) and (65) become

$$\frac{d^2 Y_i}{d\tau^2} + Y_1 + \alpha(Y_1 - Y_2) = \frac{1}{\phi} \tag{69}$$

$$\frac{d^2 Y_2}{d\tau^2} + Y_2 + \alpha(Y_2 - Y_1) = \frac{\beta}{\phi} \tag{70}$$

When the failure envelop given by Eqs. (67) and (68) is reached, sliding commences, which is governed by Eqs. (59) and (70). In some cases, the sliding of one block induces the sliding of the second block.

We first consider the symmetric case where both blocks have the same frictional behavior $\beta = 1$. An example with $\alpha = 3$ and $\phi = 1.25$ is given in Fig. 1.28. The diagonal lines converging at $Y_1 = Y_2 = 1$ are the failure envelope given by Eqs. (67) and (68). Cyclic behavior is found. Strain accumulates on the lines of unit slope when no sliding occurs. When an accumulation line intercepts the failure envelope either the first block slides (a horizontal line) or the second block slides (a vertical line). The blocks alternatively slip as shown.

We consider next an asymmetric case with $\beta = 2.5$, $\alpha = 5$, and $\phi = 1.25$. The results are illustrated in Fig. 1.29. Fully chaotic behavior is found. Curves lying outside the failure envelope are cases where both blocks are sliding at the same time. The chaotic behavior of this simple analog model is evidence that the interaction between actual faults also results in chaotic behavior.

Carlson and Langer (1989), who studied the behavior of large numbers of slider blocks, also found chaotic behavior. They obtained fractal statistics for the number of sliding events as a function of their size. Similarly statistics have been obtained by Bak and Tang (1989) and Ito and Matsuzaki (1990) using cellular automata models. In this approach, units of stress are randomly added to a network of grid points. When a point has four units the units are removed, and one unit is added to each of the four adjacent grid points. This transfer

1. FRACTAL DISTRIBUTIONS IN GEOLOGY

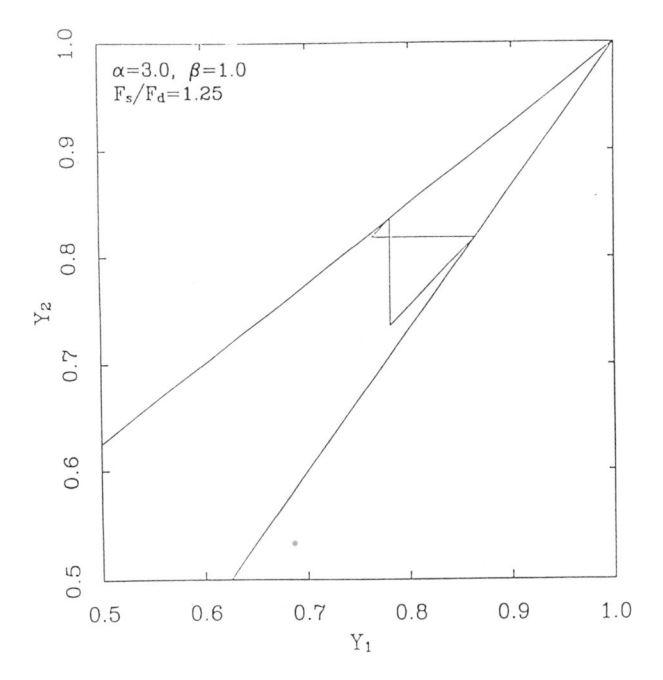

FIGURE 1.28. Behavior of the symmetrical $\beta = 1$ two-block illustrated in Fig. 1.27 with $\alpha = 3$ and $\phi = 1.25$. Cyclic behavior is obtained by alternating the slip of the two blocks.

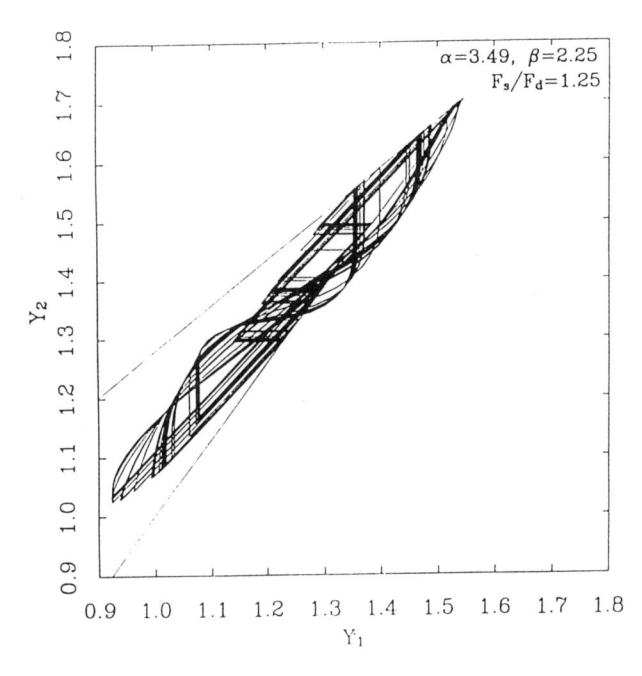

FIGURE 1.29. Behavior of an asymmetrical two-block model with $\beta = 2.25$, $\alpha = 3.49$, and $\phi = 1.25$. Chaotic behavior is obtained.

may lead to secondary failures and the process can cascade. The number of units moved is a measure of the size of the event. This is an example of a self-organized critical phenomena.

The first example of deterministic chaos was given by Lorenz (1963), who obtained numerical solutions to a set of three nonlinear total differential equations that exhibited deterministic chaos. Although other sets of equations also exhibit deterministic chaos, the Lorenz equations have been thoroughly studied, and these clearly illustrate characteristics of chaotic behavior (Sparrow, 1982). The Lorenz equations are

$$\frac{dA}{d\tau} = -\Pr A + \Pr B \tag{71}$$

$$\frac{dB}{d\tau} = -AC + rA - B \tag{72}$$

$$\frac{dC}{d\tau} = AB - bC \tag{73}$$

The Lorenz equations were derived to represent thermal convection in a fluid layer heated from below. The variables A, B, and C are coefficients in highly truncated Fourier expansions of the nondimensional stream function and temperature, and τ is the nondimensional time. The parameters are the Prandtl number Pr, the ratio of the Rayleigh number to the critical Rayleigh number r, and an aspect ratio parameter b.

A numerical solution of the Lorenz equations (71)–(73) for $\Pr = 10$, $r = 28$, and $b = 4/3$ projected on the A-B plane in the three-dimensional A-B-C phase space is given in Fig. 1.30. The unstable fixed points $A = \pm 6\sqrt{2}$, $B = \pm 6\sqrt{2}$, $C = 27$ are shown by crosses. The solution is chaotic because solutions infinitesimally close together diverge exponentially in time; fixed points behave as strange attractors. The dependence of the parameter B on time is shown in Fig. 1.31. We see that the cellular flow randomly jumps

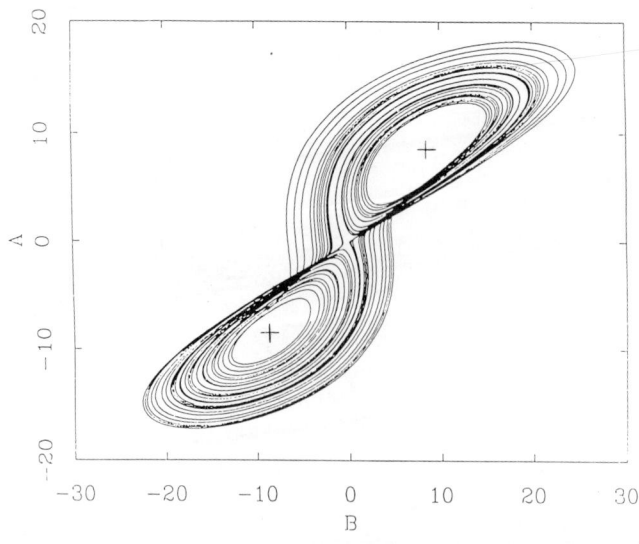

FIGURE 1.30. A solution of the Lorenz equations in Eqs. (71)–(73) with $\Pr = 10$, $r = 28$, and $b = 4/3$ in the phase plane A, B, C projected onto the A-B plane. Crosses indicate the unstable fixed points $A = \pm 6\sqrt{2}$, $B = \pm 6\sqrt{2}$, $C = 27$.

1. FRACTAL DISTRIBUTIONS IN GEOLOGY

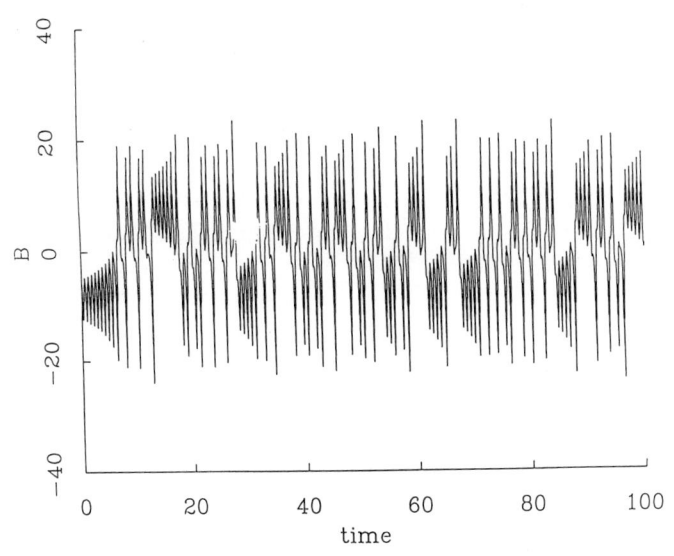

FIGURE 1.31. Dependence of the parameter B on time is given for the conditions shown in Fig. 1.30.

from clockwise to counterclockwise rotation and back. We emphasized that this random behavior is superimposed on the well-defined period of the cellular flow. These solutions exhibit fractal behavior in terms of Poincaré sections; distributions of crossing points on the A-B, A-C, and B-C planes are fractal.

Because of the very large viscosity, thermal convention in the Earth's mantle takes place at very large values of the Prandtl number—the ratio of kinematic viscosity to thermal diffusivity. For very large values of the Prandtl number, Lorenz equations do not yield chaotic solutions. This can be taken as evidence that mantle connection is not chaotic. However, using a higher order truncation, Stewart and Turcotte (1989) showed that chaotic solutions could be obtained for very large values of the Prandtl number.

Random reversals of the earth's magnetic field are very similar to reversals illustrated in Fig. 1.31. In fact a set of parameterized dynamo equations proposed by Rikitake (1958) has been shown to exhibit chaotic behavior similar to that of the Lorenz equations (Cook and Roberts, 1970).

1.12. CONCLUSIONS

Scale invariance has long been recognized as an important feature of many geological problems. A fractal distribution is the only statistical distribution that is scale-invariant. Thus it is not surprising that many geological and geophysical data sets are fractals. In this context, fractals are a method of empirically correlating a variety of quantitative observations. Fractal concepts also suggest how complex problems can be studied. Examples include

1. Length of a perimeter as a function of the length of the measuring stick
2. Perimeter length as a function of area
3. Surface area as a function of volume

4. Fraction of time increments containing an event as a function of the length of the time increment
5. Number of square boxes containing material as a function of box size
6. Number of cubic boxes containing material as a function of box size
7. Energy spectral density as a function of frequency or wave number

Sets of nonlinear total differential equations and recursion relations have been shown to yield chaotic solutions. Infinitesimal differences in initial conditions lead to order one differences in the evolving solutions. Many researchers now accept that the turbulent behavior of oceans and the atmosphere is chaotic. The conclusion is that the evolution of this system is not predictable in a deterministic sense but must be treated statistically.

For a variety of problems in geology, the basic applicable equations are not known. One example is the deformation of the Earth's crust. We do not know how to formulate fault displacements. Nevertheless we have ample evidence that the frequency-magnitude statistics of earthquakes is fractal, and we know that this involves scale-invariant behavior. Fractal behavior is also generally associated with chaotic solutions. It appears entirely reasonable to hypothesize that the deformation of the curst is a chaotic process. A conclusion is that seismicity too is not predictable in a deterministic sense but must be treated in a statistical manner.

ACKNOWLEDGMENT

The preparation of material in Chapter 1 was supported by the National Aeronautics and Space Administration under grant NAG 5-319. This is contribution 847 of the Department of Geological Sciences, Cornell University.

REFERENCES

Adler, P. M., *Int. J. Multiphase Flow* **11**, 91 (1985a).
———, *Int. J. Multiphase Flow* **11**, 213 (1985b).
———, *Int. J. Multiphase Flow* **11**, 241 (1985c).
———, *Int. J. Multiphase Flow* **11**, 853 (1985d).
———. *Phys. Fluids* **29**, 15 (1986).
Aki, K., in *Earthquake Prediction* (D. W. Simpson and P. G. Richards, eds.) (American Geophysical Union, Washington, DC, 1981), pp. 556–74.
Anderson, J. G., *Seismol. Soc. Am. Bull.* **76**, 273 (1986).
Anderson, J. G., and Luco, J. E., *Seismol. Soc. Am. Bull.* **73**, 471 (1983).
Aviles, C. A., Scholz, C. H., and Boatwright, J., *J. Geophys. Res.* **92**, 331 (1987).
Bak, P. and Tang, C., *J. Geophys. Res.* **94**, 15,635 (1989).
Barenblatt, G. I., Zhivago, A. V., Neprochnov, Y. P., and Ostrovskiy, A. A., *Oceanol.* **24**, 695 (1984).
Barton, C. C. and Hsieh, P. A., *Physical and Hydrologic-Flow Properties of Fractures*, 28th International Geological Congress Field Trip Guidebook T385 (American Geophysical Union, Washington, DC, 1989).
Bell, T. H., *Deep Sea Res.* **22**, 883 (1975).
———, *Deep Sea Res.* **26A**, 65 (1979).
Bennett, J. G., *J. Inst. Fuel* **10**, 22 (1936).
Berkson, J.M. and Matthews, J.E. *Acoustics and the Sea Bed* (N. G. Pace, ed.) (Bath University Press, Bath, England, 1983), pp. 215–23.
Biegel, R. L., Sammis, C. G., and Dieterich, J. H., *J. Struc. Geol.* **11**, 827 (1989).
Brown, S. R., *Geophys. Res. Lett.* **14**, 1095 (1987).

Brown, S. R. and Scholtz, C. H., *J. Geophys. Res.* **90**, 12,575 (1985).
Cargill, S. M., Root, D. H., and Bailey, E. H., *J. Int. Assoc. Math. Geol.* **12**, 489 (1980).
——, *Econ. Geol.* **76**, 1081 (1981).
Carlson, J. M., and Langer, J. S., *Phys. Rev.* **A40**, 6470 (1989).
Chase, C. G., *Geomorph.* **5**, 39 (1992).
Chepil, W. S. *Soil Sci.* **70**, 351 (1950)
Chinnery, M. A., *Bull. Seism. Soc. Am.* **69**, 757 (1979).
Clark, G. B., *Principles of Rock Fragmentation* (Wiley, New York, 1987).
Cook, A. E. and Roberts, P. H., *Proc. Camb. Philos. Soc.* **68**, 547–69 (1970).
Culling, W. E. H., *J. Geol.* **68**, 336–44 (1960).
——, *Trans. Jap. Geomorph. Un.* **7**, 221 (1986).
——, *Earth Surface Processes and Landforms* **13**, 619 (1988).
Culling, W. E. H. and Datko, M., *Earth Surface Processes and Landforms* **12**, 369 (1987).
Curl, R. L., *J. Int. Assoc. Math. Geol.* **18**, 765 (1986).
Curran, D. R., Shockey, D. A., Seaman, L., and Austin, M., *Impact and Explosion Cratering* (D. J. Roddy, R. O. Pepin, and R. B. Merrill, eds.) (Pergamon Press, 1977), pp. 1057–87.
Daccord, G. and Lenormand, R., *Nature* **325**, 41 (1987).
Devaney, R. L., in *The Science of Fractal Images* (H. O. Peitgen and D. Saupe, eds.) (Springer-Verlag, 1988), pp. 137–67.
Dewey, J. F., *Am. J. Sci.* **275A**, 260 (1975).
Drew, L. J., Schuenemeyer, J. H., and Bawiec, W. J., Estimation of the Future Rates of Oil and Gas Discoveries in the Western Gulf of Mexico, US Geol. Survey Prof. Paper, 1252, 1982.
Fluigeman, R. H., and Snow, R. S., *Pure Appl. Geophys.* **131**, 307 (1989).
Fox, C. G., *Pure Appl. Geophys.* **131**, 211 (1989).
Fox, C. G., and Hayes, D. E., *Rev. Geophys.* **23**, 1 (1985).
Fujiwara, A., Kamimoto, G., and Tsukamoto, A., *Icarus* **31**, 277 (1977).
Gilbert, L. E., *Pure Appl. Geophys.* **131**, 241 (1989).
Gilbert, L. E., and Malinverno, A., *Geophys. Res. Lett.* **15**, 1401 (1988).
Goodchild, M. F., *Math. Geol.* **12**, 85 (1980).
Grady, D. E., and Kipp, M. E., *Fracture Mechanics of Rock* (B. K. Atkinson, ed.) (Academic Press, London, 1987), pp. 429–75.
Gutenberg, B., and Richter, C. F., *Seismicity of the Earth and Associated Phenomena*, 2d ed. (Princeton University Press, Princeton, 1954).
Hack, J. T., Studies of Longitudinal Stream Profiles in Virginia and Maryland, US Geol. Survey Prof. Paper, 294B, 1957.
Harris, D. P., *Mineral Resources Appraisal* (Oxford University Press, 1984).
Hartmann, W. K., *Icarus* **10**, 201 (1960).
Hartmann, W. K., *Icarus* **10**, 201 (1969).
Hewett, T. A., Fractal Distributions of Reservoir Heterogeneity and Their Influence on Fluid Transport, Soc. Petrol. Eng. Paper 15386, 1986.
Hirata, T., *Pure Appl. Geophys.* **131**, 157 (1989a).
——, *J. Geophys. Res.* **94**, 7507 (1989b).
Hirata, T., Satoh, T., and Ito, K., *Geophys. J. Roy. Astro.* **90**, 369 (1987).
Huang, J. and Turcotte, D. L., *Earth Planet. Sci. Lett.* **91**, 223 (1988).
——, *J. Geophys. Res.* **94**, 7491 (1989).
——, *J. Opt. Soc. Am.* **A7**, 1124 (1990a).
——, *Geophys. Res. Lett.* **17**, 223 (1990b).
Hyndman, R. D. and Weichert, D. H., *Geophys. J. Roy. Astro.* **72**, 59 (1983).
Ito, K. and Matsukai, M., *J. Geophys. Res.* **95**, 6853 (1990).
Ivanhoe, L. F., *Oil Gas. J.* **6**, 154 (1976).
Kanamori, H. and Anderson, D. L., *Bull. Seism. Soc. Am.* **65**, 1073 (1975).
Katz, A. J. and Thompson, A. H., *Phys. Rev. Lett.* **54**, 1325 (1985).
King, G., *Pure Appl. Geophys.* **121**, 761 (1983).
King, G. C. P., *Pageoph* **124**, 567 (1986).
King, G. C. P., Stein, R. S., and Rundle, J. B., *J. Geophys. Res.* **93**, 13,307 (1988).
Krohn, C. E., *J. Geophys. Res.* **93**, 3286 (1988a).

Krohn, C. E., *J. Geophys. Res.* **93**, 3286 (1988a).
———, *J. Geophys. Res.* **93**, 3297 (1988b).
Krohn, C. E. and Thompson, A. H., *Phys. Rev.* **B33**, 6366 (1986).
Lange, M. A., Ahrens, T. J., and Boslough, M. B., *Icarus* **58**, 383 (1984).
Laverty, M., *Earth Surface Proc. Landforms* **12**, 475 (1987).
Lorenz, E. N., *J. Atmos. Sci.* **20**, 130 (1963).
Main, I. G. and Burton, P. W., *Seismol. Soc. Am. Bull.* **76**, 297 (1986).
Malinverno, A., *Pure Appl. Geophys.* **131**, 139 (1989).
Mandelbrot, B. B., *Science* **156**, 636, 1967.
———, *Proc. Nat. Acad. Sci. (USA)* **72**, 3825 (1975).
———, *The Fractal Geometry of Nature* (Freeman, San Francisco, 1982).
———, *Phys. Scripta* **32**, 257 (1985).
Mareschal, J. C., *Pure Appl. Geophys.* **131**, 197 (1989).
May, R. M., *Nature* **261**, 459 (1976).
McClelland, S. T., Summers, M., Nielson, E., and Stein, T. C., eds., *Global Volcanism 1975–1985* (Prentice Hall, Englewood Cliffs, NJ, 1989).
Molnar, P., *Seismol. Soc. Am. Bull.* **69**, 115 (1979).
Newman, W. I. and Turcotte, D. L., *Geophys. J. I.*, **100**, 433 (1990).
Nicolis, C. and Nicolis, G., *Nature* **311**, 529 (1984).
Nolte, D. D., Pyrak-Nolte, L. J. and Cook N. G. W., *Pure Appl. Geophys.* **131**, 111 (1989).
Okubo, P. G. and Aki, A., *J. Geophys. Res.* **92**, 345 (1987).
Pfeifer, P. and Obert, M., *The Fractal Approach to Heterogeneous Chemistry* (D. Avnir, ed.) (Wiley, Chichester, England, 1989), p. 11–43.
Plotnick, R. E., *J. Geol.* **94**, 885 (1986).
Rapp, R. H., *Geophys. J. I.* **99**, 449 (1989).
Rikitake, T., *Proc. Cambr. Philos. Soc.* **54**, 89 (1958).
Ross, B., *Water Resour. Res.* **22**, 823 (1986).
Sadovskiy, M. A., Golubeva, T. V., Pisarenko, V. F., and Shnirman, M. G., *Phys. Solid Earth* **20**, 87 (1985).
Sammis, C. G. and Biegel, R. L., *Pure Appl. Geophys.* **131**, 255 (1989).
Sammis, C., King, G., and Biegel, R., *Pure Appl. Geophys.* **125**, 777 (1987).
Sammis, C. G., Osborne, R. H., Anderson, J. L., Banerdt, M., and White, P., *Pure Appl. Geophys.* **123**, 53 (1986).
Scholz, C. H. and Aviles, C. A., in *Earthquake Source Mechanisms* (S. Das, J. Boatwright, and C. H. Scholz, eds.), Geophys. Mono. 37 (American Geophysical Union, Washington, DC, 1986), pp. 147–155.
Schoutens, J. E., Empirical Analysis of Nuclear and High-Explosive Cratering and Ejecta: Nuclear Geoplosics Sourcebook, vol. 55, pt. 2, sec. 4, Rep. DNA 65 O1H-4-2 (Def. Nucl. Agency, Bethesda, MD, 1979).
Sieh, K. E., *J. Geophys. Res.* **83**, 3907 (1978).
Singh, S. K., Rodriguez, M., and Esteva, L., *Bull. Seism. Soc. Am.* **73**, 1779 (1983).
Smalley, R. F., Chatelain, J. L., Turcotte, D. L. and Prevot, R., *Bull. Seism. Soc. Am.* **77**, 1368 (1987).
Smith, S. W., *Geophys. Res. Lett.* **3**, 351 (1976).
Snow, R. S., *Pure Appl. Geophys.* **131**, 99 (1989).
Sparrow, C., *The Lorenz Equations: Bifurcations, Chaos and Strange Attractors* (Springer-Verlag, New York, 1982).
Stewart, C. and Turcotte, D. L., *J. Geophys. Res.* **94**, 13,707 (1989).
Thompson, J. M. T. and Stewart, H. B., *Nonlinear Dynamics and Chaos* (Wiley, Chichester, England, 1986).
Turcotte, D. L., *J. Geophys. Res.* **91**, 1921 (1986a).
———, *Econ. Geol.* **81**, 1528 (1986b).
———, *Tectonophys.* **132**, 261 (1986c).
———, *Pure Appl. Geophys.* **131**, 171 (1989a).
———, *Tectonophys.* **167**, 171 (1989b).
Voss, R. F., in *Scaling Phenomena in Disordered Systems* (R. Pynn and A. Skjeltorp, eds.) (Plenum Press, NY, 1985), pp. 1–11.
———, Random fractal forgeries: In *Fundamental Algorithms for Computer Graphics* (R. A. Earnshaw, ed.), NATO ASI Series, vol. F17 (Springer-Verlag, Berlin, 1985), pp. 805–35.
———, in *The Science of Fractal Images* (H. O. Peitgen and D. Saupe, eds.) (Springer-Verlag, New York, 1988), pp. 21–70.
Youngs, R. R. and Coppersmith, K. J., *Bull. Seism. Soc. Am.* **75**, 939 (1985).

2

Some-Long-Run Properties of Geophysical Records

B. B. Mandelbrot and J. R. Wallis

PRESENTATION

By preparing this book Chris Barton and Paul La Pointe have earned the gratitude of all geologists and students of fractals. I continue to belong to this second group, and Chris and Paul clearly have put me in a very special debt to them.

How to express my thanks? They have suggested that I contribute a chapter to their book, an idea I found flattering but perhaps not completely appropriate. It would have been a great opportunity to write about long-run dependence and the statistical technique of R/S analysis, a topic that has fascinated me since around 1970, wide attention was drawn to it by Feder (1988). Historically the acceptance of long-run dependence had been an essential step toward formulating fractal geometry (Mandelbrot, 1982), and it has also helped give birth to an active branch of pure probability theory (Eberlein and Taqqu, 1986). From the scientists' viewpoint however, this specific chapter of fractal geometry was eventually submerged in a more general theory. As a result, it cries out for a fresh treatment.

In any event, I did not find the time to write a chapter. As an alternative, we all agreed that this book should reproduce a slightly edited version of a work that James R. Wallis and I published in *Water Resources Research*. This text may have escaped the attention of the broader community of geologists. In the process of reprinting, a few selected references later than 1969 have been added, and the bottom portions of all illustrations of R/S have been removed to take account of remarks in Taqqu (1970).

The original *Water Resources Research* paper was part of a series in that journal by the same authors, or by me alone, all of which are recommended to the reader. Abstracts of other papers in that series are appended to the present reprint (each followed by errata).

B. B. Mandelbrot

B. B. Mandelbrot • Yale University, New Haven, Connecticut, and IBM T. J. Watson Research Center, Yorktown Heights, New York.   *J. R. Wallis* • IBM T. J. Watson Research Center, Yorktown Heights, New York.

Fractals in the Earth Sciences, edited by Christopher C. Barton and Paul R. La Pointe. Plenum Press, New York, 1995.

2.1. INTRODUCTION

Chapter 2 presents and comments on some long-run properties of geophysical records and in particular tests the validity of an empirical law Harold Edwin Hurst discovered during his preliminary studies for the future Aswan High dam (Hurst, 1951, 1956; Hurst and others, 1965). We show that this law must be amended, tightened, and hedged, but its essential claim is confirmed.

Hurst's law concerns the dependence of the rescaled bridge range on the lag δ. It claims that $R(t,\delta)/S(t,\delta)$ takes the form $F\delta^H$, where F is a prefactor and the exponent H is constant, typically differing from 0.5. In this expression, $R(t,\delta)$ is the range of variation between the minimum and maximum values as a function of a particular starting time t and lag δ, while $S(t,\delta)$ is the standard deviation over the same t and δ.

Table 2.1 shows that in most cases $H \neq 0.5$. This inequality has striking consequences. If the records in question were generated by a random process such that observations far removed in time can be considered independent, $R(t,\delta)/S(t,\delta)$ would become asymptotically proportional to $\delta^{0.5}$. This means that Hurst's law would have to "break" for sufficiently large lags, but no such break has ever been observed. According to Mandelbrot and Wallis (1968), and the last section of Chapter 2, it follows that for all practical purposes, geophysical records must be considered to have an infinite span of statistical interdependence.

Fractional Gaussian noises discussed in Mandelbrot and Van Ness (1968) and Mandelbrot and Wallis (1969a), are a family of random processes specifically designed to satisfy Hurst's law. Comparing figures in Chapter 2 with those of Mandelbrot and Wallis (1969a) confirm in our opinion that geophysical records can be modeled by fractional noises. Analysis techniques are not discussed at length because they are found in Mandelbrot and Wallis (1969a), which the interested reader can consult. Definitions relative to spectra, being standard, are not repeated here; however the method of constructing the ratio R/S and pox diagram notion are described in detail.

Extreme caution is required when using R/S to analyze records where strong cycles are present. This circumstance is documented later, in particular in the case of sunspot numbers. As a result, R/S may not be the most suitable statistic when strong cycles are present. The marginal distribution of the record on the contrary has little effect on the behavior of R/S.

The records studied in Chapter 2 fall into three subject matter categories:

Recent hydrological records of streamflow, rainfall, and temperature measurements; the longest among such records are records of annual maximum and minimum stage for the Nile River.

Fossil hydrological records, such as tree ring indices, varve thickness, and other records of geological deposits; the importance of such records in hydrology are discussed in the last section of Chapter 2.

Other miscellaneous records, such as sunspot numbers, earthquake frequencies, and directions of river meanders.

Figures in Chapter 2 represent a small fraction of our present files. We have withheld many figures that would be purely repetitive and a few that require extensive discussion. The remainder of Chapter 2 is devoted to a table of rough estimates of the H-coefficient (the basic parameter of Hurst's law) and to a text devoted to three subjects: the graphical devices

TABLE 2.1. Values of H and of the Third and Fourth Moments for Some Geophysical Data[a]

Description	H	Third	Fourth
Varve Data from De Geer (1940) Swedish Time Scale			
Haileybury, Canada, +310 to −369	0.80	0.74	4.41
Timiskaming, Canada, +1321 to −588	0.96	1.03	6.33
Sirapsbacken, Sweden, +576 to +257	≈1.00	0.95	3.52
Degeron, Sweden, +499 to 0	0.97	1.10	3.65
Omnas, Sweden, +1399 to +1162	0.65	3.39	23.01
Resele, Sweden, +1399 to +1132	0.67	4.00	26.02
Hammerstrand, Sweden, +2000 to +1767	0.85	2.10	9.04
Ragunda, Sweden, +1933 to +1800	0.99	1.59	5.21
Lago Corintos, Argentina, −801 to −1168	0.89	0.85	3.58
Biano, Himalaya Mountains, −1180 to −1279	0.50	1.84	7.82
Sesko, Himalaya Mountains, −1293 to −1374	0.99	0.31	2.20
Enderit River, E. Africa, −406 to −546	0.91	1.07	4.10
Schulman (1956) Tree ring indices			
Table 28, Douglas Fir and Ponderosa Pine, Fraser River, British Columbia, 1420–1944	0.70	0.19	2.84
Table 30, Douglas Fir, Jasper, Alberta. 1537–1948	0.75	0.60	3.34
Table 31, Douglas Fir, Banff, Alberta, 1460–1950	0.65	0.32	2.76
Table 33, Douglas Fir and Ponderosa Pine, Middle Columbia River Basin, 1650–1942	0.75	0.10	3.20
Table 36, Douglas Fir, Snake River Basin, 1282–1950	0.77	0.24	3.60
Table 38, Limber Pine, Snake River Basin. 1550–1951	0.60	−0.06	3.47
Table 40, Douglas Fir, Upper Missouri River Basin, 1175–1950	≈0.50	0.33	3.14
Table 41, Limber Pine, Upper Missouri River Basin, 978–1950	0.63	0.29	3.71
Table 43, Douglas Fir, North Platte River Basin, 1336–1946	0.70	0.65	3.88
Table 44, Douglas Fir, South Platte River Basin, 1425–1944	0.62	0.21	2.77
Table 45, Douglas Fir, Arkansas River Basin, 1427–1950	0.66	0.03	2.86

(*Continued*)

TABLE 2.1. (*Continued*)

Description	H	Third	Fourth
Table 49, Mixed Species, three-year means, Colorado River Basin, 70 B.C. to 1949 A.D.	0.55	0.25	3.07
Table 50, Douglas Fir, Colorado River Basin, 1450–1950	0.64	−0.15	2.82
Table 52, Pinyon Pine, Colorado River Basin, 1320–1948	0.68	−0.43	3.32
Table 65, Ponderosa Pine, Upper Gila River, 1603–1930	0.69	−0.07	3.25
Table 66, Douglas Fir, Southern Arizona, 1414–1950	0.69	−0.03	3.11
Table 70, Douglas Fir, Upper Rio Grande, 1375–1951	0.65	0.21	2.88
Table 71, Pinyon Pine, Upper Rio Grande, 1356–1951	0.68	−0.23	2.76
Table 72, Douglas Fir, Middle Rio Grande (Guadalupe), 1650–1941	0.59	0.24	2.87
Table 73, Douglas Fir, Middle Rio Grande (Big Bend), 1645–1945	0.71	0.59	3.13
Table 75, Ponderosa Pine, southeastern Oregon, 1453–1931	0.58	0.04	3.27
Table 76, Ponderosa Pine, northeastern California, 1485–1931	0.78	0.11	3.12
Table 77, Jeffrey Pine, east central California, 1353–1941	0.72	−0.15	2.93
Table 78, Big-cone Spruce, southern California, 1385–1950	0.56	0.11	3.62
Table 79, Ponderosa Pine, southern California, 1350–1931	0.72	−0.40	2.89
Table 80, Douglas Fir, west central Mexico, 1640–1943	0.86	−0.10	2.40
Table 85, Cipres at Cerro Leon, Argentina, 1572–1949	0.91	0.41	3.41
Data from Statistical History of the United States (1965)			
Annual Precipitation			
Albany, New York, 1826–1962	0.87	0.56	3.46
Baltimore, Maryland, 1817–1962	0.75	−0.06	2.64
Charleston, South Carolina, 1832–1962	0.89	0.65	3.53
New Haven, Connecticut, 1873–1962	0.73	0.30	2.33
New York, New York, 1826–1962	0.65	0.59	2.89
Philadelphia, Pennsylvania, 1820–1962	0.81	0.21	2.85
San Francisco, California, 1850–1963	0.64	0.45	3.11
St. Louis, Missouri, 1857–1962	0.64	0.63	4.20
St. Paul, Minnesota, 1837–1962	0.67	0.49	5.15
Data from Ruth B. Simon (Personal Communication)			
Weekly derby earthquake frequencies, April 1962–June 1967	0.93	3.99	24.45

(*Continued*)

2. SOME LONG-RUN PROPERTIES OF GEOPHYSICAL RECORDS

TABLE 2.1. (Continued)

Description			H	Third	Fourth
Munro (1948) Sunspot data (1749–1948)					
Monthly sunspot frequency 1749–1948			0.96	1.04	3.88
Data from V.M. Yevdjevich (1963)					
Gota River near Sjotrop-Vanersburg, Sweden, 1807–1957		U^b	≈0.50	−0.06	2.35
		Y^c	≈0.50	0.42	2.94
Neumunas River at Smalininkai, Lithuania, 1811–1943		U^b	0.61	0.47	3.15
		Y^c	0.48	0.61	3.31
Rhine River near Basle, Switzerland, 1807–1957		U^b	≈0.50	0.14	2.80
		Y^c	≈0.50	0.23	2.89
Danube River at Orshawa, Romania, 1837–1957		U^b	≈0.50	0.27	2.26
		Y^c	≈0.50	0.22	2.53
Mississippi River near St. Louis, Missouri, 1861–1957		U^b	0.79	0.29	2.75
		Y^c	0.68	0.18	2.45
St. Lawrence River near Ogdensburg, New York, 1860–1957		U^b	0.98	−0.26	2.70
		Y^c	0.69	0.14	2.70
Professor J. C. Mann, Data from Paleozoic Era Sediments (Personal Communication)					
Wolfcampian Section, Kansas	Thickness of beds		0.75	4.60	32.27
	Lithology of beds		0.71	0.25	2.97
Virginian–Desmoinesian Section, Superior, Arizona	Lithology of beds		0.55	0.78	3.27
	Bedding type		0.67	0.19	2.21
Missourian–Atokan Section, Honacker Trail, Utah	Thickness of beds		0.70	3.51	24.81
	Lithology of beds		0.61	−0.05	2.06
	Bedding type		0.58	0.47	2.11
Data from J. de Beauregard (1968)					
Rhine River, monthly flows, 1808–1966			0.55	0.65	3.01
Loire River, monthly flows, 1863–1966			0.69	1.53	5.47
Nile River Data, Prince Omar Toussoun (1925)					
Annual maximums, 622–1469			0.84	−0.86	7.04
Annual minimums, 622–1469			0.91	0.32	3.50
Data from J. G. Speight (Personal Communication)					
Moorabool River meander azimuths			0.73	0.07	2.82

[a]The first and second moments were, in all cases, normalized to be 0 and 1, respectively.
[b]Unadjusted for overyear carryover.
[c]Adjusted for overyear carryover.

we call pox diagrams, pitfalls in the statistical estimation of H, and the significance of Hurst's findings in geophysics.

2.2. R/S AND POX DIAGRAMS

The following definition of the rescaled bridge range for an empirical record parallels the definition of the rescaled bridge range for a random process as given in Mandelbrot and Wallis (1969a).

Let $X(t)$ be a record containing T readings uniformly spaced in time from $t = 1$ to $t = T$, and let $\int X(t)$ designate $\Sigma_{u=1}^{t} X(u)$. Thus

$$\delta^{-1} \int X(\delta)$$

is the average of the first δ readings, and

$$\delta^{-1}[\int X(t + \delta) - \int X(t)]$$

is the average of readings within the subrecord from time $t + 1$ to time $t + \delta$.

The $S^2(t,\delta)$ is defined as the sample variance of the subrecord from time $t + 1$ to $t + \delta$, namely,

$$S^2(t,\delta) = \delta^{-1} \sum_{u=t+1}^{t+\delta} X^2(u) - \left[\delta^{-1} \sum_{u=t+1}^{t+\delta} X(u)\right]^2 \tag{1}$$

The bridge range $R(t,\delta)$, illustrated in Fig. 2.1, is defined as follows

$$\begin{aligned} R(t,\delta) = &\max_{0 \leq u \leq \delta} \{\int X(t + u) - \int X(t) - (u/\delta)[\int X(t + \delta) - \int X(t)]\} \\ &- \min_{0 \leq u \leq \delta} \{\int X(t + u) - \int X(t) - (u/\delta)[\int X(t + \delta) - \int X(t)]\} \end{aligned} \tag{2}$$

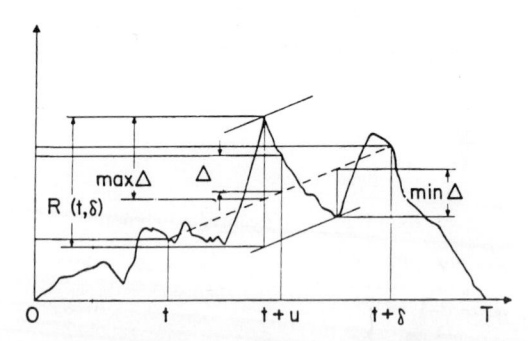

FIGURE 2.1. Construction of the sample bridge range $R(t,\delta)$ for a given sequence $X(t)$. To make the graph more compact and more legible, we have plotted (*bold line*), the bridge function of the whole sample, defined as $\Delta(t,0,T) = \int X(t) - (t/T)\int X(T)$. Replacing $\int X(t)$ with $\Delta(t,0,T)$ does not affect the values of $\Delta(u)$ or $R(t,\delta)$, two quantities defined below. Moreover, since empirical records are necessarily taken in discrete time, the function Δ should have been drawn as a series of points, but it was drawn as a line for the sake of clarity. The bridge function of the sample from $t + 1$ to $t + \delta$ is defined as

$$\Delta(u,t,\delta) = \int X(t + u) - \int X(t) - (u/\delta)[\int X(t + \delta) - \int X(t)]$$

and denoted by Δ in Fig. 2.1. The sample bridge range is defined as

$$R(t,\delta) = \max_{0 \leq u \leq \delta} \Delta(u,t,\delta) - \min_{0 \leq u \leq \delta} \Delta(u,t,\delta)$$

2. SOME LONG-RUN PROPERTIES OF GEOPHYSICAL RECORDS

The ratio $R(t,\delta)/S(t,\delta)$ is then called a rescaled bridge range. Given the lag δ, the starting t could assume $T - \delta + 1$ different values, but the resulting mass of values of $R(t,\delta)/S(t,\delta)$ is both redundant and unmanageable. The data were replaced by averages, but the resulting reduction was in our opinion too drastic. The middle ground we chose is to plot the values of $R(t,\delta)/S(t,\delta)$ in the form of what we call a pox diagram.

To construct a pox diagram, we select a limited sequence of equally spaced values for log δ and marked these on the horizontal axis of double logarithmic paper. For each δ, we select a limited number of starting points t, then plot on the vertical axis the corresponding values of $R(t,\delta)/S(t,\delta)$ so obtained. Thus above every marked value of δ, several points (enlarged into + signs) are aligned. For each δ, the sample average of the quantities $R(t,\delta)/S(t,\delta)$ is marked by a little square box. The line connecting the boxes weaves through the pox diagram.

The trend line of the pox diagram and Hurst's law. An empirical record is said to satisfy Hurst's law if, except perhaps for very small and very large values of δ, the pox diagram of $R(t,\delta)/S(t,\delta)$ is tightly aligned along a straight trendline whose slope is designated by H. Originally Mandelbrot (1965) chose the letter H to honor Hurst, but since the 1970s and 1980s, H is often related to the Hölder exponent. Theory shows that in all cases, $0 < H < 1$, and in geophysics $H > \frac{1}{2}$. The value $H = 0.5$ has a special significance because it suggests that observations sufficiently distant from each other in time are statistically independent.

Mandelbrot and Wallis (1969b) discuss the concepts of tight alignment along a trend line more precisely and define several variants (of increasingly demanding scope) of the δ^H law.

2.3. PITFALLS IN THE GRAPHIC ESTIMATION OF H

Until better estimation procedures are developed, the exponent H must be estimated graphically from something like the pox diagram. The following pitfalls in this approach exist.

2.3.1. The Initial Transient

For small values of δ, the scatter of the values of $R(t,\delta)/S(t,\delta)$ is large, and various irrelevant influences are felt. Consequently even if values of R/S for small δ are known, they should be disregarded when drawing conclusions from the pox diagram.

2.3.2. The Final Tightening of the Pox Diagram

Let the total sample run from $t = 1$ to $t = T$. For small δ, the starting points t can be so selected that the corresponding subsamples from $t + 1$ to $t + \delta$ remain nonoverlapping, whereas the number of subsamples remains large. For larger δ, the situation is not so ideal. We have the choice between very few nonoverlapping subsamples and a larger number of overlapping subsamples. Either choice narrows the pox diagram; the former does so because few points are strongly correlated and differ little from each other.

2.3.3. Graphically Fitting a Straight Trend Line

The fact that pox diagrams exhibit an initial transient and final tightening implies that when fitting a straight trend line, small and large values of δ should be given less weight than values removed from both $t = 1$ and $\delta = T$.

2.3.4. A Warning against an Error in Hurst's Initial Statement of His Law

The statement we call Hurst's law is a significant innovation. Hurst (1951) claimed that the pox diagram of R/S has a trend line going through the point of abscissa $\log 2$ and the ordinate $\log 1 = 0$, which means that Hurst took the prefactor and the exponent to be related by $F = 2^{-H}$. Hurst's claims led to the following estimation procedure: In cases where $R(t,\delta)/S(t,\delta)$ was known for a single sample from $t + 1$ to $t + \delta$, he estimated H to be

$$K(t,\delta) = \frac{\log [R(t,\delta)/S(t,\delta)]}{\log \delta - \log 2} \qquad (3)$$

In cases where $R(t,\delta)/S(t,\delta)$ was known for more than one sample, he averaged the values of the function $K(t,\delta)$ corresponding to all the points of coordinates of δ and R/S.

Actual pox diagrams have a straight trend line of slope H that fails to pass through the point of abscissa $\log 2$ and ordinate 0. Hurst's average K is thus a very poor estimate of the slope H: It tends to be too low when $H > 0.72$ and too high when $H < 0.72$. As a result, the trend line and Hurst's method may both suggest identical typical values for H, but Hurst's method greatly underestimates the variability of H around its typical value. The reader can verify the preceding statements in Figs. 2–7 of Mandelbrot and Wallis (1969a, part 2). For such processes as the stuttering noises of Mandelbrot and Wallis (1969b), the results of Hurst's method are entirely meaningless. To avoid confusion, we labeled the actual slope with the letter H, which differs from the letter K used by Hurst.

2.4. DETERMINING H FOR A SPECIFIC HYDROLOGICAL PROJECT

We found great variability in the values of H measured on R/S pox diagrams of actual records. To explain such variability may prove difficult. When designing a specific hydrological project, we must in many cases be content with an intelligent but imprecise estimated value of H. When data needed to determine H for a certain river are unavailable, related hydrological records may be usable. Regardless of the data used, H is far from 0.5, except for such obviously well-behaved rivers as the Rhine, which is analyzed in Fig. 2.9. However the difficulty of estimating H is an extremely poor justification for an indiscriminate use of short-memory models that imply that $H = 0.5$.

2.4.1. A Warning Concerning Sunspot Records and Other Cases with a Strong Periodic Element

To see the effect of strong periodic elements on a pox diagram of R/S, examine Fig. 2.2 relative to the Wolf sunspot numbers; also examine Fig. 2.3, which enlarges the portion of Fig. 2.2 near a lag of $\delta = 11$ years $= 132$ months, the wavelength of the well-known periodic element in sunspot numbers. We clearly see (1) a contraction in the scatter of the values of $\log [R(t,\delta)/S(t,\delta)]$ and (2) a break in the trend line, shaped like a horizontally laid reverse S. The partial trend lines before and after the 11-year break have equal slopes, with H much exceeding 0.5.

The contraction and break are both characteristic consequences of the presence of a strong periodic element. In addition, we observe other breaks, corresponding to each subharmonic of $\delta = 11$ years. Since however H is very large in the sunspot case, the other

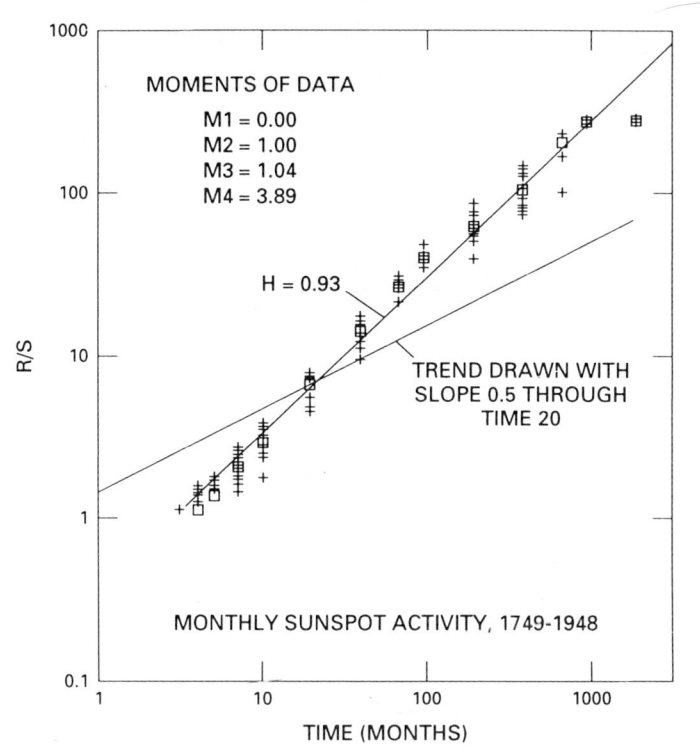

2. SOME LONG-RUN PROPERTIES OF GEOPHYSICAL RECORDS

FIGURE 2.2. Pox diagram of $\log R/S$ versus $\log \delta$ for the monthly Wolf numbers of sunspot activity, as reported by Munro (1948) (skewness = 1.04; kurtosis = 3.89). In this and later figures relative to R/S, time means time lag. In Fig. 2.2, the slope of the trend line greatly exceeds 0.5. This new observation is much more apparent than the well-known presence in sunspot numbers of a cycle of 11 years = 132 months and perhaps of a cycle of 80 years. Each of these cycles manifests itself mainly by a narrowing of the pox diagram followed by a break shaped like a horizontally laid letter S. Before drawing any conclusion from the high apparent value of H for sunspots, the reader should consult Mandelbrot and Wallis (1969b).

This pox diagram was constructed as follows: Dots (+) correspond to values of the lag δ restricted to the sequence 3, 4, 5, 7, 10, 20, 40, 70, 100, 200, 400, 700, 1000, 2000, 4000, 7000, and 9000. For every δ satisfying $\delta < 500$, we plot 14 dots, corresponding to values of t equal to 1, 100, . . . , 1400. For every δ satisfying $\delta > 500$, t was made successively equal to 1000, 2000, up to either 8000 or $(T - \delta) + 1$, whichever is smaller.

breaks are weak. Hence the presence of a periodic element complicates the picture, but it does not hide the Hurst phenomenon. Note also that the 80-year cycle reported by Willett (1964), which does not appear on the spectral pox diagram of the sunspot data (see Fig. 2.4), has been detected at the extremity of Fig. 2.2.

In other examples, the basic one-year cycle and its subharmonics overwhelm the long-run interdependence expressed by Hurst's law, and the value of H cannot be inferred graphically from the pox diagram of R/S built with raw data. The presence of several periodic elements of different wavelengths makes the situation even more complicated. For example, let several cyclic components be added to a fractional noise with a medium-sized value of H. The breaks corresponding to each period and its subharmonics merge together and yield a pox diagram with a much reduced apparent slope. The slope may fall below 0.5, and the diagram may even appear horizontal. For enormous values of δ, a trend line of slope

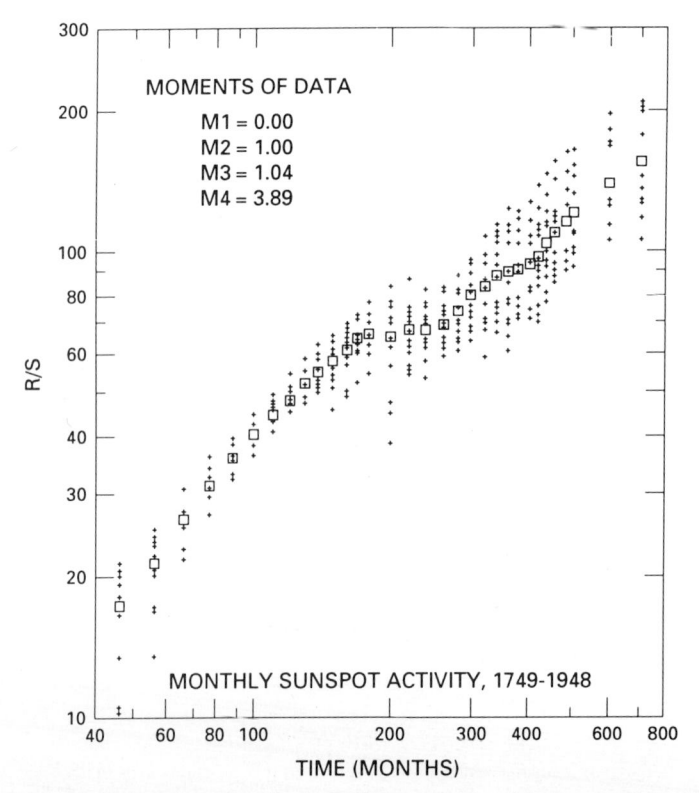

FIGURE 2.3. Detail of portion of Fig. 2.2. The preceding caption describes how we observed, near the lag corresponding to the 11-year cycle (132 month) cycle, a narrowing followed by break, which is characteristic of strongly cyclic phenomena. It was imperative to blow up this narrowing to examine details. In the resulting pox diagram, values of δ range from 50–180, uniformly spaced by increments of 10, then from 180–500 uniformly spaced by increments of 20, with two additional values at 600 and 700. For each δ, values of t were uniformly spaced with increments of 159. Several breaks are clearly visible here.

H ultimately reestablishes itself, but as we have argued in many other contexts, an asymptotic behavior is of little use either in engineering or science unless it is very rapidly attained. Examples where strong subharmonics coexist with multiple cycles require lengthy discussion, and these are not presented in Chapter 2.

In summary, the pox diagrams of R/S are less useful in the presence of strong periodic elements. Fortunately most periodic elements in natural records are fairly obvious, and these can be removed; the search for hidden periodicities is usually fruitless. In the following section, draw R/S pox diagrams of the corrected records remaining after periodic elements are removed.

No clear cut cyclic effect is present in data analyzed in Figs. 2.5–2.16.

2.4.2. Significance of Hurst's Law in Geophysics

Among the classical dicta of the philosophy of science is Descartes's prescription "to divide every difficulty into portions that are easier to tackle than the whole." This advice

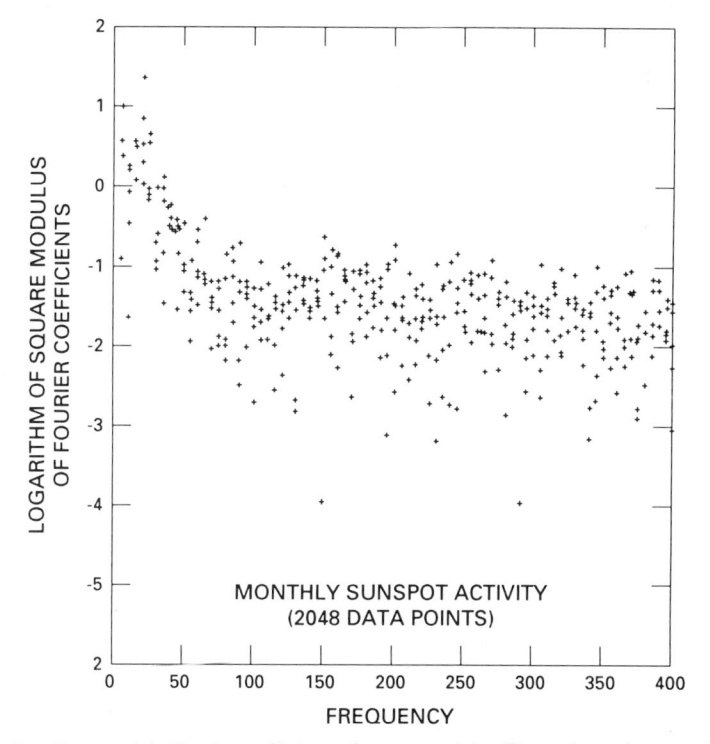

2. SOME LONG-RUN PROPERTIES OF GEOPHYSICAL RECORDS

FIGURE 2.4. Pox diagram of the Fourier coefficients of sunspot activity. The total sample was reduced to 2048 pieces of data to enable us to use a Fast Fourier Transform computer program. The frequency is measured in number of cycles per 2048 pieces of data. Frequencies are grouped together in fives, with the hth group made of the frequencies $5h$, $5h$ --1, $5h$ --2, $5h$ --3, and $5h$ --4, where h takes all positive integer values. Values of the squared Fourier moduli corresponding to frequencies in a group are all plotted above the abscissa, corresponding to the highest frequency in such a group, namely, $5h$. This procedure introduces a local smoothing that eliminates many apparent cycles and makes it possible to some extent to avoid spectral window processing of Fourier coefficients (see Mandelbrot and Wallis, 1969a, part 2). Curiously the well-known 11-year period of sunspot numbers does not appear as a sharp peak in Fourier coefficients but continues the tendency of the diagram to rise sharply as k decreases. Using spectral windows on the same data, as in Granger and Hatanaka (1964, p. 66), leads to essentially the same conclusion reached by this pox diagram.

has been extraordinarily useful in classical physics, because boundaries between distinct subfields of physics are not arbitrary. They are intrinsic in the sense that phenomena in different fields interfere little with each other and each field can be studied alone before a description of the mutual interactions is attempted.

Subdivision into fields is also practiced outside of classical physics. Consider, for example, atmospheric science. Students of turbulence examine fluctuations with time scales of the order of seconds or minutes; meterologists concentrate on days or weeks; specialists whom we may call macrometerologists concentrate on periods of a few years; climatologists deal with centuries; and finally paleoclimatologists are left to deal with all longer time scales. The science that supports hydrological engineering falls somewhere between macrometerology and climatology.

The question then arises whether or not this division of labor is intrinsic to the subject

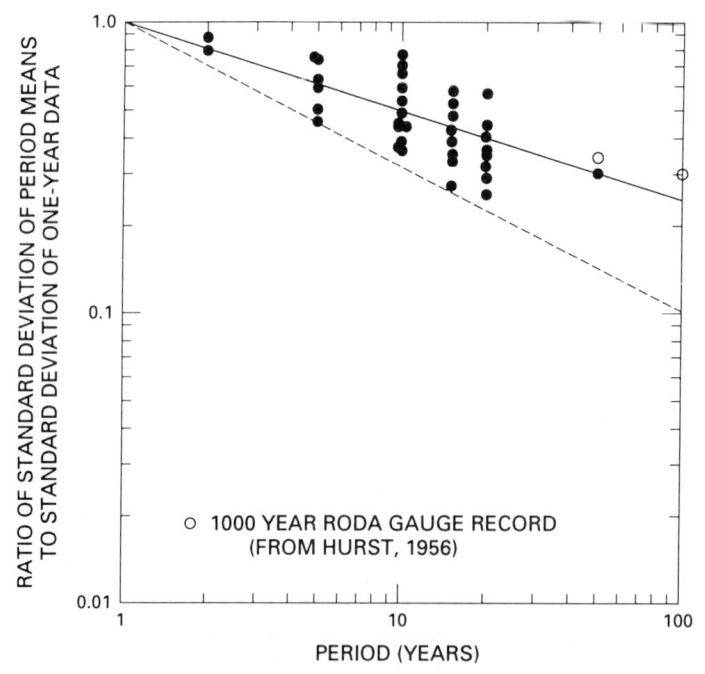

FIGURE 2.5. A superposition of several variance–time diagrams. To draw the variance–time diagram of a record, select a sequence of lags δ (plotted as abscissas). For each δ, average δ successive pieces of data in a record, then compute the variance of these δ-year averages around the overall average of each sample. The quantity plotted on the ordinate is the ratio of this variance for δ plotted on the abscissa and δ = 1. Several variance–time diagrams are superposed on Fig. 1.5, which does not represent new calculations but is reproduced with the author's permission from Langbien (1956), who used miscellaneous records from Hurst (1956). Superposition has eliminated much of the information in the data, and Fig. 1.5 shows only that every one of the superposed variance–time diagrams is located above the dashed line, which has slope $H - 0.5$. That line represents the behavior to be expected if observations placed at very distant instants of time were independent. The variance decreases less rapidly than expected because the R/S pox diagram has a trendline of slope $H > 0.5$ and the Fourier pox diagram rises steeply for low frequencies. The last two points to the right are for Hurst's data concerning the 1000-year record of Rhoda gauge readings.

matter. In our opinion, it is not, in the sense that it does not seem possible when studying a field in the preceding list to neglect its interactions with others. We fear therefore that dividing the study of fluctuations into distinct fields is mainly a matter of convenient labeling, but it is hardly more meaningful than either classifying bits of rock into sand, pebbles, stones, and boulders or enclosed water-covered areas into puddles, ponds, lakes, and seas.

Take the example of macrometerology and climatology. Although they can be defined formally as the sciences of fluctuations on time scales respectively smaller and longer than one lifetime, macrometerology and climatology should not be considered as really distinct unless and until they have actually been shown to be ruled by processes that can be separated from each other by actual experiment. In particular, there should exist at least one duration λ for a record, with λ of the order of magnitude of one lifetime, that is both long enough for macrometerological fluctuations to be averaged out and short enough to keep

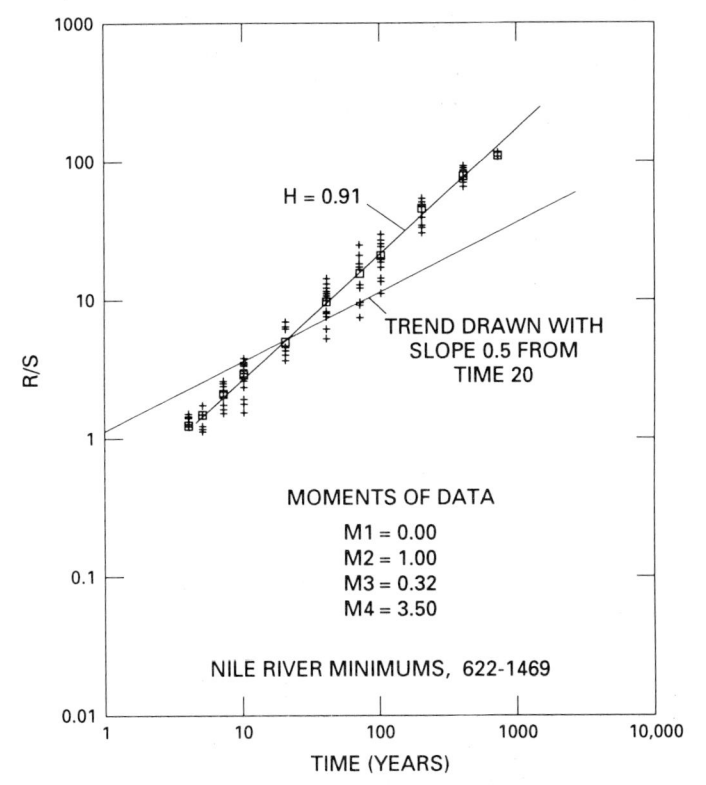

2. SOME LONG-RUN PROPERTIES OF GEOPHYSICAL RECORDS

[Plot: R/S vs TIME (YEARS), log-log. Labels: H = 0.91; TREND DRAWN WITH SLOPE 0.5 FROM TIME 20; MOMENTS OF DATA M1 = 0.00, M2 = 1.00, M3 = 0.32, M4 = 3.50; NILE RIVER MINIMUMS, 622-1469]

FIGURE 2.6. Pox diagram of $\log R/S$ versus $\log \delta$ for the Nile minimums, as reported in Toussoun (1925) (skewness = 0.32, kurtosis = 3.50). This pox diagram was constructed as in Fig. 2.2. The Nile is interesting for three reasons: It is the site of the biblical story of Joseph, which suggested to Mandelbrot and Wallis (1968) the term Joseph effect; it is the river whose flow Hurst's investigations were designed to help regularize; and it is the object of the longest available written hydrological records. The R/S pox diagram of Nile maximums is not plotted because it is extremely similar to the diagram shown in Fig. 2.6.

clear of climate fluctuations. We examine how the existence of such a λ would affect spectral analysis and R/S analysis.

Unfortunately both spectral and R/S analyses possess limited intuitive appeal. Let us therefore first discuss a more intuitive example of the kind of difficulty encountered when two fields gradually merge into each other. We summarize the discussion in Mandelbrot (1967a) of the concept of the length of a seacoast or a riverbank, see also Mandelbrot (1982). To measure a coast with increasing precision, start with a very rough scale and add increasingly finer detail. For example, walk a pair of dividers along a map and count the number of equal sides of length G of an open polygon whose vertices lie on the coast. When G is very large, the length is obviously underestimated. When G is very small, the map is extremely precise; the approximate length $L(G)$ accounts for a wealth of high-frequency details that are surely outside the realm of geography. As G is made very small, $L(G)$ becomes meaninglessly large. Now consider the sequence of approximate lengths corresponding to a sequence of decreasing values of G. The $L(G)$ may increase steadily as G decreases, but the zones in which $L(G)$ increases may be separated by one or more

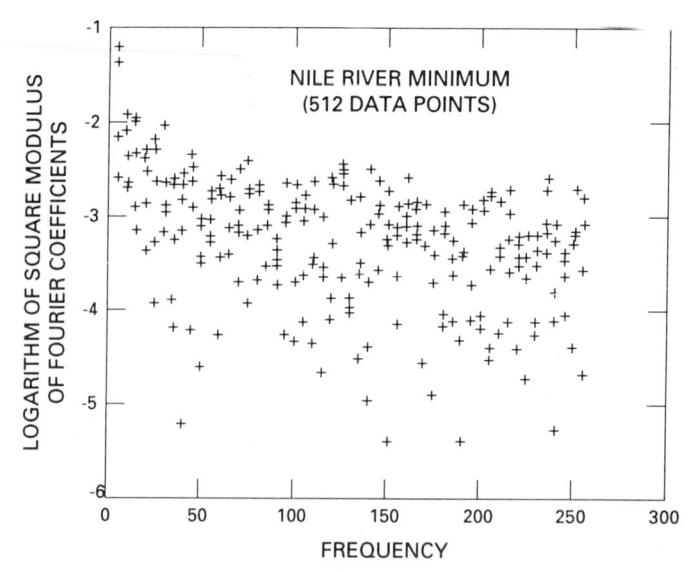

FIGURE 2.7. Spectral pox diagram of Nile minimums. The rise of the diagram, as k decreases, confirms that the value of H is large, as observed in fig. 2.6.

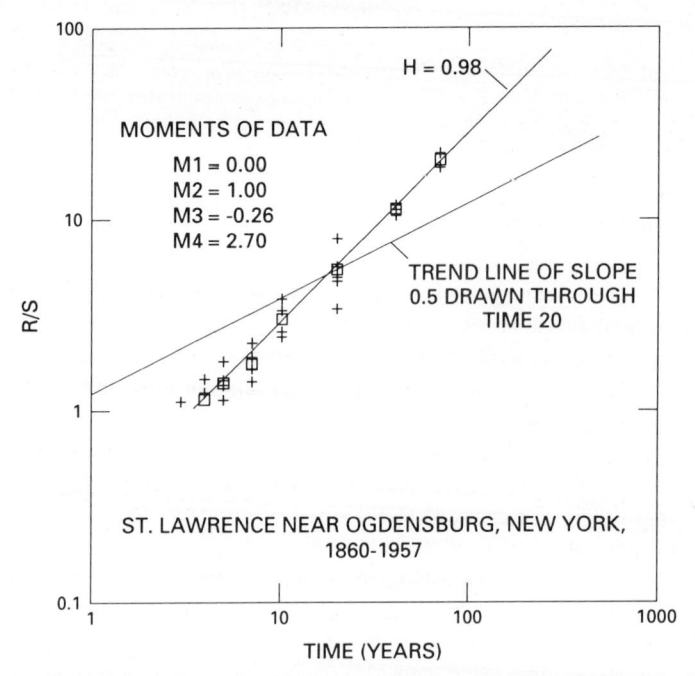

FIGURE 2.8. Pox diagram of $\log R/S$ versus $\log \delta$ for the annual flow of the St. Lawrence River. Data are from Yevjevich (1963) (skewness = -0.26, kurtosis = 2.7). This pox diagram was constructed as in Fig. 2.2. Note observed that the estimate of H in Table 2.1 would be substantially modified if carry over were taken into account.

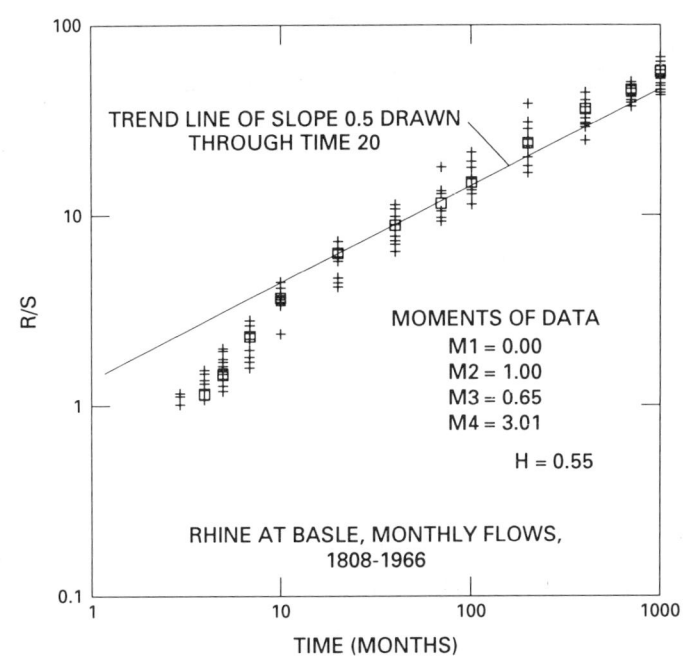

2. SOME LONG-RUN PROPERTIES OF GEOPHYSICAL RECORDS

FIGURE 2.9. Pox diagram of $\log R/S$ versus $\log \delta$ for the annual flow of the Rhine River. Data are from de Beauregard (1968) (skewness = 0.65; kurtosis = 3.01). This pox diagram was constructed as follows: Log δ was restricted to the values 3, 4, 5, 7, 10, 20, 40, 70, 100, 200, 400, 700, 1000, 2000, 4000, 7000, and 9000. For every δ, we selected 15 values of δ, spaced uniformly over the available samples. The Rhine provides a rare example of river flow in which the asymptotic slope of the trend line is very nearly 0.5. This means that low-frequency effects equaling $H > 0.5$ are absent or weak and high-frequency effects are not overwhelmed. In our opinion, dealing with low-frequency effects (when present) takes precedence over dealing with high-frequency effects to which traditional statistical models are exclusively devoted. In the case of the Rhine, it appears legitimate to use models restricted to high-frequency effects.

"shelves" where $L(G)$ is essentially constant. To define clearly the realm of geography, we think it is necessary for a shelf to exist for values of G near λ, where features of interest to the geographer satisfy $G \gg \lambda$ and geographically irrelevant wiggles satisfy $G \ll \lambda$. If a shelf exists, we can call $G(\lambda)$ a coast length. A shelf indeed has been observed for many coasts marked by sand dunes and also for many man-regulated coasts (say, the Thames in London). In many cases however, there is no shelf in the graph of $L(G)$, and the concept of length must be considered entirely arbitrary. Similar comments have been made in other contexts, such as the distinction between domains of relevances of economics and the theory of speculation in Mandelbrot (1963).

After this preliminary, let us return to the distinction between macrometeorology and climatology. It can be shown that to make these fields distinct, the spectral density of the fluctuations must have a clear-cut dip in the region of wavelengths near λ, with large amounts of energy on both sides. This dip would be the spectral analysis counterpart of the shelf in coast length measurements. But in fact no clear-cut dip is ever observed.

Similarly from the viewpoint of R/S analysis, macrometeorology and climatology are not distinct sciences unless the pox diagram of R/S exhibits distinct macrometeorology and

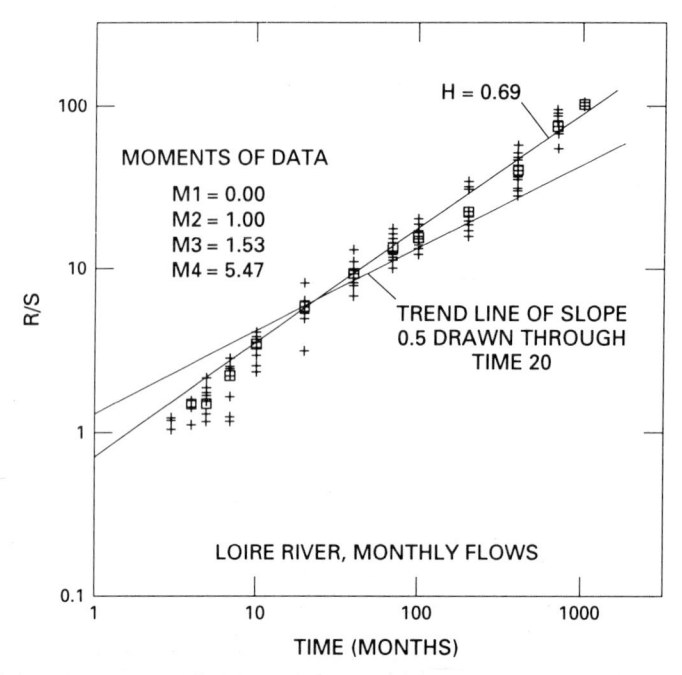

FIGURE 2.10. Pox diagram of log R/S versus log δ for the annual flow of the Loire River. Data are from de Beauregard (1968) (skewness = 1.33, kurtosis = 5.47). This pox diagram was constructed as in Fig. 2.9. The fact that the value of H is much larger for the Loire than for the Rhine fully confirms the intuitive feeling of French literary geographers about the respective degrees of irregularity of these two rivers. Whenever possible we should substitute the value of H for the geographer's intuitive descriptions. The latter may however be available in cases where the former is unattainable. To make such intuitive knowledge viable, it is useful to establish a correlation between H and intuition in cases where both are available.

climatology regimes, with an intermediate regime near $\delta = \lambda$, where a variety of configurations is conceivable. To distinguish between macrometerology and climatology, the intermediate regime of the R/S diagram must be clearly marked, with a straight trend line of slope 0.5; this region is the R/S analysis counterpart of the shelf in coast length measurements. The narrower the separating regime, the less clear is the distinction between macrometerology and climatology. At the limit, our two fields cannot be distinguished unless the R/S diagrams are very different for $\delta < \lambda$ and $\delta > \lambda$. For example, Hurst's law may apply for $\delta < \lambda$ with an exponent $H_1 \neq 0.5$ and for $\delta > \lambda$ with an exponent H_2 differing from both 0.5 and H_1.

When we wish to determine whether or not such distinct regimes are in fact observed, short hydrological records of 50 or 100 years are of little use. Much longer records are needed; thus, we followed Hurst in searching for very long records among the fossil weather data exemplified by varve thickness and tree ring indices. However even when the R/S pox

FIGURE 2.12. Pox diagram of log R/S versus log δ for a dendrochronological series from Schulman (1956) (skewness = 0.24, kurtosis = 3.60). Other records in Schulman yield extremely similar pox diagrams. However, measured values of H vary greatly between different trees in the same general geographical area. This variability confirms that long-run characteristics of tree growth are highly sensitive to microclimatic circumstances.

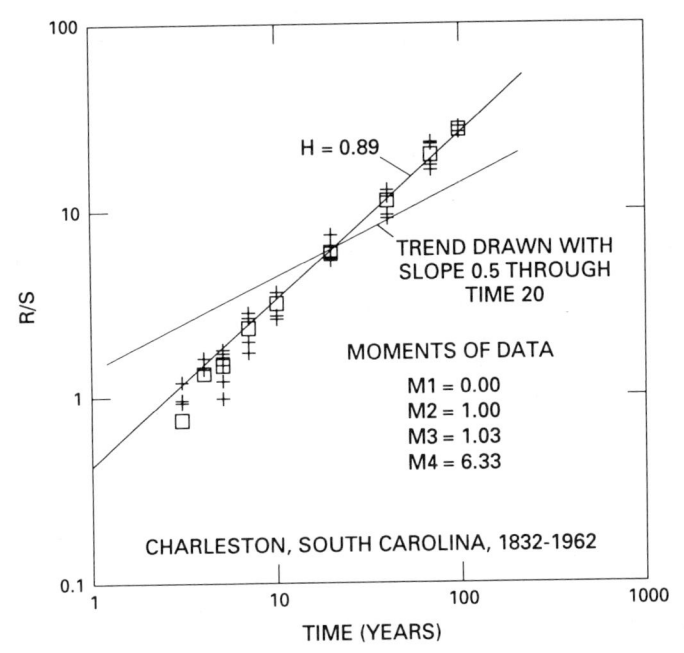

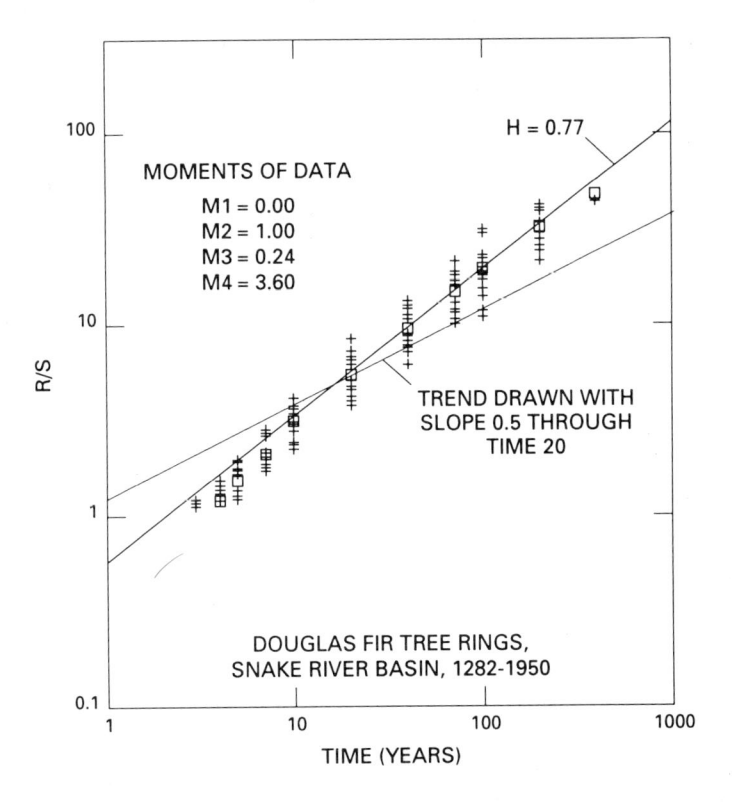

2. SOME LONG-RUN PROPERTIES OF GEOPHYSICAL RECORDS

FIGURE 2.11. Pox diagram of log R/S versus log δ for precipitation in Charleston, South Carolina. Data are from the US Bureau of the Census (1965) (skewness = 0.65, kurtosis = 3.5). Other precipitation records from the same source yield diagrams essentially indistinguishable from this example.

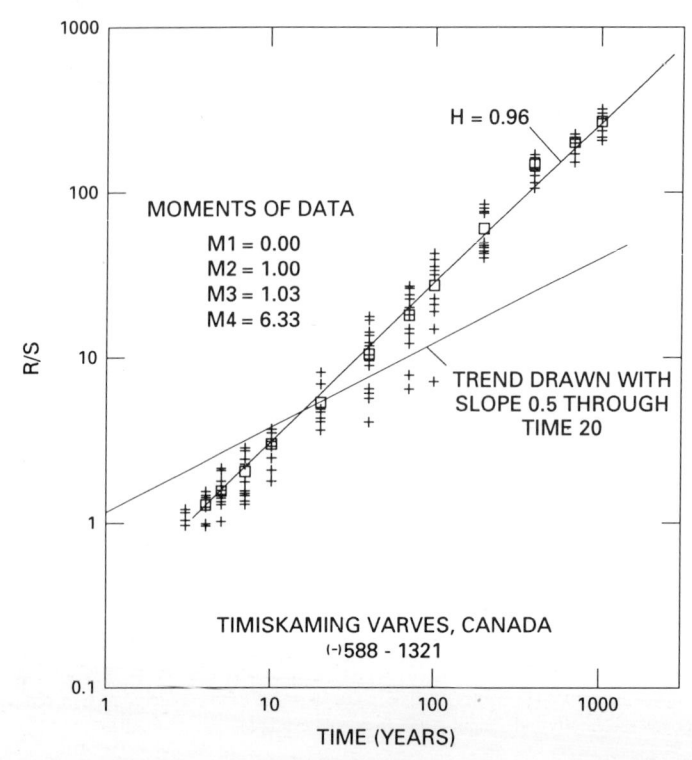

FIGURE 2.13. Pox diagram of log R/S versus log δ for the thickness of varves in Timiskaming, Canada. Data are from De Geer (1940) (skewness = 1.03, kurtosis = 6.33). Other records in de Geer yield extremely similar pox diagrams. The present example was selected because its length is unusual even among varve records. Every time an unusually long record becomes available, we rush to check whether the slope of the R/S pox diagram returns to the value of 0.5 corresponding to asymptotic independence. So far we are aware of no example of such a return. (Returns due to very strong cyclic effects are a different matter.)

diagrams are so extended, they still do not exhibit the kind of break that identifies two distinct fields.

We can now return to the claim made at the beginning of Chapter 2: The span of statistical interdependence of geophysical data is infinite. By this we mean only that this span is longer than the longest records so far examined. This gives to the term "infinity" a physical definition rather than the well-known mathematical definition (namely, "something" larger than any real number). For a detailed discussion of physical infinity, see Mandelbrot (1963). Among the related works that have been published since the original issue of Chapter 2, see Mandelbrot (1982).

In summary, even though distinctions between macrometerology and climatology or between climatology and paleoclimatology are unquestionably useful in ordinary discourse, they are not intrinsic to the subject matter, and it is imprudent to let them influence hydrological design.

One last word about why H is so well-defined for so many geological data? Why is the Rhine well-behaved with $H = 0.5$, while the typical river yields $H > 0.5$? We can state the problem, but we have no solution to offer.

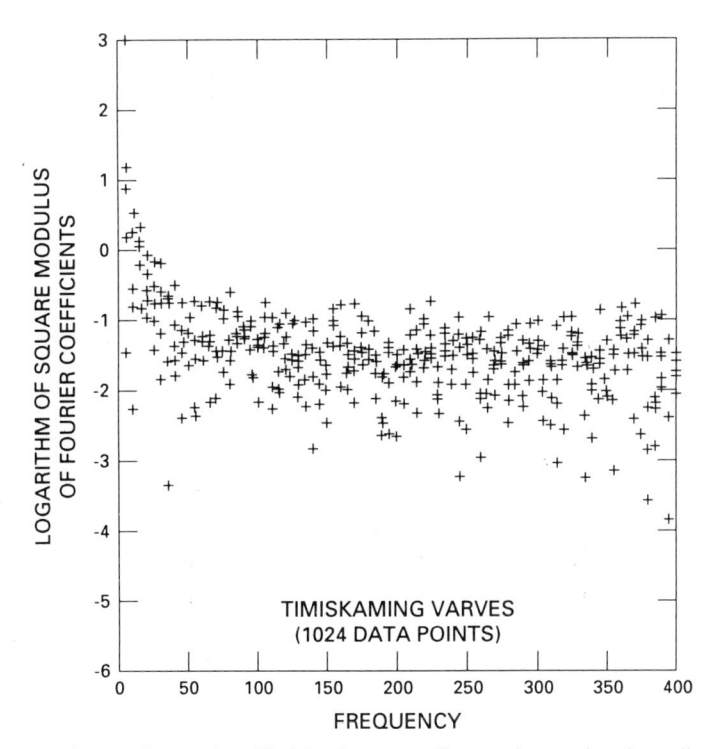

2. SOME LONG-RUN PROPERTIES OF GEOPHYSICAL RECORDS

FIGURE 2.14. Fourier pox diagram from Timiskaming varves. Frequencies are plotted on a linear scale of the abscissas. The extremely steep low-frequency rise of this diagram confirms the high value of H noted in Fig. 2.13.

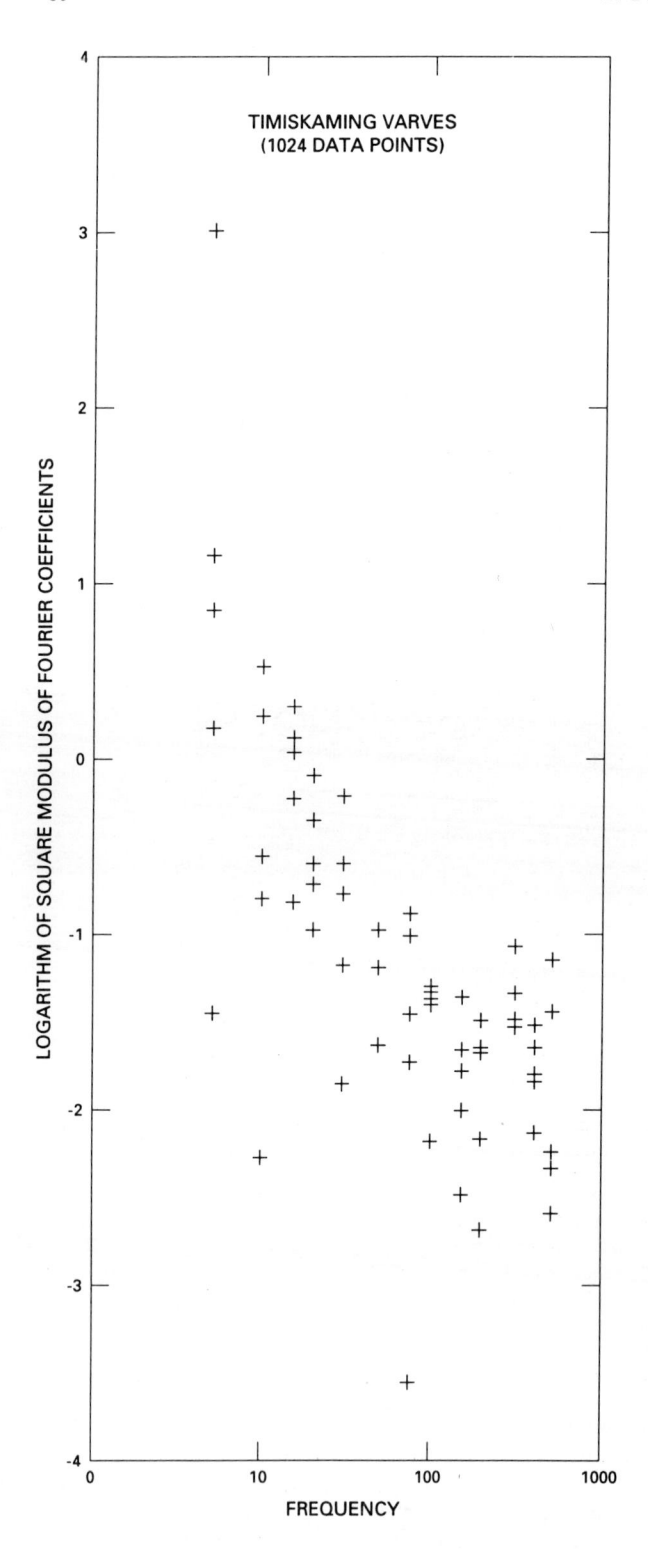

FIGURE 2.15. Alternative plot on double-logarithmic coordinates of the Fourier pox diagram for Timiskaming varves. For clarity only a small number of Fourier coefficients is plotted. Above the abscissa k, where k is one of the frequencies 5, 10, 15, 20, 30, 75, 100, 150, 200, 300, 400, 500, 700, we plotted Fourier coefficients corresponding to the frequencies k, $k-1, k-2, k-3$, and $k-4$. There is some inconclusive evidence of a dip in spectral density in the neighborhood of frequency 200, corresponding to a wavelength of 5 years.

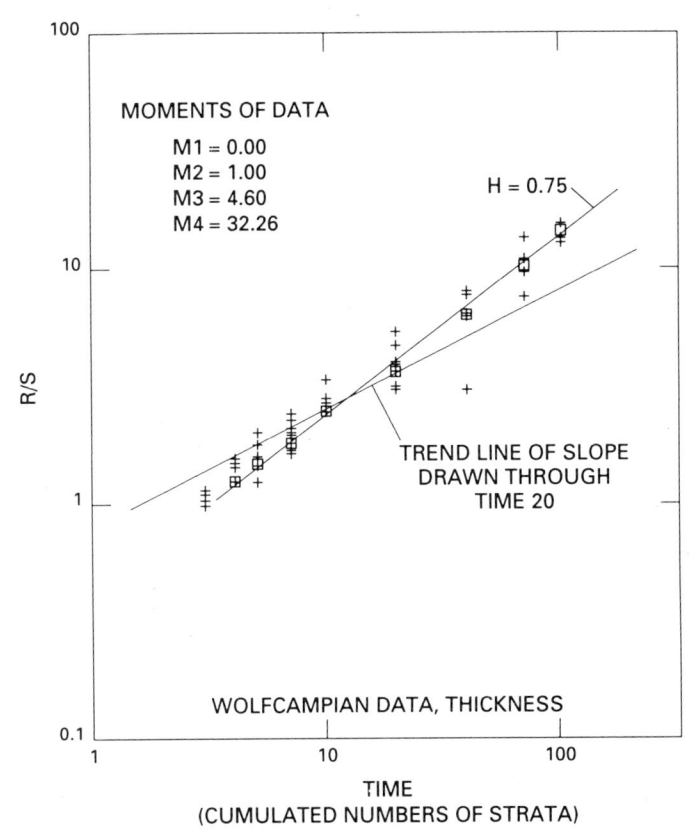

2. SOME LONG-RUN PROPERTIES OF GEOPHYSICAL RECORDS

FIGURE 2.16. Pox diagram of $\log R/S$ versus $\log \delta$ for the thickness of bedding during the Wolfcampian era in Kansas, which spanned approximately 21 million years. Data are from a private communication from Dr. G. C. Mann (skewness = 4.60, kurtosis = 32.26; this record is highly non-Gaussian, personal communication). Such records differ from hydrological records in that time is measured in cumulated numbers of strata, where the nth item in the record is the thickness of the nth stratum.

REFERENCES

deBeauregard, J., *Electricité de France* (1968).
De Geer, G., *Geochronologia Suecica Principles* (Almquest and Wiksells, Stockholm, 1940).
Eberlein, X. and Taqqu, M. S., *Dependence in Probability and Statistics* (Birkhauser, Boston, 1986).
Feder, J., *Fractals* (Plenum Press, New York, 1988).
Granger, C. W. J. and Hatanaka, M., *Spectral Analysis of Economic Time Series* (Princeton University Press, Princeton, 1964).
Hurst, H. E., *Trans. Am. Soc. Civ. Eng.* **116**, 770 (1951).
———, *Proc. Inst. Civ. Eng.*, 519 (1956).
Hurst, H. E., Black, R. P. and Simaika, Y. M., *Long-Term Storage, an Experimental Study* (Constable, London, 1965).
Langbein, W. B., Contribution to the discussion of Hurst, *Proc. Inst. Civ. Eng.*, 565 (1956).
Mandelbrot, B., *J. Polit. Econ.* **71**, 421 (1963).
Mandelbrot, B., *Comptes-rendus de l'Académie des Sciences (Paris)* **260**, 3274 (1965).
———, *Science* **155**, 636 (1967a).

———, *Encyclopédie de la Pléiade*, Logique et Connaissance Scientifique (J. Piaget, ed.) (Gallimard, Paris, 1967b).
———, *Water Resour. Res.* **7**, 543 (1971).
———, *Zeit Wahr* **31**, 271 (1975).
———, *The Fractal Geometry of Nature* (W. H. Freeman, New York, 1982).
———, *Multifractals: Noise, Turbulence, and Aggregates* (Springer-Verlag, New York, 1991).
Mandelbrot, B. and Van Ness, J. W. *SIAM Review* **10**, 422 (1968).
Mandelbrot, B. and Wallis J. R., *Water Resour. Res.* **4**, 909 (1968).
———, *Water Resour. Res.* **5** (1969a).
———, *Water Resour. Res.* **5**, 967 (1969b).
Munro, E. H., *Terr. Magn.* **53**, 241 (1948).
Schulman, E., *Dendroclimatic Changes in Semiarid America* (University of Arizona Press, Tucson, 1956).
Taqqu, M., *Water Resour. Res.* **6**, 349 (1970).
Toussoun, O., *Mem. Inst. Egypt* 8–10 (1925).
US Bureau of the Census, *The Statistical History of the United States from Colonial Times to the Present* (Fairfield Publishers, Stanford, 1965).
Willett, H. C., *Weather and Our Food Supply* (Iowa State University, Ames, 1964), p. 123–51.

NOTATIONS

H	Hurst's constant (also called Holder constant); the asymptotic slope of the plot of $\log(R/S)$ versus $\log \delta$.
K	Slope of $\log R/S$ versus $\log \delta$ using the algorithm suggested by Hurst; this K was used as an estimate of H, but ordinarily it underestimates the value of $H - 0.72$.
$L(G)$	Length of a coastline as approximated by a G-sided polygon with equal sides and vertices lying on the coastline.
λ	Arbitrary time boundary separating the domains of applicability of two scientific discipline, both of which study a single set of phenomena but use different time scales.
$R(t,\delta)$	Sample bridge range for lag δ.
δ	Time lag.
$S^2(t,\delta)$	Variance of δ values of $X(t + 1) \ldots X(t + \delta)$ around their sample average.
t	Time
T	Total available sample size.
$X(t)$	Record containing T readings uniformly space in time from $t = 1$ to $t = T$.
$\int x(t)$	Equal to $\Sigma_{u=1}^{t} x(u)$.

APPENDIX: ABSTRACTS AND ERRATA FOR RELATED PAPERS BY THE SAME AUTHORS

Notation: In these errata, the letters L and R stand for left and right column, respectively; line -13 stands for line 13 from the bottom of the column.

Remark: In all the papers by Mandelbrot and Wallis, the values of R/S for small lags are incorrect and must be disregarded (Taqqu, 1970). These values were not taken into account; hence the papers' conclusions remain unaffected.

Noah, Joseph, and Operational Hydrology (Mandelbrot and Wallis, 1968)

Abstract: By the Noah effect, we designate the observation that extreme precipitation can be very extreme indeed, and by Joseph effect, the finding that a long period of unusual

2. SOME LONG-RUN PROPERTIES OF GEOPHYSICAL RECORDS

(high or low) precipitation can be extremely long. Current models of statistical hydrology cannot account for either effect and must be superseded. As a replacement, self-similar models appear very promising. They account particularly well for the remarkable empirical observations of Harold Edwin Hurst. The present paper introduces and summarizes a series of investigations on self-similar operational hydrology.

In the papers that the present work introduces and summarizes, the complex interplay between the random variable $R(\delta)/S(\delta)$, its sample distribution, its expectation, and its variance, is described in detail. Therefore the present work postpones all qualification and handles $R(\delta)/S(\delta)$ quite casually. Depending on the context, the same letters $R(\delta)/S(\delta)$ stand for this random variable itself, its expectation, or a sample estimate.

Errata

p. 911 L, line 12, replace 0.5 by $s^{0.5}$.
p. 913 R, line -13, replace $\text{expl}(-|s|/s)$ by $\exp(-|s|/s_2)$.
p. 913R line -11, replace $(1+|s|_2/S_3)^{-2}$ by $(1+|s|/S_3)^{-2}$.
p. 914 L, line -14, replace ΔX^* by $\Delta X^* =$.
p. 915 L, line -3 replace $l:f$ by $1:f$.
p. 916 L, line 28, replace $C\sqrt{s}$ by $K\sqrt{s}$, with some (positive and finite) constant.
p. 916 L, line -5, replace X(u) by C(u).
p. 916 R line 17, replace C a constant by G a random variable independent of t and s.
p. 916 R, line -8, replace by $C_H(s) = Q[(s-1)^{2H} - 2s^{2H} + (s+1)^{2H}]$.
p. 916 R, line -6, replace by $C_H(s) = [2H(2H-1)Q]s^{2H-2}$.
p. 916 R, line -9, replace $s \geqslant$ by $s \geqslant 1$.

Computer Experiments with Fractional Gaussian Noises (Mandelbrot and Wallis, 1969a)

Abstract of Part 1, Averages and variances: Fractional Gaussian noises are a family of random processes such that the interdependence between values of the process at instants of time very distant from each other is small but nonnegligible. It has been shown by mathematical analysis that such interdependence has precisely the intensity required for a good mathematical model of log run hydrological and geophysical records. This analysis is not illustrated, extended, and made practically usable with the help of computer simulations. In Part 1, we stress the shape of the sample functions and relations between past and future averages.

Abstract of Part 2, Rescaled ranges and spectra: Continuing the preceding paper, we report on computer experiments on the rescaled range and the spectrum of fractal Gaussian noise.

Abstract of Part 3, Mathematical appendix: The present appendix is devoted to mathematical considerations designed to fill in the innumerable logical gaps left in the preceding paper. Even though most proofs are merely sketched or omitted, the notation remains heavy, and some readers may wish to skip to Formulas 1 and 2, which define Type 1 functions.

Mandelbrot and Van Ness (1968) contains additional mathematical details and references.

Errata

Frieze 8 should follow Friezes 1 to 7 along the tops of succeeding pages.
p. 228 L, line 8, replace [1968a] by [1968].

pl. 233 L, line 6, add = (equal sign) so that it reads $\epsilon(\Delta X^*)^2 = \epsilon(\Delta_H)B^2 = C_H s^{2H}$.
p. 233 R, line 11, read $\sqrt{C_H^H}$.
p. 234 R, line −10, add ϵ to $\epsilon S^2(t,s) = \epsilon X^2 - \epsilon(s^{-1}\Delta X^*)^2$.
p. 234 R, line −3 erase ⊥.
p. 236 L, replace $(P + F)^{2F}$ by $(P + F)^{2H}$.
p. 239, interchange the figures on this and the next page. The captions do not move.
p. 240, replace this figure by the figure on the last page.
p. 240 R, line 2, add − (minus sign) before the second term of Δ_{FP}/S_p.
p. 240 L, line −1, delete the short paragraph beginning with "A corollary of Studen's result . . ."
p. 241 L, line 5, after the table, erase ⊥.
p. 242 L, line −1, replace Σ^t_{u+1} by $\Sigma^t_{u=1}$.
p. 243, caption of Figure 1, line −1, replace (in both places) $0 < u$ by $0 \leq u$.
p. 243 R, line −17, replace *the* by *either*.
p. 256 R, line 4, replace | by 1.
p. 263 R, line 5, read "that s > 2 and 1 ⩽ . . ."

Robustness of the Rescaled Range R/S in the Measurement of Non-Cyclic Long-Run Statistical Dependence (Mandelbrot and Wallis, 1969b)

Abstract: The rescaled range $R(t,\delta)/S(t,s)$ is shown by extensive computer simulation to be a very robust statistic for testing the presence of noncyclic long-run statistical dependence and for estimating its intensity. Processes examined in the paper include extraordinarily non-Gaussian processes with huge skewness and/or kurtosis (that is, third and/or fourth moments).

A Fast Fractional Gaussian Noise Generator (Mandelbrot, 1971)

Abstract: By design, fast fractional Gaussian noises (ffGn) have the following characteristics: The number of operations to generate them is relatively small, the long-run statistical dependence that they exhibit is strong and has the form required for self similar hydrology. Their short-run properties, as expressed by correlations between successive or nearly successive yearly averages, are adjustable (within bounds) and can be fitted to the corresponding short-run properties of records. Extension to the multidimensional (multisite) case should raise no essential difficulty. Finally, their definition, as sums for Markov–Gauss and other simple processes, fits the intuitive ideas that climate can be visualized as either unpredictably inhomogeneous or ruled by a hierarchy of variable regimes.

Erratum

p. 545 R, replace $\dfrac{1 - P^{-(1-H)}H(2H - 1)}{\Gamma(3 - 2H)}$ by $1 - \dfrac{B^{-(1-H)}H(2H - 1)}{\Gamma(3 - 2H)}$

3

Some Remarks on the Numerical Estimation of Fractal Dimension

S. A. Pruess

3.1. INTRODUCTION

Fractal geometry is currently of major interest in many fields. A commonly occurring problem involves determining the fractal dimension. Hausdorff (1919) rigorously defined the concept of fractional dimension, and further theoretical work was done by Besicovitch and others (see Falconer, 1985 or 1990, for an extensive list of references). Few applications were made using the concept of fractals until the early 1970s when Mandelbrot began his work. In the past ten years, many researchers (e.g., Grassberger, 1983; Barton and Larson, 1985; Sreenivasan and Meneveau, 1986; and Hunt and Sullivan, 1989) from a wide variety of disciplines have devised algorithms for estimating fractal dimensions.

While there are many different approaches to estimating fractal dimensions, Chapter 3 concentrates on those related to box counting, as described in the following section. The main reason for this choice is that this is often the first technique tried on physical data, probably because it seems so simple a concept. Unfortunately there are some limitations to box-counting methods that are not so widely known as they should be. A second reason for concentrating on box-counting methods is that many alternatives for physical data are based on similar mathematical limiting processes and may suffer from similar difficulties.

In Chapter 3, I examine some of the commonly used methods to explore their mathematical bases and find their weaknesses. Section 3.2 presents mathematical background helpful in understanding the algorithms; Section 3.3 contains output for some test examples, and Section 3.4 gives some suggestions for using these methods.

S. A. Pruess • Department of Mathematics, Colorado School of Mines, Golden, Colorado.

Fractals in the Earth Sciences, edited by Christopher C. Barton and Paul R. La Pointe. Plenum Press, New York, 1995.

3.2. MATHEMATICAL BACKGROUND

While the material in this section may appear abstract, it does shed light on the behavior of the algorithms, especially why they may yield misleading results. The following definitions of Hausdorff measure and Hausdorff dimension are a combination of those in Hausdorff (1919) and Falconer (1990); the text of Barnsley (1988) is also a fine reference. For simplicity, I restrict the discussion to subsets of the plane.

The diameter of a set V, denoted by diam (V), is defined as the diameter of the smallest circle containing all points in V. If a given set S is contained in the union of sets $\{V_i\}$ and each V_i has diameter less than δ, then we say $\{V_i\}$ is a δ-cover of S. For a δ-cover $\{V_i\}$ of S and a nonnegative number d define

$$H_d(\delta) = \inf_{V_i \cap S \neq \phi} \sum \text{diam}\,(V_i)^d \qquad (1)$$

The notation inf stands for infimum, the greatest lower bound (essentially the minimum). The Hausdorff d-dimensional measure of S is denoted by μ_d and defined to be

$$\mu_d = \lim_{\delta \to 0} H_d(\delta) \qquad (2)$$

This limit does exist, but it may be infinite, since H_d increases as δ decreases. It is not difficult to argue (see Falconer, 1990 for details) that the function $H_d(\delta)$ is a nonincreasing function of δ, i.e., as $\delta \to 0$, values of $H_d(\delta)$ cannot decrease. This is in sharp contrast to its algorithmic approximations described in the following paragraphs.

The definition of fractal or Hausdorff–Besicovitch dimension, denoted by D, as given by Hausdorff and studied by Besicovitch, is that quantity uniquely characterized by

$$\mu_d = \begin{cases} \infty & \text{if } 0 < d < D \\ 0 & \text{if } D < d < \infty \end{cases} \qquad (3)$$

It is convenient (and nearly always valid) to assume that when $d = D$, the limit exists, and is a finite positive number. In this case, the value μ_D is known as the *Hausdorff measure* of S. The main uses for these abstract concepts (either to topologists or scientists) are (1) to characterize numerically the idea of size of a given set S or (2) to compare the sizes of two given sets. In the latter case, the set with the larger dimension D is considered larger; if they both have the same D, then the one with the largest measure μ_D is considered larger.

There appears to be two potential difficulties with implementing these definitions as an algorithm: How do we compute the limit $\delta \to 0$ in Eq. (2) and the infimum in Eq. (1)? Essentially both of these calculations involve limiting processes, and these must be replaced by some finite sample as an approximation. This sampling produces errors that can be very difficult to analyze mathematically and may lead to significantly inaccurate values in trying to calculate D or μ_D. Examples are given in the next section.

While the definition in Eq. (1) says to minimize over *all* δ-covers, most algorithms use only coverings by uniform V_i, e.g., circles or squares of a fixed size. In fact nearly all published algorithms based on coverings use squares (usually called boxes). First the set S is embedded in some larger square that is then subdivided into smaller subsquares of constant diagonal δ. Fortunately for all but pathological sets S, this simple choice for V_i produces the correct D and μ_D as long as the infimum and limit as $\delta \to 0$ are carefully chosen. Falconer (1990) provides more mathematical details relating the mathematical D to the so-called box dimension, corresponding to choosing boxes for the V_i.

3. SOME REMARKS ON THE NUMERICAL ESTIMATION OF FRACTAL DIMENSION

For uniformity of presentation, I assume that the grid of boxes is generated as follows. First place the set S in the first quadrant of a Cartesian coordinate system. Next choose a diameter δ and let $h = \delta/\sqrt{2}$, which is the length of the side of each box. The ijth box has vertices $[(i-1)h, (j-1)h]$, $[(i-1)h, jh]$, $[ih, (j-1)h]$, and (ih, jh) for $1 \leq i \leq m$ and $1 \leq j \leq n$, with m and n sufficiently large that S is covered. Because there is a uniform covering, Eq. (1) becomes

$$H_d(\delta) = \inf \sum_{V_i \cap S \neq \phi} \delta^d$$

$$= \inf[N(\delta)\delta^d] \quad (4)$$

where $N(\delta)$ is the number of boxes that overlap points of S. Note that the infimum is still present, since I have not said how S is to be placed in the first quadrant and mathematically it is necessary to minimize over all possible orientations (rotations and shifts). Of course in an algorithm, this cannot be done; at best only at few characteristic orientations can be considered. In any case, I can assume that the limiting process $\delta \to 0$ removes this source of error, so that

$$H_d(\delta) = N(\delta)\delta^d \quad (5)$$

and

$$\mu_D = \lim_{\delta \to 0} H_D(\delta)$$

$$= \lim_{\delta \to 0} N(\delta)\delta^D$$

Then

$$\log \mu_D = \lim_{\delta \to 0} [\log N(\delta) + D \log \delta]$$

or

$$0 = \lim_{\delta \to 0} \{\log N(\delta) - D \log 1/\delta - \log \mu_D\} \quad (6)$$

Rearranging this yields

$$D = \lim_{\delta \to 0} \frac{\log N(\delta) - \log \mu_D}{\log 1/\delta} \quad (7)$$

$$= \lim_{\delta \to 0} \frac{\log N(\delta)}{\log 1/\delta} \quad (8)$$

Equation (8) is often called the fractal equation.

Two standard existing methods are now described for estimating the Hausdorff dimension of a set S in the plane R^2. The first, described in Mandelbrot (1982), has been applied by various authors (e.g., Sreenivasan and Meneveau, 1986). Let

$$D_1(\delta) = \frac{\log N(\delta)}{\log \delta^{-1}} \quad (9)$$

then, for any fixed δ, this yields an estimate for D. It is not always made clear that only approximations are being generated and it is crucial to take $\delta \to 0$ in accord with Eq. (8). Since in practice the limiting process must stop somewhere, it is important to consider the error made. One part of the error is clearly the neglected term in Eq. (7), viz.,

$$-\frac{\log \mu_D}{\log \delta^{-1}}$$

While it is true that this term can be made arbitrarily small by taking δ arbitrarily small, the rate of convergence can be excruciatingly slow. In fact, if $\mu_D = 2$, to make this term smaller than 0.001 requires $\delta < 10^{-301}$; this necessitates a very large number of boxes! Of course when $\mu_D = 1$, this term is not present, so convergence may be faster.

The second algorithm I formally call the box count method, and it is based on an idea that was developed independently by many authors (e.g., Grassberger, 1983, and Barton and Larsen, 1985) from a description by Mandelbrot (1982). With $N(\delta)$ as in (4), plot $N(\delta)$ versus the inverse of δ on a log/log scale. The slope of the resulting function provides an estimate $D_2(\delta)$ for D. Algorithmically a finite set of δ-values are chosen, which results in a discrete set of points in the log/log plane. The dimension D is usually estimated by the slope of the best least squares line for this data. Mathematically let

$$x = \log \delta^{-1}$$

and

$$y = \log N(\delta) \tag{10}$$

then the slope of y as a function of x provides an estimate $D_2(\delta)$ for D (the antilog of the intercept is an estimate for μ_D). The motivation for this is also Eq. (5), which in terms of x and y states that

$$0 = \lim_{\delta \to 0} [y - Dx - \log \mu_D]. \tag{11}$$

Again the mathematics suggest that D is the limiting slope as $\delta \to 0$ of y as a function of x, though in practice, only a finite number of sample points are used and a single slope D_2 is produced. The error incurred is quite difficult to analyze, but it has recently been studied in the impressive work of Hunt (1990) under certain distributional assumptions.

The final algorithm considered is a straightforward implementation of Eqs. (1)–(3). This algorithm, recently proposed by Samuel (1988), is called the Hausdorff algorithm. Choose an initial bracket [DL, DU] containing D, e.g., [0, 2]; then for a given sequence $\delta_1 > \delta_2 > \ldots > \delta_m$ determine if the sequence

$$m_{d,j} = N(\delta_j)\delta_j^d$$

goes to zero or infinity when $d = (DL + DU)/2$. One way of determining this is to fit a least squares line to (δ_j, m_{dj}), then see if the line decreases or increases. If it decreases (limit presumed to be zero), then set DU := (DL + DU)/2; otherwise (limit presumed to be infinity) set DL := (DL + DU)/2. This can be iterated until convergence. Since decisions are based on a finite set of δ values, again only an estimate, denoted D_3, is produced for D. This algorithm yields incorrect results only if an incorrect judgment is made concerning the limit of $m_{d,j}$ as $j \to \infty$. Intuitively since the only possible limits are infinity or zero, the choice of which limit is correct for a given $m_{d,j}$ should not be difficult. In fact as d approaches the exact dimension D for a fixed set of δs, the sequence $m_{d,j}$ becomes nearly constant. Hence this algorithm is often no more accurate than the preceding one for a fixed choice of box sizes.

Although the Hausdorff–Besicovitch theory is used to some extent in each of the preceding methods, there is some ambiguity in estimates for the fractal dimension due

3. SOME REMARKS ON THE NUMERICAL ESTIMATION OF FRACTAL DIMENSION

mainly to choices of δ. Reading the literature can produce further confusion, since it is often not made clear that only approximations to D are being produced.

Several possibilities have been suggested in an effort to mimic the effects of the infimum in Eq. (4). Samuel (1988) rotated the data through several angles to seek alternative estimates for D; this has been generalized in Barton and others (1989) to offer an option for minimizing the count N over several rotations for each choice of δ. However my experience is that these rotations are not necessary if δ is taken sufficiently small. Of course in practice, for a fixed set of δs, some rotations may help. Another possibility is to use a Monte Carlo approach where for each δ a box of random position and orientation is generated and tested to see if it overlaps the data set S. This is done repeatedly, thereby generating an estimate for the probability of overlapping S with a random box of size δ. This probability is easily translated into $N(\delta)$ for use in the algorithms. As is typical with Monte Carlo algorithms, experience indicates that this approach is rather slow even for moderate accuracy; however, it does have the advantage that some smoothing occurs, which as we see in the next section is very desirable. For computational details on a Monte Carlo algorithm, see Hunt and Sullivan (1989).

Some important consequences of the preceding theory to the algorithms are

1. $N(\delta)$, and consequently $H_d(\delta)$, are step functions of δ, because $N(\delta)$ is integer-valued.
2. N in Eq. (5) is not necessarily a decreasing function of δ; hence $\log N$ is not necessarily an increasing function of $\log \delta^{-1}$. This is in sharp contrast to the case when Eq. (4) is used; i.e., the infimum is taken over all covers.
3. Sampling N at only a finite number of δ values may produce poor estimates for its slope.
4. To produce reasonable estimates for D may require δ to be very small to mimic the limit in Eq. (8) or (11).

3.3. SOME SAMPLE RESULTS

To test algorithms, it is a common practice to apply them initially to data with known answers. Unfortunately in the case of fractals, all the interesting examples with known dimension require infinite amounts of data (infinite iterations of some initiator-generator algorithm). Stopping at a finite number of iterations changes the fractal dimension (to the topological dimension) if we take $\delta \to 0$ as mathematical theory requires. We can talk about a fractal dimension over a range of δ, but then the exact value is unknown.

The first example considered is a straight line. Let the line be quite general connecting the vertex (x, y) to (u, v). The advantage of such a simple example is that analysis is straightforward and there is no question the dimension $D = 1$. For the box method discussed in the preceding section, it is often simpler to refer to the side length h instead of the diagonal length δ. Even for such a simple example, the exact formula for the box count N as a function of h is rather complicated. It can be shown that

$$N = -\left[\frac{-u}{h}\right] - \left[\frac{-v}{h}\right] - \left[\frac{x}{h}\right] - \left[\frac{y}{h}\right] \quad (12)$$

is more or less correct (depending on whether or not the line intersects interior grid vertices and the convention used for counting in that case). The notation [x] stands for the greatest integer less than or equal to x.

Even though Eq. (12) applies to only a single straight line, it typifies more general cases, since data sets in the plane can be considered as (or are usually digitized as) collections of straight lines. For a set of disjoint lines when δ is sufficiently small, the formula for N consists of sums of terms like Eq. (12). For general sets of lines, this sum must be modified to account for overlap.

As an example, consider the line from (2/5, 4/7) to (3/4, 2/3). The graph of log N versus log δ^{-1} is shown in Fig. 3.1. While the slope may approach 1 as $\delta^{-1} \to \infty$, a poor choice of δ values can produce a rather different approximation. In fact Table 3.1 displays the results of estimates for the fractal dimension using the algorithms described in the preceding section for two sets of six indicated δ values. Note the poor accuracy of D_1, although it converges to the correct value. Several authors (e.g., Grassberger, 1983) have tried to accelerate or extrapolate these values but with little success. The reason for this failure is that all such acceleration schemes assume some kind of regular behavior as $\delta \to 0$, which is invalid here because of Consequence (1). Actually a single line is much better behaved than general data, since more line segments produce even more irregularity in $N(\delta)$.

The box sizes used for Table 3.1 were chosen arbitrarily; in general a good strategy chooses points so that the log of δ is equally spaced. Moreover since Fig. 3.1 shows more scatter in the data when δ is large, it is recommended to make the maximum δ used successively smaller over several sets of box sizes if possible. (A weighted least squares

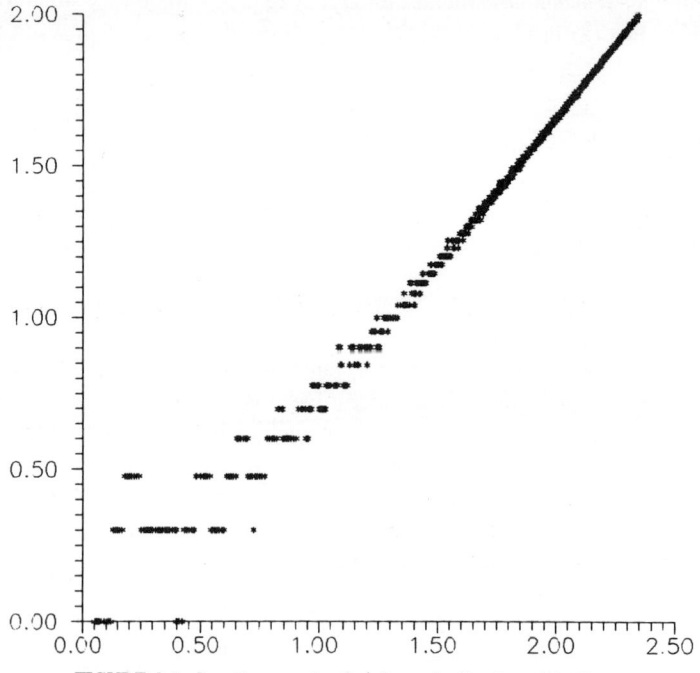

FIGURE 3.1. Log N versus log δ^{-1} for an inclined straight line.

3. SOME REMARKS ON THE NUMERICAL ESTIMATION OF FRACTAL DIMENSION

TABLE 3.1. Estimates for
the Fractal Dimension of a Straight Line
(Exact $D = 1$)

δ	$N(\delta)$	$D_1 = \text{Log} N/\text{Log}\delta^{-1}$
0.24531	3	0.782
0.10711	6	0.802
0.04406	12	0.796
0.02020	24	0.814
0.00959	48	0.833
0.00463	96	0.849
$D_2 = 0.869$		
$D_3 = 0.856$		
0.18813	2	0.415
0.09431	5	0.682
0.04677	11	0.783
0.02338	20	0.784
0.01172	38	0.818
0.00584	78	0.850
$D_2 = 1.028$		
$D_3 = 1.060$		

method seems like it might help, but this has never led to much improvement for me.) When 21 box sizes are equally spaced logarithmically in (0.001, 0.01), we obtain $D_2 = 0.99582$ and $D_3 = 0.99679$, while for 21 δs in (0.0001, 0.001) we obtain $D_2 = 0.99923$ and $D_3 = 0.99908$. Note that the error appears on the order of the largest δ, a fact not difficult to prove for a single straight line. In any case, for physical data, there is nearly always a resolution limit (the length of the smallest line segment) that establishes a minimum value for δ.

A more interesting example is the middle thirds (or triadic) Cantor dust. This is defined most easily in terms of generations: At the zeroth generation begin with the interval (0,1); at each succeeding generation, delete the middle third. This set has fascinated mathematicians for a hundred years, and it is not difficult to prove (e.g., see Falconer, 1985) that $D = \log 2/\log 3 = 0.63093$, and $\mu_D = 1$ in the limit. However in practice we must stop at a finite number of generations and hope that this is a good approximation. Unfortunately the mathematics breaks down, since taking $\delta \to 0$ must surely produce $D = 1$ for any particular generation. Figure 3.2, which displays $\log N$ versus $\log \delta^{-1}$ for the sixth generation of Cantor dust with one hundred δ values in (0.00001, 0.9), demonstrates this. The stepped nature of $\log N$ is quite evident for the larger δ (smaller $\log \delta^{-1}$) and a pronounced upward bending occurs around $\log \delta^{-1} = 3$ or $\delta = 0.001$ (the minimum line segment length is 0.0014). This is easy to explain: If δ is very small, then the individual line segments are seen, so $D \approx 1$. This phenomenon is a common error, since the mathematics requires $\delta \to 0$; of course as pointed out earlier, stopping at a finite generation has rendered the mathematics irrelevant to the limiting Cantor set.

Because of the large scatter in the graph (typical of most physical data sets), calculated values for N and consequently for D_2 and D_3 are highly sensitive to relatively minor changes in the initial box size. For example for the eighth generation, when $h_{\text{max}} = 0.00304$ and

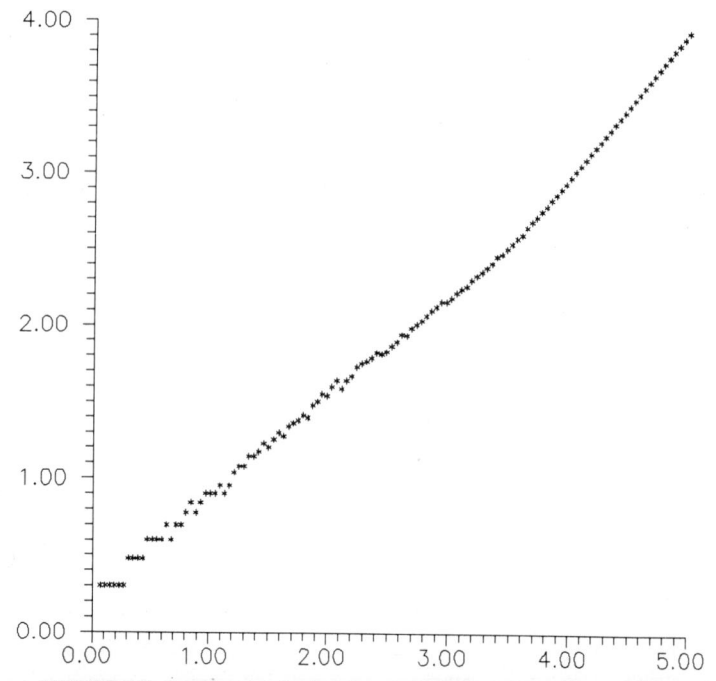

FIGURE 3.2. Log N versus log δ^{-1} for triadic Cantor dust (sixth generation).

$h_{min} = 0.000305$, $D_2 = 0.63628$ and $D_3 = 0.63801$; for the same h_{max} and $h_{min} = 0.000303$, $D_2 = 0.63816$ and $D_3 = 0.64233$. Hence a change of 0.000002 in the smallest h (which produced equally minor changes in all the hs) led to changes in the D estimates of 0.002–0.004. This indicates that some smoothing is probably desirable; e.g., calculate several sets of approximations and average these. Table 3.2 contains means and population standard deviations for several generations. For each generation in Table 3.2, 41 δ values were used with equally spaced logs in eight different intervals generated as small random perturbations of the interval [2*min length segment, sqrt(min length segment)]. Convergence to $D = 0.63093$ certainly appears to be slow and erratic. It is unknown how much of the

TABLE 3.2. Estimates for the Fractal Dimension of Cantor Dust
($D = 0.63092975$ at Generation ∞)

Generation	D_2 Mean	D_2 Standard Deviation	D_3 Mean	D_3 Standard Deviation
4	0.65929	0.02132	0.65225	0.02161
5	0.65429	0.01318	0.65468	0.01594
6	0.63759	0.00974	0.63567	0.00826
7	0.63892	0.00718	0.63978	0.00989
8	0.63595	0.00593	0.63652	0.00729
9	0.63460	0.00327	0.63383	0.00573
10	0.63203	0.00197	0.63164	0.00263

3. SOME REMARKS ON THE NUMERICAL ESTIMATION OF FRACTAL DIMENSION

difference from this "exact" D is due to the finite number of generations used and how much is due to the choice of a finite set of δ. Similar uncertainty occurs for other standard examples, such as Sierpinski triangles and the triadic or quadric Koch curves.

The standard deviations and the differences between the output of the two algorithms for the same δs, suggest that only the first two or three digits are accurate. To attain more accuracy (in theory) would require smaller values for δ, but we are already getting close to the minimum line length. As a last comment on this Cantor dust, note that the "right" set of δs is $\{1, \frac{1}{3}, \frac{1}{9}, \frac{1}{27}, \ldots\}$, since this yields the exact $D = \log 2/\log 3$ and $\mu_D = 1$ as long as we do not go below the minimum line segment length.

As a final example, consider the 8025 line segments from a digitized data set (see Fig. 3.3) of Barton and Larsen (1985) representing fracture patterns in a geological outcrop on Nevada's Yucca Mountain. The vertices of the segments vary over the interval (0,20), and the average length is 0.0055, with quite a wide variation in length occurring. Figure 3.4 displays a plot of $\log N$ versus $\log h^{-1}$ for h in the interval (0.003, 0.5). The stepped nature of $\log N$ is clear for small $\log h^{-1}$, and there is a bend near $\log h^{-1} = 1.8$ (or $h = 0.016$), which may indicate that individual lines are being resolved. Assuming the fractal dimension is sought over the range (0.01, 0.10) where the data appears to be linear, values were calculated using eight sets of 41 step sizes over small perturbations of this range. The mean $D_2 = 1.79350$ with standard deviation 0.00249, and mean $D_3 = 1.80005$ with standard

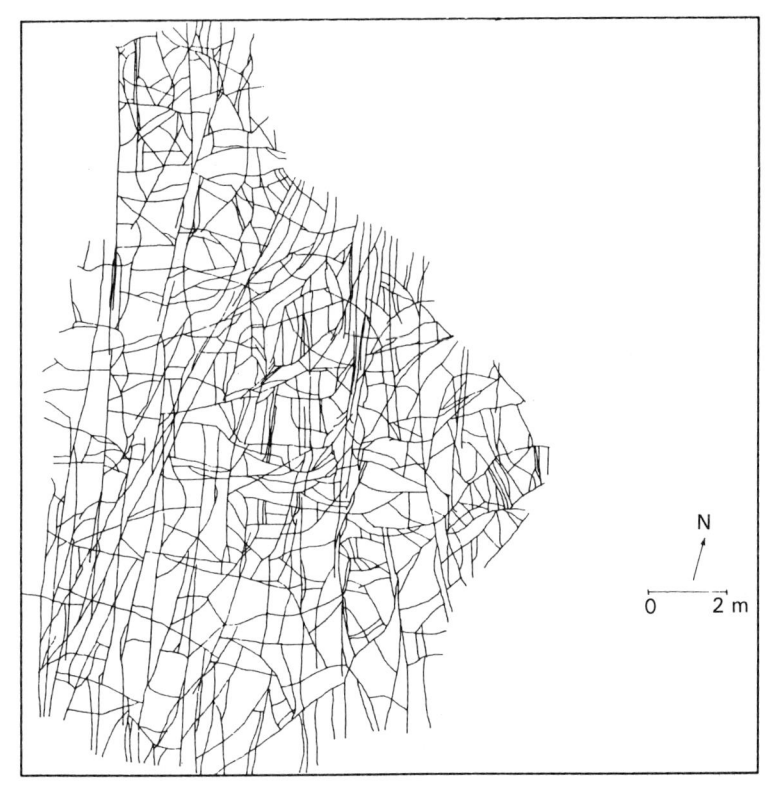

FIGURE 3.3. Fracture pattern from Yucca Mountain, Nevada.

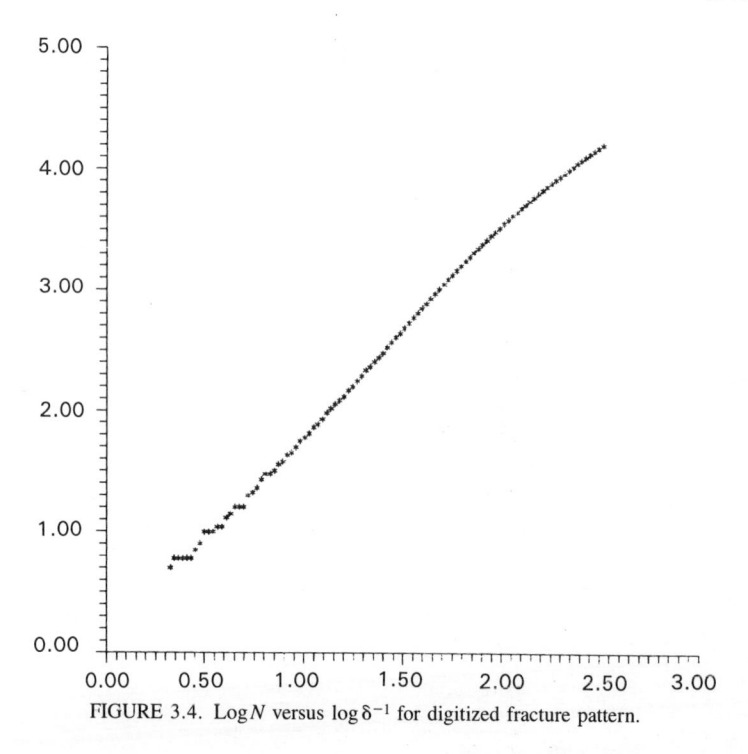

FIGURE 3.4. Log N versus log δ^{-1} for digitized fracture pattern.

deviation 0.00289. Using the larger range (0.0029, 0.05) produced a mean $D_2 = 1.57419$ and mean $D_3 = 1.62135$. The fact that they differ by 3% in their answers indicates something is awry (each standard deviation was only about 0.004). The D_1 values ranged from 1.781 for the larger box sizes to 1.664 for the smaller h, which is fairly inconclusive. This example demonstrates the desirability of examining a plot of log $N(\delta)$ versus log δ^{-1} before deciding on the box sizes to use.

3.4. CONCLUSIONS

From the mathematical motivation of Section 3.2 and the examples in Section 3.3, several conclusions can be drawn. While the mathematics indicates that box sizes should be taken arbitrarily small, this is not realistic, since there is usually a limiting resolution in the data or there are computer time and storage limitations. Hence errors due to finite sampling are unavoidable.

For data sets consisting of a finite number of line segments in a plane, too large a δ yields $D \approx 2$ (space-filling), while too small a δ produces $D \approx 1$ when individual line segments are resolved. In some applications, the fractal dimension is observed to change over different scales, since different physical phenomena are being observed. However the researcher must be careful that an observed change in dimension is not due to a poor choice of box sizes!

The likelihood of slow convergence renders the first algorithm (D_1) useful for only rough approximations to D. The other two algorithms exhibit similar behaviors, and it is probably useful to calculate both D_2 and D_3 for a given set of δs just for comparison. For two-dimensional data, the expense of computing both D_2 and D_3 is minor compared to the cost of producing N.

At a minimum, algorithms should (1) sample at a reasonable choice of δ values and (2) provide some statistical information about the accuracy and sensitivity of the output. By reasonable I mean not so large that data fill all the boxes and not so small that individual lines are resolved. These should probably be determined interactively rather than automatically; a graph of $\log N(\delta)$ versus $\log \delta^{-1}$ is extremely helpful. Perturbations should then be done on these box sizes to produce enough approximations to D to yield meaningful statistics. Finally for any published estimates for D, the range of δs or hs used should be reported, since estimates are often so sensitive to this choice.

REFERENCES

Barnsley, M., *Fractals Everywhere* (Academic Press, San Diego, 1988).
Barton, C., and Larsen, E., in *Fundamentals of Rock Joints* (Ove Stephannson, ed.) (Proceedings of the International Symposium on Fundamentals of Rock Joints, Bjorkliden, Sweden, 1983), pp. 77–84.
Barton, C., Schutter, T., and Samuel, J., *DIMENSION—A Computer Program That Computes the Fractal Dimension of Lines or Points in a Plane*, US Geological Survey Open-file Report, 1989.
Falconer, K., *The Geometry of Fractal Sets* (Cambridge University Press, New York, 1985).
———, *Fractal Geometry* (Wiley, New York, 1990).
Grassberger, P., *Phys. Lett.* **97A**, no. 6, 224 (1983).
Hausdorff, F., *Math. Ann.* **79**, 157 (1919).
Hunt, F., *SIAM J. App. Math.* **50**, 307 (1990).
Hunt, F., and Sullivan, F., in *Nonlinear Semigroups, PDEs and Attractors*, Lecture Notes in Mathematics, vol. 1394 (Springer-Verlag, New York, 1989).
Mandelbrot, B., *The Fractal Geometry of Nature* (W. H. Freeman, San Francisco, 1982).
Samuel, J., "A Method for Estimating the Hausdorff Dimension of a Planar Line Pattern," M.S. thesis, Colorado School of Mines, 1988.
Sreenivasan, K., and Meneveau, C., *J. Fluid Mech.* **173**, 357 (1986).

4

Measuring the Dimension of Self-Affine Fractals: Example of Rough Surfaces

S. R. Brown

4.1. INTRODUCTION

Many geophysical records are in the form of time or spatial series, where a parameter is observed sequentially at discrete intervals in time or space. Examples include the roughness of natural fractures in rock, land and seafloor topography, the spatial series comprising geophysical well logs, and some geophysical time series (some hydrologic records, for example). A description of the geometry of these functions is essential to studies of the mechanical and transport properties of fractures in rock (Brown and Scholz, 1986; Brown, 1989; scattering of acoustic waves from the sea floor (Fox and Hayes, 1985); detecting fractures by acoustic methods (deBilly and others, 1980); characterizing the spatial variation of permeability in oil and gas reservoirs and aquifers (Hewett, 1986); and evaluating the persistence of some time-dependent geophysical phenomena, including stream flows (Mandelbrot and Wallis, 1969). In each of the examples just cited, a class of fractals known as self-affine have been used to describe the spatial or time series. The distinction between self-affine fractals and the more familiar self-similar fractals is discussed in a subsequent section.

In the example of surface roughness, fractals are very useful. Recent work, discussed later, shows that the topography of natural surfaces in rock is adequately described by a self-affine fractal model. The fractal model of surface topography includes scaling or size-dependent properties; therefore the concept of surface roughness becomes much more general when traditional measures of surface roughness, such as root mean square (rms) roughness (the standard deviation of surface height) are combined with the fractal dimen-

---

S. R. Brown • Geomechanics Division, Sandia National Laboratories, Albuquerque, New Mexico.

Fractals in the Earth Sciences, edited by Christopher C. Barton and Paul R. La Pointe. Plenum Press, New York, 1995.

sion. For this reason, the mathematical formulation of fractals provides both a realistic and useful model of rough surfaces. The fractal model also allows realistic rough surfaces to be generated by computer (see Peitgen and Saupe, 1988), allowing many physical properties to be studied by numerical simulation, such as the mechanical and transport properties of a single fracture. This approach allows fracture models to be developed that are soundly based on the principles of physics, and successful models of this type can be more reliably extrapolated to other conditions than purely empirical descriptions.

To assist the reader in using fractals to study these types of geophysical records, three methods for estimating the fractal dimension of a one-dimensional self-affine time or spatial series are compared: the box-counting method, the divider method, and the spectral method. During the following discussion, some limitations of these methods are emphasized. For purposes of illustration, I center the discussion around the topography of fracture surfaces in rock, beginning with a justification for the study of fractures concluding with a summary of experimental observations on surface roughness.

4.2. FRACTURES IN ROCK

Fractures of all sizes, ranging from microcracks to joints and faults, are well-known for their effects on mechanical and transport properties of rock. Mechanical properties, such as bulk elastic constants and shear strength, are strongly affected by the presence of fractures (Goodman, 1976; Barton and Choubey, 1977; Walsh and Grosenbaugh, 1979; Brown and Scholz, 1986). Fractures also control the hydraulic conductivity of crystalline and tight sedimentary rock (Gangi, 1978; Kranz and others, 1979; Walsh and Brace, 1984; Brown, 1989). These effects arise from the fact that surfaces composing a fracture are rough and mismatched at some scale. The shape, size, number, and strength of contacts between surfaces control the mechanical properties. The separation between surfaces, or the aperture, determines the transport properties.

Walsh and Grosenbaugh (1979) show that the normal stiffness of a fracture varies approximately as the inverse of the rms asperity height. For interlocked surfaces, relationships between the rms surface slope and the peak shear strength of a joint have been suggested by Tse and Cruden (1979) following the experimental work of Barton and Choubey (1977). Therefore if roughness varies with the characteristic surface size, such as the diameter of a circular sample of a surface, so must the mechanical and transport properties. This remark must be tempered by the fact that mechanical and transport properties of fractures depend not only on the topography of the individual surfaces, but also on how well the two surfaces are correlated (Brown and others, 1986). Studying the scaling properties of individual surfaces however provides the groundwork for understanding scaling properties of fractures.

Scaling properties of natural rock surfaces have been studied in detail (Brown and Scholz, 1985; Scholz and Aviles, 1986; Power and others, 1987). In these studies, surface heights were digitized at equally spaced intervals along a line to produce a profile. These data were analyzed by computing the power spectral density function.

The power spectral density is computed essentially by breaking a time series, in this case the profile, into a sum of sinusoidal components—each with its own wavelength, amplitude, and phase. The squared amplitude of each component is referred to as its power, and a plot of power versus wavenumber or inverse of wavelength is referred to as the power spectrum. The phase indicates the position of the first peak of each sinusoidal component relative to all others. The phase spectrum is a plot of the phase as a function of wave number. Phase spectra for rough surfaces are typically random; that is, there is no consistent relation between phase and wave number. Excellent introductions to spectral analysis are given by Bendat and Piersol (1971) and Bath (1974).

Comparison of power spectra shows that all natural rock surfaces, including bedding planes, tension cracks (joints), and frictional wear surfaces (faults), are remarkably similar. To a first approximation, all surfaces have a power spectral density functions $G(k)$ of the form:

$$G(k) = Ck^{-\alpha} \qquad (1)$$

where $k = 2\pi/\lambda$ is the wave number and λ is the wavelength or distance along the profile. The proportionality constant C varies among surfaces. The power α usually falls in the range of $2 < \alpha < 3$. Sayles and Thomas (1978) found similar behavior for numerous other random surfaces. Since the power spectrum of rough surfaces are power laws with a negative exponent, then long wavelength features have higher amplitudes and contribute more to the overall roughness than short wavelength features. This power law form of the power spectrum can be interpreted in terms of fractal geometry.

4.3. SCALING, FRACTALS, AND CROSSOVER LENGTH

Mandelbrot (1983) suggested that fractals are useful mathematical models of rough surfaces, and indeed this has some physical basis (Termonia and Meakin, 1986). In the present context, a fractal is a particular mathematical model of irregular geometry whose scaling properties are described by the fractal dimension D. The fractal dimension can range between topologic and Euclidian dimensions. For example, a profile of a rough surface is topologically a line (dimension 1), but is defined in Euclidian two-space, and the fractal dimension falls between 1 and 2. In this sense, the fractal dimension is a measure of how much space a particular function fills.

Two classes of fractals are distinguished, self-similar and self-affine. One familiar example of a self-similar fractal is Brownian motion in the plane of a microscope slide (Mandelbrot, 1983, Plate 13). If we trace on graph paper the path of the particle through time at two different magnifications, the two drawings look statistically the same (i.e., they have the same statistical moments). Since a simple change in magnification left the complexity of the curve unchanged, this process is self-similar; in this case, the fractal dimension is D = 2. If we define a coordinate system on the microscope slide and graph the x-position of the particle as function of time, then a self-affine fractal of dimension D = 1.5 is obtained. To be precise, if we refer to this function as $B(t)$ and the time axis is rescaled by two positive numbers h and g, then the functions $h^{D-2}B(ht)$ and $g^{D-2}B(gt)$ are statistically the same, whereas $B(ht)$ and $B(gt)$ are not. To obtain statistically equivalent graphs, the position axis must be scaled differently than the time axis; thus $B(t)$ is merely self-affine. The important distinction between self-similar and self-affine functions is this particular anisotropy of scaling (Mandelbrot, 1983; Feder, 1988). As we will see, the distinction between self-similar and self-affine fractals has some bearing on the method used to estimate the fractal dimension.

Self-affine fractals can be used as models of rough surfaces, since they have power spectral density functions of the form Eq. (1) when they are in the form of linear profiles. Figures 4.1 and 4.2 show examples of a self-affine fractal and its power spectral density function. When $2 \leq \alpha < 3$, the fractal dimension D can be estimated from the relation $D = 2.5 - \alpha/2$, corresponding to $1 < D \leq 1.5$ (Berry and Lewis, 1980; Mandelbrot, 1983; Voss, 1985). The justification for the relationship between D and the power spectrum is based on nonrigorous mathematical derivations and empirical evidence. In the same vein, this relationship is further supported in Chapter 4 by comparing the spectral method with two other methods.

The use of the spectral method is not without its problems. The preceding expression relating α and D may not be reliable with $D > 1.5$ (Brown and Scholz, 1985). Time or spatial series always have the property that they are single-valued (i.e., they do not have overhangs), and this property is a requirement for the power spectrum to be defined. When $D > 1.5$, it becomes possible for a self-affine fractal to be multivalued. Sampling this function as a discrete time or spatial series will miss the overhangs and as a consequence the fractal dimension computed by the spectral method will be too small. Additionally the power law form of the power spectrum is not a sufficient condition for determining whether or not a particular function is fractal. The function must also have a random phase spectrum. We can easily construct functions with nonfractal characteristics from a power law distribution of Fourier components having nonrandom phase (Hough, 1989).

The power spectrum $G(k)$ provides a useful description of the surface roughness if we consider the moments of the power spectrum, defined as

$$m_n = \int_{k_0}^{\infty} k^n G(k) \, dk \qquad (2)$$

where m_n is the nth moment and $k_0 = 2\pi/\lambda_0$ is the wave number corresponding to the profile length λ_0. In practice the upper limit of integration is the Nyquist cutoff corresponding to a wavelength of twice the sample interval. By combining the derivative theorem and Parseval's theorem for the Fourier transform (Bath, 1974) we find that m_0 is the variance (mean square value) of heights on the profile, m_2 is the variance of slopes, and m_4 is the variance of curvatures. When $G(k)$ represents a self-affine fractal, then Eq. (2) gives

$$m_0 = \sigma^2 = \kappa \lambda_0^{2(2-D)}$$

where κ depends on the constant C in Eq. (1). This is the scaling law for the rms surface height σ. Since this fractal model of surface topography includes the scaling or size-dependent properties, the concept of surface roughness becomes much more general than when using only the traditional measures of surface roughness, such as rms height σ.

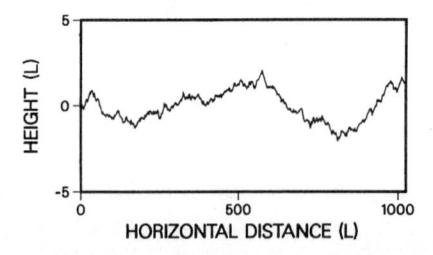

FIGURE 4.1. Self-affine fractal with fractal dimension $D=1.5$, generated by using the algorithm of Fournier and others (1982). Height and horizontal distance are expressed in arbitrary length units (L). This function contains a total of 1025 points.

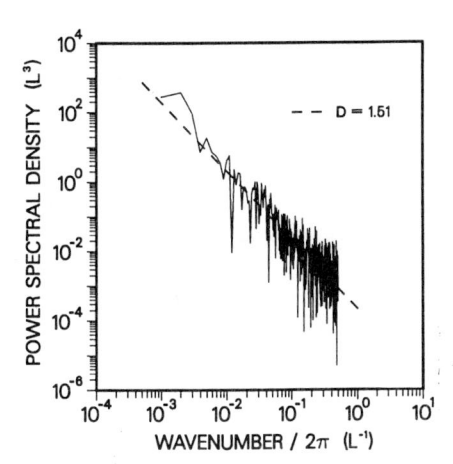

4. DIMENSION OF SELF-AFFINE FRACTALS: EXAMPLE OF ROUGH SURFACES   81

FIGURE 4.2. Power spectral density function for the self-affine fractal in Fig. 4.1. The wave number and power are expressed in terms of arbitrary length units (L). The dashed curve is a least squares fit for wave number/$2\pi < 3 \times 10^{-1}$. The slope gives the fractal dimension $D = 1.51$.

For the case $D = 1.5$, Sayles and Thomas (1978) refer to κ as the topothesy, and they tabulate its value for various surfaces. Wong (1987) defines a different constant, the crossover length b, such that the standard deviation of heights is

$$\sigma = b\left(\frac{\lambda_0}{b}\right)^{2-D}$$

thus $\kappa = b^{2D-2}$. The crossover length is in fact the same generalized topothesy suggested by Berry and Hannay (1978) in a comment to Sayles and Thomas (1978). One important property of the crossover length is that when $\lambda = b$, then $\sigma = b$.

4.4. ESTIMATING THE FRACTAL DIMENSION

The definition of the fractal dimension is tied to the measure or size of a set of points in space, where the measure is the sum of the sizes of small elements used to cover the set. For example, a surface may be covered by small circular disks to produce an estimate of the total area. The fractal dimension D describes how the total size of the set depends on the size of the covering elements. An excellent discussion of this subject is given by Feder (1988). The calculation of the fractal dimension from this precise definition is difficult because the method requires the most efficient covering of the set and does not assume a priori a particular shape of the covering elements. For this reason, less general alternative methods of estimating the fractal dimension must be used in practice (Mandelbrot, 1983, 1985).

I discuss three of the many possible methods for estimating the fractal dimension. Of all methods in common use, perhaps that closest to the definition of fractal dimension is the box-counting method (Mandelbrot, 1983; Feder, 1988). This is performed by laying a regular grid of boxes of characteristic size r over the curve and counting the number of boxes intersected by the curve. This is repeated for a large number of box sizes. The number of boxes filled is plotted as a function of the total number of boxes in the grid on a log-log plot. If linear, the slope of this curve is related to D. Two other methods of estimating the dimension of a self-affine fractal have also been used. The spectral method is discussed in the preceding section. The other method is known alternatively as the ruler, compass, or divider method (Mandelbrot, 1983; Feder, 1988). This is performed conceptually by opening a pair of dividers to some distance r and walking them along the profile to estimate its total length. The total number of steps (the total length of the line) is plotted as a function of r on a log-log plot. If linear, the slope of this curve is related to D.

Each of these three methods has some limits to its applicability. Problems with the spectral method are discussed in the preceding section. The spectral method is expected to be valid only for self-affine fractals in the form of time or spatial series. Both the box and divider methods always give a good approximation to the correct value of D for self-similar fractals. Both methods have also been applied in some form to self-affine fractals (Aviles and others, 1987; Carr and Warriner, 1987; Okubo and Aki, 1987; Turk and others, 1987). Both the box and divider methods give the correct fractal dimension for self-affine fractals only under certain conditions (Mandelbrot, 1985; Wong, 1987), and indeed some values reported from the divider method seem to be quite low [for example, compare values $D \approx 1$ from the divider method obtained by Aviles and others (1987) to values $D \approx 1.1–1.5$ from the spectral method obtained from the same data by Scholz and Aviles (1986)].

In Chapter 4, I compare the box, divider, and spectral methods for estimating the fractal dimension of self-affine functions. I begin by recounting an evaluation of the divider method given previously by Wong (1987), which assumes that the spectral dimension is correct. Following the same logic, I then extend this derivation to the box method. Some problems in applying these methods are made clear. Finally these problems are further illustrated through an example.

4.4.1. Divider Method

Suppose we have a self-affine fractal with a nominal length λ_0 digitized at discrete intervals of length r (see Fig. 4.3). Vertical fluctuations over the distance r are on average equal to the standard deviation of heights σ (defined as the average deviation from the mean). There are λ_0/r segments in all, and the total length of line λ as a function of r is approximately

$$\lambda = \frac{\lambda_0}{r}(r^2 + \sigma^2)^{1/2} \qquad (3)$$

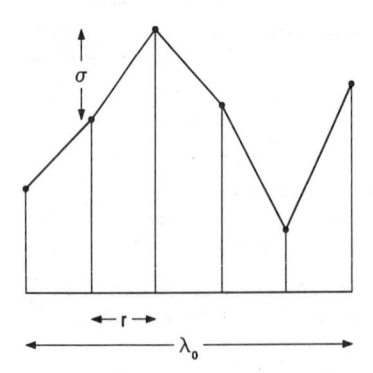

FIGURE 4.3. Definition of terms used in evaluating the divider method for estimating the fractal dimension of a self-affine fractal, where σ is the standard deviation of surface height (average deviation from the mean) over a length scale r and λ_0 is the total profile length.

4. DIMENSION OF SELF-AFFINE FRACTALS: EXAMPLE OF ROUGH SURFACES

Recalling that for a self-affine fractal over distance r, the standard deviation of heights is

$$\sigma = b\left(\frac{r}{b}\right)^{2-D}$$

(Wong, 1987), then Eq. (3) becomes

$$\lambda = \lambda_0\left[1 + \left(\frac{r}{b}\right)^{2(1-D)}\right]^{1/2} \quad (4)$$

The behavior of this equation is shown in Fig. 4.4. Since the fractal dimension $D > 1$, then for $r \ll b$ we obtain

$$\lambda \approx \lambda_0\left(\frac{r}{b}\right)^{1-D}$$

Thus $\log(\lambda)$ versus $\log(r)$ has slope $1 - D$. When $r \gg b$, then $\lambda \approx \lambda_0$. In this case, calculating D by using the slope of the r-λ curve gives $D \approx 1$. The crossover length b is interpreted as the horizontal sampling interval above which the divider method breaks down. Apparently to obtain meaningful results from the divider method, we must have data digitized at a scale much smaller than the crossover length.

Even without prior knowledge of b however, a simple solution to this problem exists. From the definition of b and Eq. (4), we notice that if the ordinate of the self-affine fractal is multiplied by a constant greater than 1 to increase the standard deviation of heights σ, then the effective crossover length can be increased relative to the sample interval. Thus for a given range of divider lengths r, the limit $r \ll b$ can be reached and the correct fractal dimension obtained without actually changing the sample interval. A computer program was written to demonstrate this procedure, and the results of calculations for the self-affine fractal of Fig. 4.1 are shown in Fig. 4.5. The value $D = 1.50$ obtained for high magnifications agrees well with that obtained by the spectral method. The crossover length need not be known ahead of time. We can simply magnify the ordinate of the function repeatedly by various factors until a stable estimate of D is obtained. In fact whenever the divider method is used, the stability of the results should be tested in this manner.

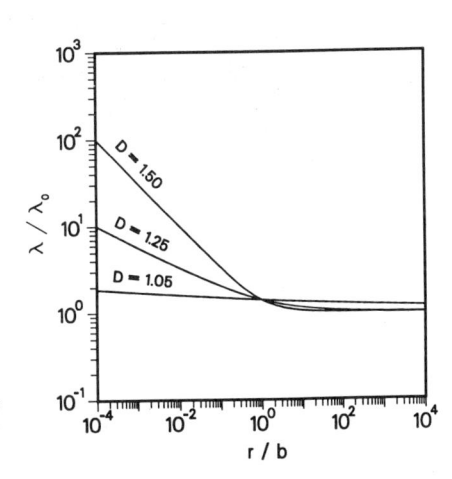

FIGURE 4.4. Graph of Eq. (4) for several values of the fractal dimension D. For $r \ll b$, the slope of the curves if $1 - D$, but for $r \gg b$, the slope is always close to zero.

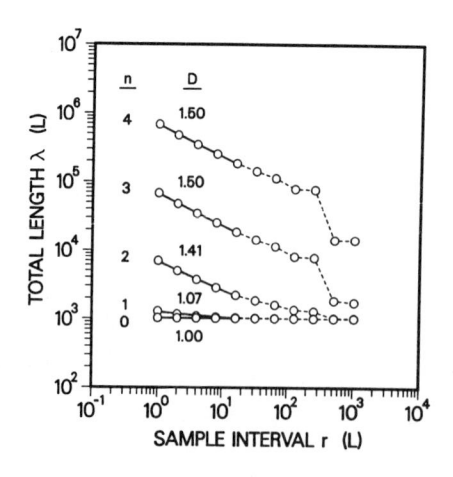

FIGURE 4.5. The divider method applied to the self-affine fractal in Fig. 4.1. The sample interval and total length are expressed in arbitrary length units (L). The label n represents the power of 10 (10^n) by which the ordinate of the fractal was magnified prior to applying the method. The label D represents the resulting fractal dimensions obtained from a least squares fit to the left-most five points on each curve. The value $D = 1.50$ obtained for high magnifications agrees well with that obtained from the spectral method.

The arguments just made were based on a modified version of the divider method, where the sample interval r was held constant at each step and the opening of the dividers varied. This is not quite the same as walking a pair of dividers with fixed openings along the function. The same conclusions hold for the true divider method, where divider lengths are held constant and the sample interval varies. While the results are not shown here, a computer program implementing the true divider method was written and run for the self-affine fractal of Fig. 4.1. The results are nearly identical to Fig. 4.5.

The crossover length b can be estimated from the power spectral density function. From Eqs. (1) and (2) and the definition of b, the constant of proportionality in Eq. (1) is found to be

$$C = (4 - 2D)(2\pi)^{4-2D} b^{2D-2}.$$

Plots of the power spectral density are normally presented in terms of the reciprocal wavelength ($1/\lambda$) rather than the wave number k. In this case, Eq. (1) can be written as

$$G\left(\frac{1}{\lambda}\right) = (4 - 2D) b^{2D-2} \left(\frac{1}{\lambda}\right)^{-(5-2D)} \quad (5)$$

By plotting $\log(G)$ versus $\log(1/\lambda)$, D can be estimated from the slope of the resulting straight line. The crossover length b can be calculated from D and the intercept of the straight line. From Fig. 4.2, the crossover length is $b \approx 10^{-3}$. The rms roughness σ is proportional to $b^{1/2}$ when D=1.5. The maximum value of r used in Fig. 4.5 to determine D is 16. Therefore to make r small, say $r < 0.1b$, the ordinate of Fig. 4.1 must be magnified by at least 400 [$= (10 \cdot 16 \cdot 1000)^{1/2}$]. After this magnification, the divider method gives the correct fractal dimension (see Fig. 4.5). Using published plots of $G(1/\lambda)$, the crossover length is found to be $b \approx 10^{-8} - 10^{-5}$ m for natural joints (Brown and Scholz, 1985) and $b \approx 5 - 100$ m for the 1906 section of the San Andreas fault (Scholz and Aviles, 1986).

4.4.2. Box-Counting Method

The box-counting method is perhaps the most familiar method of calculating the fractal dimension. It is implemented by covering the function with a grid of identical square boxes, then counting the number of boxes intersecting the function as a function of the box

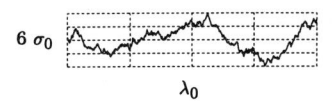

4. DIMENSION OF SELF-AFFINE FRACTALS: EXAMPLE OF ROUGH SURFACES

FIGURE 4.6. Illustration of the grid used in the derivation of the box-counting method. The function completely fills the outer rectangle, which has length λ_0 and height $\approx 6\sigma_0$. The outer rectangle is divided into n^2 boxes, each with the same aspect ratio as the outer rectangle.

size (see Mandelbrot, 1983, or Feder, 1988, for example). A version fo the box-counting method for self-affine fractals can be derived in a manner similar to the divider method. In this derivation, we find that the problem with the crossover length does not arise. To begin suppose we have a self-affine fractal with a nominal length λ_0. As before vertical fluctuations over the distance r are on average equal to the standard deviation of heights $\sigma = \sigma(r)$. Assume that the height distribution is nearly Gaussian. For a Gaussian distribution, 99.7% of the heights fall with $\pm 3\sigma$ of the mean value. Therefore we assume that local peak–valley distances of heights on the surface is of the order 6σ. Likewise total peak–valley distances over the entire function length λ_0 is approximately $6\sigma_0$, where $\sigma_0 = \sigma(\lambda_0)$ is the standard deviation of heights of the entire function. Actual rough surfaces for example have finite maximum and minimum values, so the chance of finding a value outside this 6σ range is extremely small.

Draw a rectangle around the function, which has length λ_0 (along the abscissa) and height $6\sigma_0$ (along the ordinate), so that the rectangle is completely filled. Divide the large rectangle into n^2 boxes with the same aspect ratio as the large rectangle; i.e., each box is $r = \lambda_0/n$ long by $6\sigma_0/n$ high (see Fig. 4.6).

We can now calculate the number of small boxes it takes to cover the function. Along each of the n intervals of length r along the abscissa, there are approximately

$$\frac{6\sigma(r)}{(6\sigma_0/n)}$$

boxes filled (see Fig. 4.7). Summing over all n intervals along the abscissa, there are a total of N boxes filled, where

$$N(n) \approx \frac{6\sigma(r)}{6\sigma_0/n} \cdot n = \frac{\sigma(r)}{\sigma_0} \cdot n^2 \qquad (6)$$

Recalling that for a self-affine fractal over distance r the standard deviation of heights is $\sigma = b(r/b)^{2-D}$ (Wong, 1987), then Eq. (6) becomes:

$$N(n) \approx \frac{\sigma(r)}{\sigma_0} \cdot n^2 = \frac{b(r/b)^{2-D}}{b(\lambda_0/b)^{2-D}} \cdot n^2 = \left(\frac{r}{\lambda_0}\right)^{2-D} \cdot n^2 \qquad (7)$$

and since $\lambda_0/r = n$, then

$$N(n) \approx n^{D-2} \cdot n^2 = n^D \qquad (8)$$

FIGURE 4.7. Each of the n intervals along the abscissa have width $r = \lambda_0/n$. The height of each of the n rectangles in this interval is $6\sigma_0/n$. Over this interval of length r, the function has a total peak valley range of approximately $6\sigma(r)$, where $\sigma(r)$ is the standard deviation of heights over the length r.

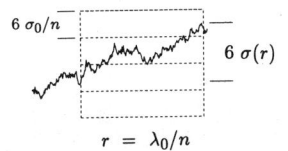

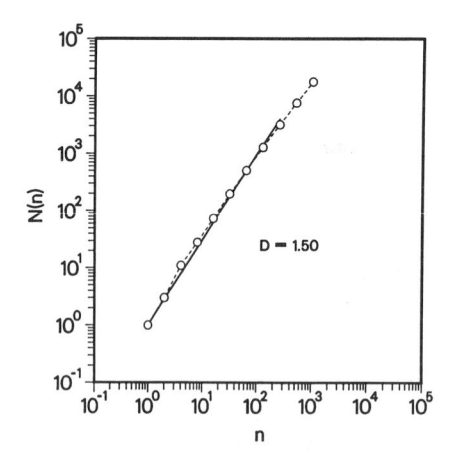

FIGURE 4.8. The box-counting method applied to the self-affine fractal in Fig. 4.1. The fractal dimension of this function is given by Eq. (8). The value shown was obtained from a least squares fit to the left portion of the curve, as indicated by the solid line (≤256 boxes). As indicated by Eq. (8), the curve was required to pass through the point (1,1). The total length of the self-affine function is 1025, thus the last two points ($n = 512$, $n = 1024$) were ignored in the least squares fit of Eq. (8), since they represent at most two sample points per box. The value $D = 1.50$ agrees well with that obtained from the other two methods.

Notice that in this particular construction of the box-counting method where boxes have the same aspect ratio as the entire data record, the crossover length b cancels out in the final expression. This is not true in general for other aspect ratios, say, for square boxes, and we would have to pay attention to the box size relative to the crossover length. The choice of this particular aspect ratio for boxes is equivalent to magnifying the ordinate of the function to overcome the crossover length problem just discussed for the divider method.

A computer program was written to demonstrate this version of the box-counting method, and results of calculations for the same self-affine function considered before in Fig. 4.1, are shown in Fig. 4.8. The value $D = 1.50$ obtained by the box-counting method agrees well with that obtained by the other two methods.

4.5. CONCLUSIONS

The scaling properties of self-affine fractals are described by two parameters, the fractal dimension and the crossover length. An analysis using these parameters allows three methods for estimating the fractal dimension of a self-affine spatial series to be compared: the box-counting, divider, and the spectral. These methods yield the same results if the sampling interval over which the function is measured is smaller than the crossover length. For sample intervals greater than the crossover length however, the box-counting method and the divider method always give fractal dimensions close to 1. This demonstrates that care must be taken when estimating the dimension of a self-affine fractal. Simple modifications to both the box-counting method and the divider method allow the correct fractal dimension to be obtained without prior knowledge of the crossover length.

ACKNOWLEDGMENTS

A substantial portion of Chapter 4 was published previously as Brown (1987) and Brown (1988). I thank Po-zen Wong, Simon Cox, Brian Bonner, and Sue Hough for helpful discussions. This work was performed at Sandia National Laboratories supported by the US Department of Energy under contract number DE-AC04-76DP00789.

REFERENCES

Aviles, C. A., Scholz, C. H., and Boatwright, J., *J. Geophys. Res.* **92**, 331 (1987).
Barton, N., and Choubey, V., *Rock Mech.* **10**, 1 (1977).
Bath, M., *Spectral Analysis in Geophysics* (Elsevier, New York, 1974).
Bendat, J. S., and Piersol, A.G., *Random Data Analysis, and Measurement Procedures* (Wiley, New York, 1971).
Berry, M. V., and Hannay, J. H., *Nature* **273**, 573 (1978).
Berry, M. V., and Lewis, Z. V., *Roy. Soc. London Proc. Ser. A* **370**, 459 (1980).
Brown, S. R., *Geophys. Res. Lett.* **14**, 1095 (1987).
———, *Geophys. Res. Lett.* **15**, 286 (1988).
———, *J. Geophys. Res.* **94**, 9429 (1989).
Brown, S. R., Kranz, R. L., and Bonner, B. P., *Geophys. Res. Lett.* **13**, 1430 (1986).
Brown, S. R., and Scholz, C. H., *J. Geophys. Res.* **90**, 12,575 (1985).
———, *J. Geophys. Res.* **91**, 4939 (1986).
Carr, J. R., and Warriner, J. B., in *Proceedings, 28th US Symposium on Rock Mechanics*, Tucson (A. A. Balkema, Boston, 1987) pp. 73–80.
deBilly, M., Cohen-Tenoudji, F., Quentin, G., Lewis, K., and Adler, L., *J. Nondestr. Eval.* **1**, 249 (1980).
Feder, J., *Fractals* (Plenum Press, New York, 1988).
Fournier, A., Fussell, D., and Carpenter, L., *Commun. ACM* **25**, 371 (1982).
Fox, C.G., and Hayes, D. E., *Rev. Geophys.* **23**, 1 (1985).
Gangi, A. F., *Int. J. Rock Mech. Min. Sci. Geomech. Abstr.* **15**, 249 (1978).
Goodman, R. E., *Methods of Geological Engineering in Discontinuous Rocks* (West Publishing, New York, 1976).
Hewett, T. A., in *Proceedings, 61st Society of Petroleum Engineers Annual Technical Conference*, Paper No. SPE 15386 (Society of Petroleum Engineers, Richardson, TX, 1986).
Hough, S. E., *Geophys. Res. Lett.* **16**, 673 (1989).
Kranz, R. L., Frankel, A. D., Engelder, T., and Scholz, C. H., *Int. J. Rock Mech. Min. Sci. Geomech. Abstr.* **16**, 225 (1979).
Mandelbrot, B. B., *The Fractal Geometry of Nature* (W. H. Freeman, San Francisco, 1983).
———, *Physica Scripta* **32**, 257 (1985).
Mandelbrot, B. B., and Wallis, J. R., *Water Resour. Res.* **5**, 321 (1969).
Okubo, P. G., and Aki, K., *J. Geophys. Res.* **92**, 345 (1987).
Peitgen, H-O., and Saupe, D., *The Science of Fractal Images* (Springer-Verlag, New York, 1988).
Power, W. L., Tullis, T. E., Brown, S. R., Boitnott, G. N., and Scholz, C. H., *Geophys. Res. Lett.* **14**, 29 (1987).
Sayles, R. S., and Thomas, T.R., *Nature* **271**, 431 (1978).
Scholz, C. H., and Aviles, C.A., in *Earthquake Source Mechanics*, Geophysical Monograph Series, Vol. 37 (S. Das, J. Boatwright, and C. H. Scholz, eds.) (American Geophysical Union, Washington, DC, 1986), pp. 147–55.
Termonia, Y., and Meakin, P., *Nature* **320**, 429 (1986).
Tse, R., and Cruden, D. M., *Int. J. Rock Mech. Min. Sci. Geomech. Abstr.* **16**, 303 (1979).
Turk, N., Greig, M. J., Dearman, W. R., and Amin, F. F., in *Proceedings, 28th US Symposium on Rock Mechanics, Tucson* (A. A. Balkema, Boston, 1987) p. 1223–236.
Voss, R.F., in *Scaling Phenomena in Disordered Systems* (R. Pynn and A. Skjeltorp, eds.) (Plenum Press, New York, 1985), pp. 1–11.
Walsh, J. B., and Brace, W. F., *J. Geophys. Res.* **89**, 9425 (1984).
Walsh, J. B., and Grosenbaugh, M. A., *J. Geophys. Res.* **84**, 3532 (1979).
Wong, P., in *Physics and Chemistry of Porous Media II*, AIP Conference Proceedings 154 (J. R. Banavar, J. Koplik, and K. W. Winkler, eds.) (American Institute of Physics, NY, 1987), pp. 304–18.

5

Review of the Fractal Character of Natural Fault Surfaces with Implications for Friction and the Evolution of Fault Zones

W. L. Power and T. E. Tullis

5.1. INTRODUCTION

Surface roughness has been a topic of considerable interest within the engineering community for many years, primarily because surface roughness affects the frictional properties and wear of sliding surfaces, the properties of electrical contacts, and the stiffness of machine parts (Thomas, 1982). Most engineering studies however concentrate on surfaces that can be described using Euclidean geometry rather than the type of surfaces described by fractal geometry. The reason is that almost all man-made surfaces, and many natural ones, mimic simple Euclidean geometric forms, such as spheres, cylinders, planes, or their derivatives. Euclidean shapes have great practical importance; for example, ball bearings function well because they are close approximations to spheres, while a mirror provides an undistorted reflection because it is a close approximation to a plane.

Most studies of surface roughness relied on Euclidean models for surface description. In this approach, the surface is thought of as the sum of two parts—first a simple Euclidean shape that describes the overall form of the object and second a random or stochastic departure from the Euclidean ideal (see Fig. 5.1a). Because of the nature of most surface preparation techniques, many roughness features of engineering interest fall within a well-defined and limited size range. For these surfaces, height distributions and simple statistical

---

W. L. Power and T. E. Tullis • Department of Geological Sciences, Brown University, Providence, Rhode Island; *present address of W.L.P.*: CSIRO Exploration and Mining, Nedlands, Western Australia, Australia.
Fractals in the Earth Sciences, edited by Christopher C. Barton and Paul R. La Pointe. Plenum Press, New York, 1995.

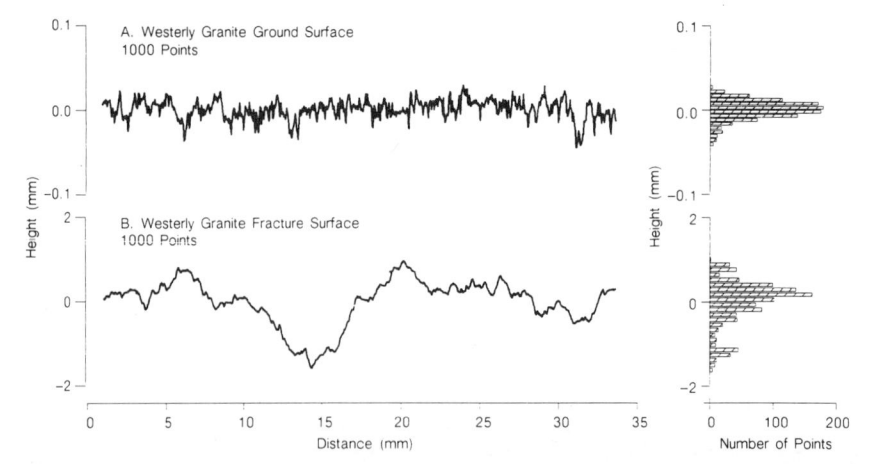

FIGURE 5.1. Example of rock surfaces illustrating the differences between Euclidean and fractal models. (a) Ground surface typical of rock surfaces used in many laboratory studies. This surface can be described as the sum of a simple, analytically describable Euclidean shape (a plane) and a stochastic departure from the Euclidean ideal usually defined as roughness. Roughness is commonly described using height distributions (*right*) or one-parameter descriptions, such as the rms roughness (see Fig. 5.2). Ground surfaces are commonly prepared in two stages: First a flat surface is machined on the sample; second the resulting surface is roughened at a small scale using loose abrasive. (b) A rock fracture surface that can be described by a fractal surface model. Fractal surfaces have roughness at a wide range of scales. Euclidean shapes cannot describe surfaces of this type because the stochastic and random character of the surface extends to the largest scale observed. Height distributions from such surfaces typically have a bi- or trimodal character whose details depend on the particular area of the surface profiled. (Reprinted with permission from the American Geophysical Union.)

parameters, such as the rms roughness or centerline average roughness, are used to characterize the surface (Myers, 1962; Thomas, 1982).

Fractal models provide a more realistic framework for the description of many natural surfaces, because fractal geometry does not rely on Euclidean shapes. In contrast to man-made surfaces, roughness features of many natural surfaces occur over a broad range of scales or over a broad bandwidth. Two concepts from fractal geometry, self-similar and self-affine scaling, provide the means of developing new surface models that are particularly well-suited to natural surfaces. In Chapter 5 we use the term self-similar to denote statistically self-similar, as defined by Mandelbrot (1983):

> A bounded random set S is statistically self-similar, with respect to the ratio t and an integer M, when S is the union of M nonoverlapping subsets, each of which is of the form $t(S_M)$ where the M sets S_M are congruent in distribution to S. (pp. 349–50)

We also use the term self-affine to denote statistically self-affine as Mandelbrot (1983, p. 350) defines it.

Mandelbrot's definitions of self-similar and self-affine scaling can be applied to a set of surface profile data consisting of a large number of measurements of the height of a surface y_i, arranged along a single linear direction x. If a surface profile is statistically self-similar, when magnified isotropically, a portion of the profile appears statistically the same as the entire profile. Another way of stating this is that the appearance of the surface is independent of scale. Thus if both an elephant and an ant were to walk on a self-similar surface, they would observe the same type of topography relative to their respective body

5. REVIEW OF THE FRACTAL CHARACTER OF NATURAL FAULT SURFACES

sizes. For a surface profile, statistically self-affine means that a portion of the profile appears statistically the same as the entire profile only if different magnification factors are used for the directions parallel and perpendicular to the surface. For self-affine profiles, asperity curvature and slope change with observation scale. For a self-affine surface, an elephant and an ant would observe different topographies, but roughness would still be present at all scales.

If we blindly apply the same description methods used for Euclidean surfaces to fractal surfaces, problems result. For example, height distributions for fractal surfaces are poorly defined, and they depend strongly on the profile length or scale of observation (see Figs. 5.1b and 5.2). This means that for a fractal surface, common roughness parameters, such as the rms roughness, depend on the scale of observation (Sayles and Thomas, 1978; Scholz and Aviles, 1986; Power and Tullis, 1991a). To solve this problem, it is necessary to describe fractal models using at least two parameters. For example, one parameter can be used to describe the amplitude or steepness of the surface roughness at one particular scale, while the other can be used to describe how the roughness varies with scale (Power and Tullis, 1991a).

5.2. TECHNIQUES FOR OBSERVING AND CHARACTERIZING FRACTAL SURFACES

In many applications, the goal of statistical surface description is to provide surface parameters that are simpler and easier to use than a complete deterministic description of a surface. To determine the best-fit fractal model to a given set of real surface data, two

FIGURE 5.2. Root mean square roughness as a function of observed profile length for the Westerly Granite surface shown in Fig. 5.1. The ground surface has a fairly constant rms roughness for all observation scales and profile lengths greater than about 1.5 mm. This is because at scales longer than about 1.5 mm, the ground surface is essentially flat. The fracture surface does not have a constant rms roughness because the surface is fractal. The calculated rms roughness depends strongly on the amount of the surface profile used in the calculation. (Reprinted with permission from the American Geophysical Union.)

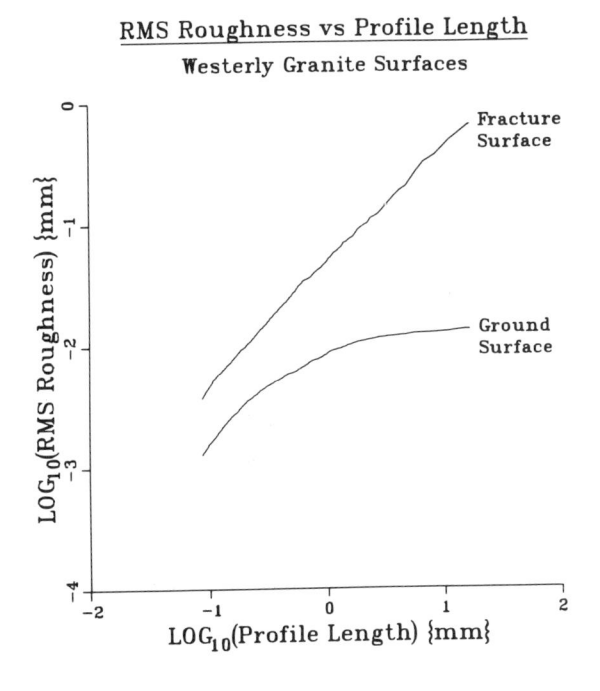

methods are in common use; these are the divider method and the spectral method. Both methods are easily adapted for use with surface profile data (Power and Tullis, 1991a). We use spectral methods in Chapter 5, because they are derived from time series analysis, a long-standing and well-established technique. Divider (or ruler) methods have been described by Mandelbrot (1983), and the relationships between divider methods and spectral methods have been explored by Berry and Lewis (1980), Scholz and Aviles (1986), Aviles and others (1987), Brown (1987), Hough (1989), Fox (1989), Carr (1989a, 1989b), and Power and Tullis (1991a).

In the spectral approach, amplitude spectra or power spectra are computed from surface profile data using methods described in many texts (Priestley, 1981; Bracewell, 1986; Bendat and Piersol, 1986, pp. 361–403). The sequence of operations used in Chapter 5 is summarized in Fig. 5.3, and described in more detail by Power and others (1988)

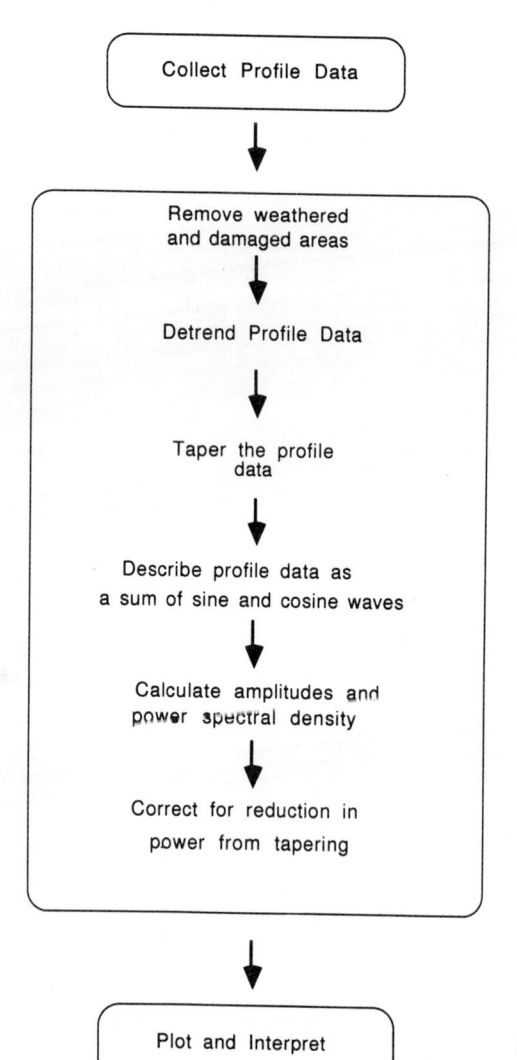

FIGURE 5.3. Sequence of operations used to calculate power spectra from surface profile data.

5. REVIEW OF THE FRACTAL CHARACTER OF NATURAL FAULT SURFACES

and Power and Tullis (1991a). The statistical character of surfaces, as opposed to the statistical character of profiles, can be easily derived from surface profile data (Nayak, 1971; Adler and Firman, 1981; Brown and Scholz, 1985; Goff and Jordan, 1988).

Spectra calculated from ideally self-similar or self-affine surfaces have power spectral density functions $G(f)$ of the form

$$G(f) = Cf^{-\alpha} \tag{1}$$

where f is the spatial frequency and α and C are constants (Berry and Lewis, 1980; Brown and Scholz, 1985; Scholz and Aviles, 1986; Brown, 1987). To determine the two parameters α and C that describe the fractal model, it is useful to recast Eq. (1) as follows

$$\log G(f) = \log C - \alpha \log(f) \tag{2}$$

and then plot $\log G(f)$ as a function of $\log(f)$ (see Fig. 5.4). The slope of the power spectrum α describes how the surface roughness changes with scale, while the intercept $\log C$ describes the steepness of the surface topography and specifically the amplitude to wavelength ratio of each Fourier component. Spectra from ideally self similar surface profiles have power spectra with α equal to exactly 3 (the slope of -3 on a plot of power spectral density versus spatial frequency), while power spectra from self-affine surfaces have power spectra with α other than 3 (Power and Tullis, 1991a).

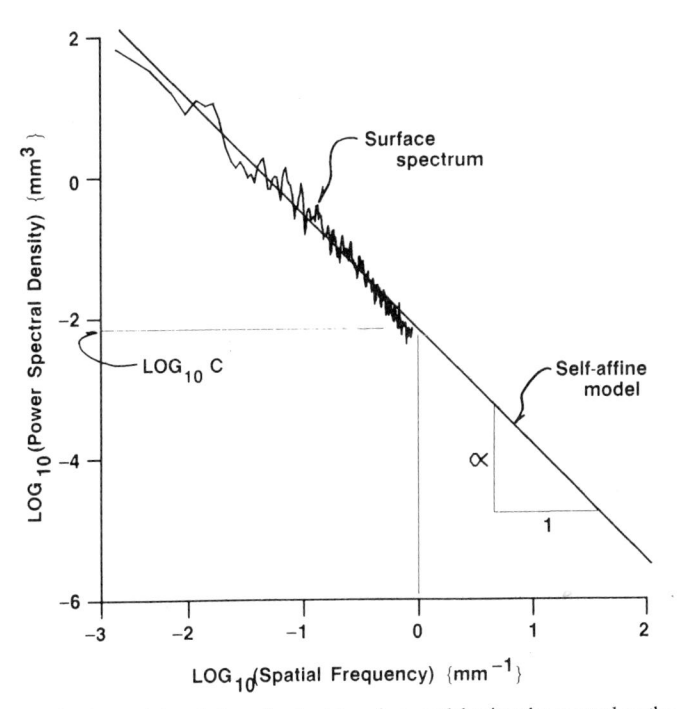

FIGURE 5.4. Determination and description of a fractal surface model using the spectral method. Estimates of the power spectral density at discrete spatial frequencies (here shown connected with the irregular line and labeled surface spectrum) are fit with a linear fractal model (*solid line*). The model is described using two constants, the slope α and the intercept $\log(C)$. See text and Eq. (2). (Reprinted with permission from the American Geophysical Union.)

5.3. OBSERVATIONS OF NATURAL ROCK SURFACES

A variety of studies have measured the roughness of natural rock surfaces (Rengers, 1970; Schneider, 1974; Weissbach, 1978; Swan, 1983; Reeves, 1985; Brown and Scholz, 1985; Power and others, 1987; Carr and Warriner, 1987; Turk and others, 1987; Carr, 1989c; Maerz and Franklin, 1990). Most quantitative studies have involved measurements using stylus profilimeters of the type described by Brown (1984) and Keller and Bonner (1985). In addition to surface profilimetry, measurements have been made using photogrammetric methods, laser cameras, video cameras or shadow profilimetry (Carr, 1989c; Maerz and Franklin, 1990), slit island analysis (Mandelbrot and others, 1984), or resin replication techniques (Gentier and others, 1989). Although a wide variety of techniques can be used, any single technique or measurement instrument typically gives information over only a limited range—usually 1–2 orders of magnitude. Some studies have increased the bandwidth or range by applying different instruments or different techniques at different scales (Brown and Scholz, 1985; Power and others, 1987; Maerz and Franklin, 1990).

Profile data from natural fault surfaces reveal a number of important characteristics. First natural fault surfaces are difficult to characterize because their roughness is spatially heterogeneous (Power and others, 1987). Surface irregularities, such as large-scale striations, tool marks, and mullion structures (see Fig. 5.5) cause the surfaces to have strongly anisotropic roughness. Fault surfaces with features of this type have been described by Means (1987), Hancock and Barka (1987), and Will and Wilson (1989), as well as many others. In most cases, the anisotropy of the surfaces is at least partly due to the greater propensity toward smoothing through wear in the direction of relative motion of the two fault blocks (Brown and Scholz, 1985; Power and others, 1987).

Natural fault surfaces are approximately self-similar from wavelengths as small as 10 μm (10^{-5} m) to wavelengths as large as 40 m, a 6.5 order of magnitude range (see Fig. 5.6). Measurements at a larger scale of the mapped trace of the San Andreas fault (see Fig. 5.7) suggest that the approximately self-similar or self-affine fractal character continues to scales as large as tens to hundreds of kilometers. Fault surface data (see Fig. 5.6) suggest that at some scales, fault surfaces have self-affine character ($\alpha \neq 3$), but over the entire wavelength range, the character is remarkably close to self-similar ($\alpha = 3$). The

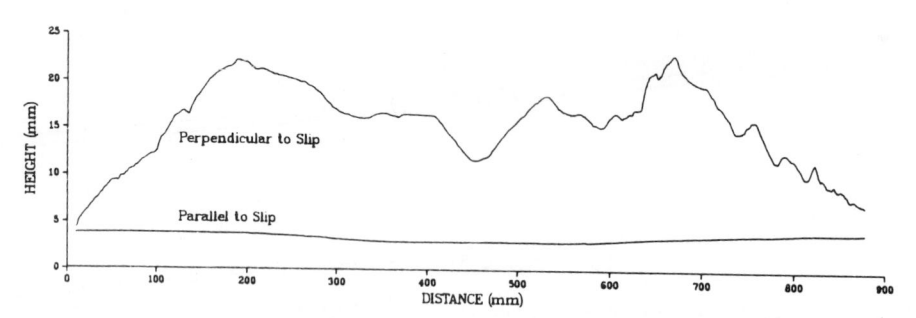

FIGURE 5.5. Illustration of the strong directional anisotropy of fault surfaces. Typically profiles measured parallel to the slip direction have overall amplitude to wavelength ratios that are 1–2 orders of magnitude lower than profiles measured perpendicular to the slip direction (Power and others, 1987; Power and Tullis, 1991a). (Reprinted with permission from the American Geophysical Union.)

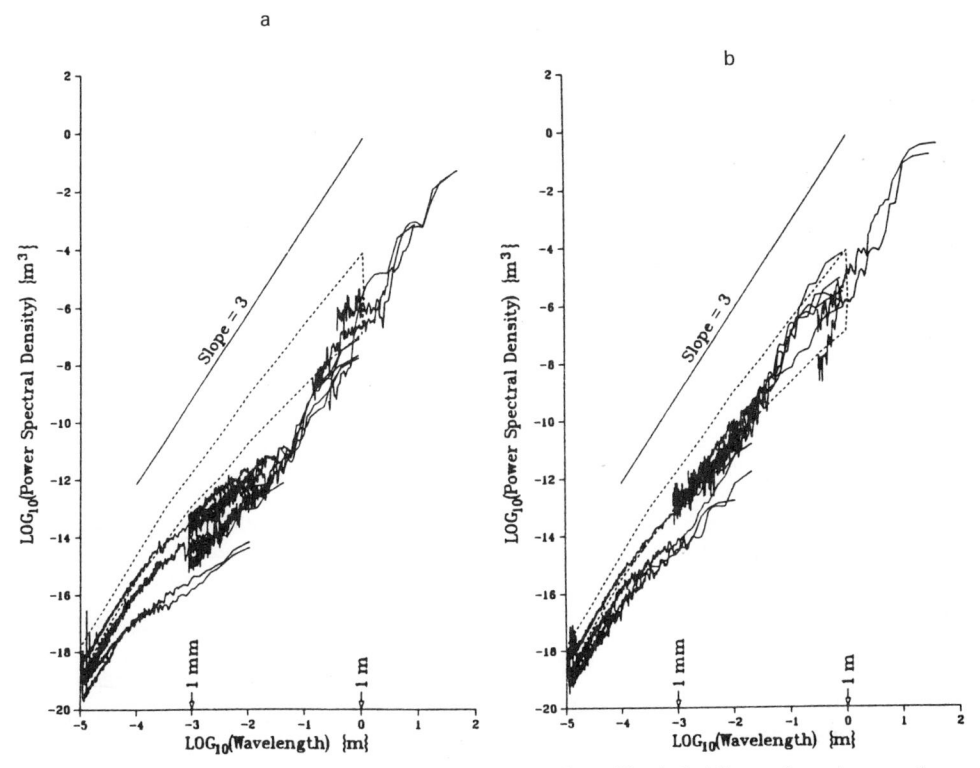

FIGURE 5.6. Power spectra from natural fault and fracture surfaces. The dashed line encloses the area where power spectra from natural fracture surfaces (Brown and Scholz, 1985) would fall if plotted on these diagrams. Fault surface spectra calculated from profiles measured both (a) parallel to the slip direction and (b) perpendicular to the slip direction are available (Powers and others, 1988). (Reprinted with permission from the American Geophysical Union.)

degree of departure from ideal self-similarity evident in Fig. 5.6 can be quantified by saying that the power level varies from ideal self-similarity by about a factor of 100, which means that the overall amplitude to wavelength ratio varies by about a factor of 10 (Power and Tullis, 1991a). We emphasize however that no natural surface is exactly self-similar or self-affine. The fault surfaces we describe as approximately self-similar may be broadly equivalent to rock fracture surfaces that Brown and Scholz (1985) describe as self-affine. We chose the terminology approximately self-similar at least partly because our measurements span 6.5 orders of magnitude in scale in comparison to other studies that typically involve smaller bandwidths (for example, Brown and Scholz, 1985—5 orders of magnitude; Maerz and Franklin, 1990—4.5 orders of magnitude).

5.4. DISCUSSION AND IMPLICATIONS

In almost all fractal systems, fractal behavior is confined between upper and lower boundaries in scale, outside of which the system is either nonfractal or characterized by different scaling parameters (Mandelbrot, 1983; Burrough, 1984). Although the fractal

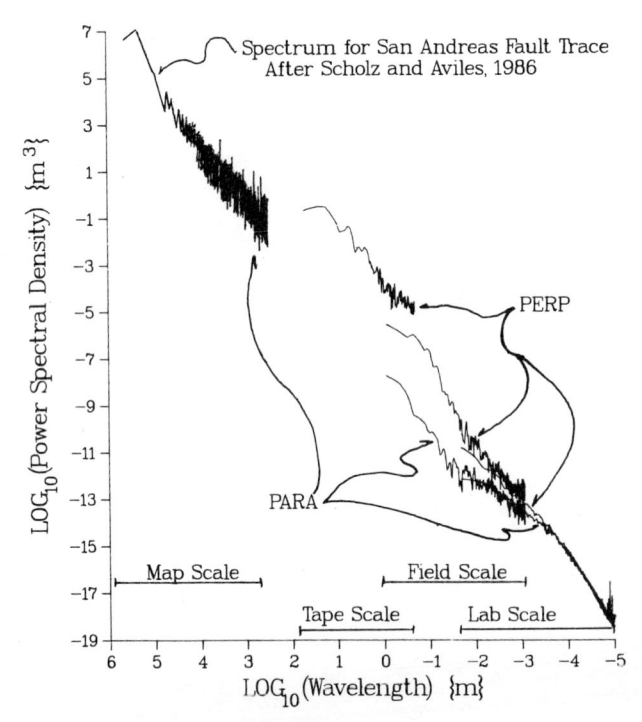

FIGURE 5.7. Selected power spectra for fault surfaces over 11 orders of magnitude in wavelength. Abbreviations follow: para—spectra from profiles measured parallel to the slip direction; perp—spectra from profiles measured perpendicular to the slip direction. Spectrum from the San Andreas fault trace is from Scholz and Aviles (1986); All other spectra are from Power and others (1987). (Reprinted with permission from the American Geophysical Union.)

character of a surface by itself does not provide conclusive evidence that particular physical mechanisms were active in surface formation, whatever mechanisms were active must be consistent with the observed geometry. Fractal geometry and the concepts of self-similar and self-affine scaling can be useful in studying surfaces in two ways. First, scales where abrupt changes in the geometry occur may signal the onset of a new or different surface generation mechanism at a smaller or larger scale. Second, ideas or concepts of self-similar or self-affine fractal scaling may have a different predictive capability than Euclidean geometries, thereby providing a new perspective. In Chapter 5, we concentrate primarily on the second path by inferring the mechanical evolution of fractal or self-similar surfaces as opposed to Euclidean surfaces.

5.4.1. Wear and Development of Gouge and Breccia

A number of authors, including Otsuki (1978) and Robertson (1983) suggested that the thickness of gouge and breccia along natural fault zones increases approximately linearly with displacement. The suggestion that linear relationships are correct for natural fault zones is controversial (Robertson, 1983; Wilder, 1984; Waterman, 1984; Robertson, 1984; Scholz, 1987; Hull, 1988; Power and others, 1988; Blenkinsop, 1989). Certainly some of the controversy stems from the realization that fault zones from different depths in the crust should not behave similarly enough so that a single relationship (linear or nonlinear) is widely applicable. However with data from a single area (where fault zones presumably formed and evolved under similar conditions) and with data that all come from fault zones in

5. REVIEW OF THE FRACTAL CHARACTER OF NATURAL FAULT SURFACES

one rock type, the cluster along a linear trend is often much stronger (Robertson, 1987). Another source of controversy may be that many if not all field-oriented geoscientists are personally acquainted with a particular fault zone that has a considerably thinner or thicker gouge zone than a simple linear relationship like that shown in Fig. 5.8 would predict. In contrast to natural fault zones, experimental rock friction studies have found conflicting or nonlinear relationships between displacement and gouge and breccia thickness (Jackson and Dunn, 1974; Engelder, 1974; Yoshioka, 1986; Power and others, 1988). Experimental observations of gouge and breccia formation during large-displacement rock friction experiments (see Fig. 5.9) generally result in the development of a thin layer of gouge material in the initial stages of sliding, which does not grow thicker once enough gouge material has accumulated to insulate the two surfaces from one another (Power and others, 1988; Yund and others, 1990).

Despite the controversial nature of thickness and displacement relationships for brittle faults (particularly natural ones), it is useful to consider the implications of the fractal

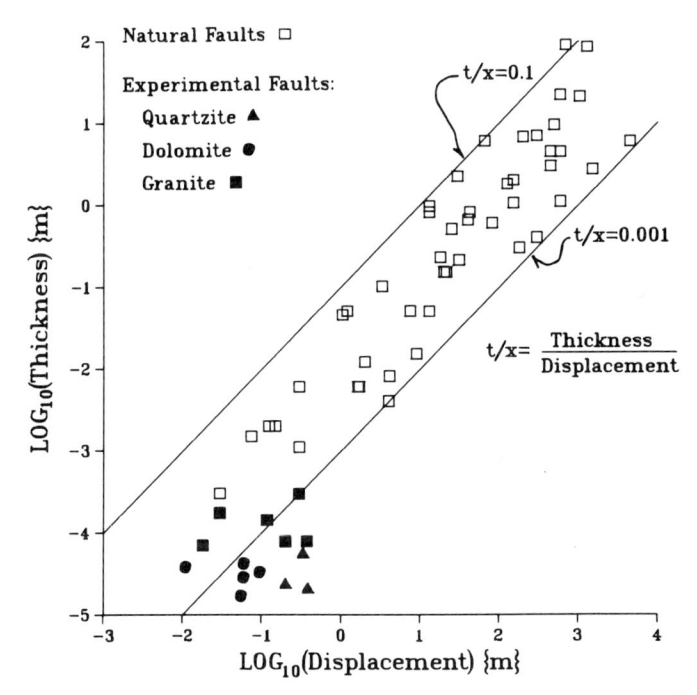

FIGURE 5.8. Wear-zone thickness versus displacement for natural faults (modified from Otsuki, 1978; Robertson, 1983; and Hull, 1988) and large displacement rock friction experiments (Power and others, 1988). A log-log plot is used for this data because of the extremely wide range of thicknesses and displacements. On a log-log plot, a strictly linear relationship (constant thickness to displacement ratio) plots as a line with slope 1. A line with any other slope represents a power law relationship. Much data from natural faults fall between thickness to displacement ratios (t/x) 0.1–0.001, suggesting an approximately linear relationship between thickness and displacement. Experimental data for large-displacement experiments have considerably different t/x ratios from natural faults. The difference between experimental and natural faults probably arises because the scaling of surface roughness of natural rock fractures differs from that of ground surfaces used in most rock friction experiments. See text and Power and others (1988) for further discussion. (Reprinted with permission from the American Geophysical Union.)

Experimental Data – Granite

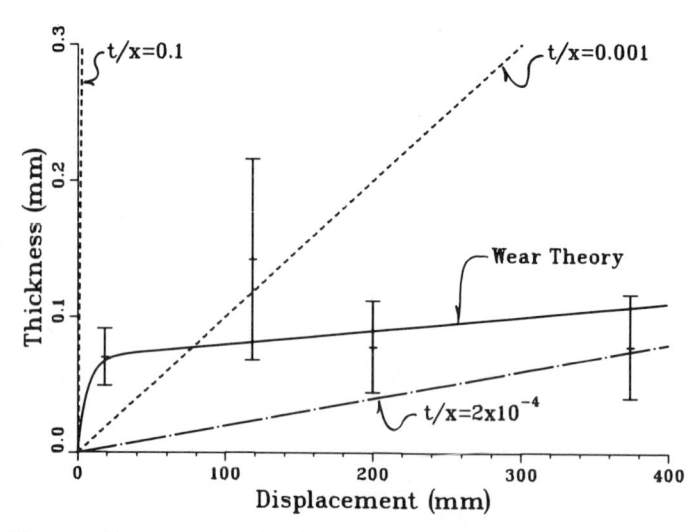

FIGURE 5.9. Wear-zone thickness as a function of displacement for rotary shear experiments done with Westerly granite. Error bars shown for the four data points represent one standard deviation. The line labeled wear theory was constructed using a nonlinear, running-in wear theory described by Queener and others (1965) and Power and others (1988). For comparison with Fig. 5.9, we also plot three linear thickness to displacement ratios. All four experiments were done at the same normal stress (75 MPa). The velocity of sliding varied during the course of each experiment (Tullis and Weeks, 1986; Blanpied and others, 1987). (Reprinted with permission from the American Geophysical Union.)

character of surface roughness on the development and evolution of gouge and breccia, here termed wear material. A simple thought experiment, accompanied by illustrations (see Fig. 5.10 and 5.11), shows potential differences between the mechanical evolution of faults that develop from fractal and Euclidean surfaces. We consider two cases; first two blocks of rock separated by Euclidean surfaces (see Fig. 5.10) and second two blocks of rock separated by approximately self-similar surfaces (see Fig. 5.11). The first case (Euclidean

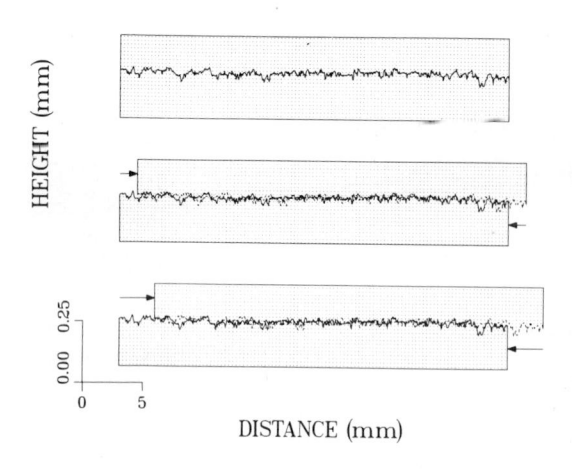

FIGURE 5.10. Development of a fault zone from Euclidean surfaces (see also Fig. 5.11). Both Figs. 5.10 and 5.11 show the result of sliding the blocks a small amount if no deformation of surfaces nor dilation between surfaces is allowed. For real faults, interpenetration areas become pulverized, forming wear material that fills the voids. The amount of interpenetration provides a crude estimate of the amount of gouge that forms between surfaces. Once a thin layer of gouge insulates Euclidean surfaces from one another, no more gouge material has to be formed, because at large wavelengths, the surfaces are essentially flat.

5. REVIEW OF THE FRACTAL CHARACTER OF NATURAL FAULT SURFACES

FIGURE 5.11. Development of a fault zone from two self-similar rock surfaces (see also Fig. 5.10). For a self-similar surface, mismatch between surfaces increases continuously with slip. The self-similar character of natural faults and fractures can explain the approximately linear relationship between wear-zone thickness and displacement observed for natural faults. See text or Power and others (1988) for further explanation. (Reprinted with permission from the American Geophysical Union.)

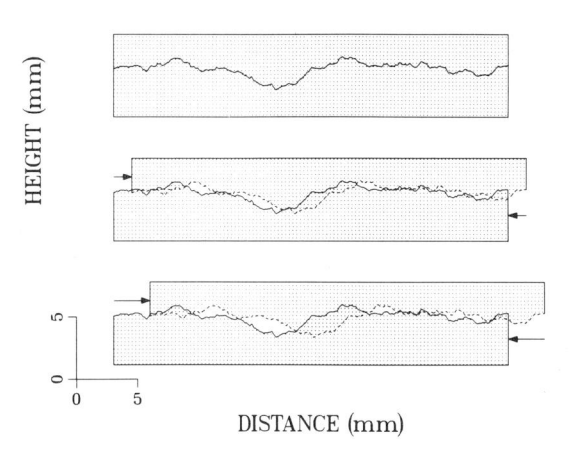

surfaces) is typical of surfaces used in many experimental studies of rock friction and joint closure. This surface is rough at small scales, but at large scales, it is essentially flat (see Fig. 5.10a). In the second case, surfaces are approximately self-similar, which is a fairly close approximation to many rock fracture surfaces. This surface possesses roughness at all scales up to and including the dimension of the block illustrated in Fig. 5.11a. We make the following assumptions: (1) Motion normal to the fault surfaces (dilation) is minimal; (2) density differences between the wall rock and the wear material can be neglected; and (3) all motion is parallel to a single slip direction (the problem is two-dimensional). In such a thought experiment, if surfaces are rough, they must interpenetrate one another to allow relative displacement. In the natural case, interpenetration is impossible. However the degree of interpenetration, or volume of mismatch, allows a quantitative estimate of the amount of gouge material produced during sliding (Power and others, 1988).

Relative sliding of Euclidean surfaces causes the gouge thickness to increase nonlinearly with displacement. Because Euclidean surfaces are rough at small scale, the first increments of sliding produce mismatches between the two surfaces (see Fig. 5.10b). During the initial sliding stages, these mismatches are worn away, creating gouge material. Because surfaces are essentially flat at large scale, progressively larger displacements do not increase the mismatch between surfaces. Also because at large scale, surfaces are essentially flat, a small amount of gouge between the surfaces can effectively insulate one surface from the other, thereby preventing the formation of more gouge material.

Sliding on the self-similar fracture surface causes a different sequence of events than sliding on the Euclidean surface. For the self-similar surface, a small amount of sliding isolates small irregularities in one surface from their initial positions in the opposite surface (see Fig. 5.11b). Larger displacements increase the mismatch because larger and larger irregularities become isolated from their original positions in the opposite surface (see Fig. 5.11c). For self-similar surfaces, the amount of mismatch increases linearly with slip, suggesting that the approximately linear relationship between thickness and displacement on natural fault surfaces results from the self-similar character of natural rock surfaces (Power and others, 1988).

Figure 5.11c illustrates one reason why some fault zones do not follow linear thickness–displacement relationships. Although considerable slip has occurred between the surfaces (see Fig. 5.11c), there are still discrete points where the two surfaces are in contact. If the surfaces are self-similar, a strictly linear relationship between wear-zone thickness and displacement is not observed. Instead we would expect a triangular relationship where the maximum or average gouge thickness increases linearly with displacement (see Fig. 5.12), but very small or even zero thicknesses could be observed near contact points between surfaces. Thus the fractal character of surfaces may explain why very thin wear zones that do not fit a linear thickness and displacement relationship are sometimes observed. Caution should be used when measuring the thickness of wear zones along natural faults. In particular thickness should be measured both near and far from points of contact between surfaces, and measurements should be scattered over a distance at least a few times the total slip distance, so that both points of contact and intervening wear material are adequately sampled (Wilder, 1984; Power and others, 1988).

Some faults should not follow linear thickness and displacement relationships because they do not develop from self-similar or approximately self-similar rock surfaces. One prominent and important large-scale example is provided by the Earth's major subduction zones. Subduction zones should not follow linear thickness and displacement relationships because the large-scale form of the Earth and the sea floor is essentially spherical, a classic Euclidean shape. Although the sea floor is rough, and approximately fractal at intermediate scale, direct observation reveals that at the largest scales, the sea floor has a deterministic shape (Goff and Jordan, 1988). Thus thickness and displacement relationships for subduction zones should be decidedly nonlinear, like the situation observed in experimental studies performed with Euclidean surfaces. Another example where fault zones may be essentially Euclidean is large overthrust faults that initiate along bedding surfaces. Although few studies have characterized the roughness of bedding surfaces, it is likely that they are Euclidean at large scale because they develop under the control of the Earth's gravity field.

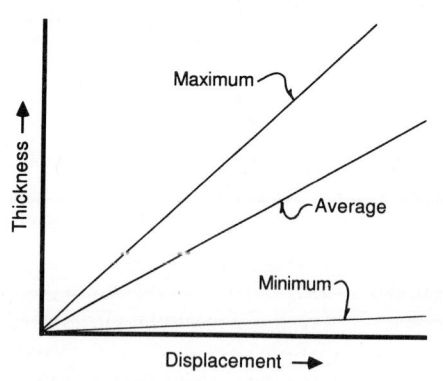

FIGURE 5.12. A triangular relationship between the gouge and breccia zone thickness and displacement. If fault zones are bounded by self-similar surfaces, there should always be some places along a fault zone where intact rock from both sides of the fault zone is in direct contact (see text and Figs. 5.11a–c). At contact points, the thickness of the gouge and breccia should be approximately zero. The thickness of the gouge and breccia in areas distant from contact points should increase approximately linearly with displacement (Power and others, 1988). The combination of the evolution of new contact spots along the surfaces and the approximately linear increase of the thickness of gouge material distant from direct contact points should result in an approximately triangular relationship between wear-zone thickness and displacement for natural fault zones that develop from approximately self-similar surfaces.

5.4.2. Design of Rock Friction Studies

An important development in the study of fault zones involves using rate- and state-variable constitutive descriptions of frictional strength (Dieterich, 1979; Ruina, 1983; Gu and others, 1984). Rate- and state-variable constitutive descriptions provide a more realistic description of frictional behavior than constant static and dynamic friction coefficients. Rate- and state-variable descriptions have been extremely successful in predicting and describing frictional properties of laboratory-scale samples (Dieterich, 1979, 1981; Tullis, 1986a; Tullis and Weeks, 1986; Blanpied and others, 1987), and innovative applications of various aspects of constitutive descriptions to natural fault zones have begun in earnest (Dieterich, 1978; Cao and Aki, 1986; Tse and Rice, 1987; Marone and Scholz, 1988; Tullis, 1988; Scholz, 1988; Lorenzetti and Tullis, 1989). The rate- and state-variable description of the frictional surface, when coupled with a knowledge of the elastic properties of the loading system (either a testing machine or the wall rock surrounding a fault surface), can be used to predict the stability of sliding on the frictional surface (Dieterich, 1978). Sliding is stable if the stiffness of the loading is greater than a critical stiffness

$$K_{cr} = \frac{-(a-b)\sigma}{D_c} \quad (3)$$

where $(a - b)$ is a measure of the change in steady-state frictional resistance with slip rate, σ is normal stress, and D_c is a characteristic distance over which frictional resistance evolves following changes in stress or slip velocity (Dieterich, 1978, 1979, 1981; Ruina, 1983; Gu and others, 1984). The characteristic decay distance D_c in constitutive friction laws is important because it controls the relative size of the volume of rock where we would expect to be able to monitor changes in stress, strain, and velocity prior to an earthquake instability (Dieterich, 1986; Tse and Rice, 1987; Tullis, 1988; Lorenzetti and Tullis, 1989).

Much of the progress in developing rate- and state-variable constitutive friction laws has come solely through the study of Euclidean surfaces. There are two main reasons why experimental studies have concentrated on Euclidean surfaces: First the general aim of much experimental work is to isolate individual aspects of the inherently complex situations observed in nature, so that individual components of natural systems can be studied in a simple enough setting to be well understood. Second fractal surfaces are extremely difficult to perform experiments on because measuring the friction between two surfaces requires many discrete points of contact between the surfaces. Thus the frictional behavior can be thought of as the averaged response of many contacts. With approximately self-similar surfaces like those illustrated in Fig. 5.11c, very few contact spots occur between the surfaces. In experiments conducted with self-similar surfaces, the frictional strength may be dominated by one or a few transient contacts, leading to unpredictable or irreproducible behavior. For example, Jaeger (1971, p. 108) concluded that the frictional behavior of natural joints (which were probably approximately self-similar) was much less regular than the frictional behavior of ground surfaces.

Progress toward understanding the effects of surface roughness on rock friction has been significant, but more work remains to be done, particularly with regard to understanding effects of surface roughness and scaling surface roughness on rate- and state-variable constitutive friction laws. A variety of studies (Byerlee, 1967, 1970, 1978; Ohnaka, 1973, 1975) discussed the importance of roughness on the initial, maximum, and residual friction (which have been defined by Byerlee, 1978). The effect of fractal or approximately self-similar roughness on these friction values is less clear. Byerlee (1978) concludes that friction is independent of roughness at high normal stress but highly dependent on roughness at low normal stress. The strong dependence of frictional strength on surface roughness at low normal stress is a well-calibrated aspect of geotechnical engineering (Barton and others, 1985; Barton, 1986).

The effect of roughness on the rate- and state-variable constitutive descriptions of friction is less well understood. The roughness of Euclidean surfaces (which were generally characterized by their rms roughnesses) has been shown to have a clear effect on the value of D_c, the characteristic decay distance (Dieterich, 1979; Okubo and Dieterich, 1984; Tullis, 1986a, 1986b, 1991), with rougher surfaces leading to larger D_c values. The effect of self-similar or self-affine roughness on D_c does not appear to have been studied experimentally, but Scholz (1988) and Power and Tullis (1991b) have made initial attempts at estimating the values of D_c for fractal surfaces. There is a great need for more study of the effect of roughness on the values of rate- and state-variable constitutive parameters as well as the initial, maximum, and residual friction.

5.4.3. Fault Zone Evolution

Careful experimental study of the development of gouge and breccia between Euclidean rock surfaces has revealed two distinct phases in the development of a "mature gouge layer" between the rock surfaces (Power and others, 1988; Yund and others, 1990; Cox, 1990). In the first stage, wear material develops rapidly, probably at least partly because the surfaces are in direct contact with one another. In the second phase, the accumulation of wear material insulates the surfaces from one another. At large slip distances, a single, relatively long-lived slip surface usually develops within the gouge material (Yund and others, 1990).

Many experimental studies conclude that this somewhat more stable and evolved (and by inference flatter) slip surface develops preferentially near one side of the gouge zone, near undeformed rock (Logan and others, 1979; Marone and others, 1989). While this may be true for experimental studies, direct application of this result to natural fault zones will probably be difficult, because if a flat surface develops within a fault zone and boundary surfaces of that fault zone are self-similar or self-affine surfaces, it is impossible for the flat surface to follow the edge of the fault zone exactly for its entire length. Due to the approximate self-similarity of rock surfaces that bound natural fault zones, individual slip surfaces accommodating all the deformation may be only transient features.

A number of different mechanisms for initiating fault zones from fractures have been proposed (Segall and Pollard, 1983; Martel and others, 1988; Cox and Scholz, 1988), but implications of different mechanisms for the character of fault surface roughness has generally not been explored. Ultimately it may be possible to use fault-initiation mechanisms to predict the roughness of initial fault surfaces. The character of the initial surfaces could then be coupled with appropriate models for wear and evolution and finally compared to measurable natural fault surface roughness.

5.4.4. Earthquake Source

A number of recent studies have considered the effects of large-scale roughness features of fault zones on the earthquake process. Studies by Aki (1984), King and Nabelek

(1985), and Sibson (1985) stress the importance of large-scale roughness features, such as asperities, fault bends, or fault jogs, as the initiation and termination points of dynamic earthquake ruptures and the generation of characteristic earthquakes. Wesnousky (1989) presented a correlation between the geometrical complexity of fault zones and their seismic character. He concludes that the most evolved and simple fault zones generate the largest earthquakes. The development of single, discrete fault surfaces, which have been termed principal slip surfaces by Sibson (1986), are clearly an essential element of the earthquake source. The development of principal slip surfaces from self-similar, fractal surfaces by linking networks of surface fractures and wear is a promising topic for future research.

5.5. CONCLUSIONS

Direct measurement of natural fault surfaces and rock fractures through surface profilimetry reveals that roughness occurs on these rock surfaces at scales as small as 10 μm and as large as 40 m and greater. Fractal models based on the concepts of self-similar and self-affine scaling provide useful yardsticks for describing natural rock surfaces. Spectral methods provide a clear and concise method for analyzing surface profile data. Spectra calculated from self-similar surfaces have slopes of -3 on log-log plots of power spectral density versus spatial frequency; spectra calculated from self-affine surfaces have slopes different than -3. Over the 6.5 order of magnitude range where observations are the most complete, natural rock fractures and fault surfaces have approximately self-similar roughness.

The approximately self-similar character of natural fault and rock fracture surfaces has important implications for the development and evolution of fault zones. In particular the self-similar character of natural rock surfaces can be used to explain the approximately linear relationship between wear zone thickness and displacement observed for many natural fault zones. In contrast to the natural situation, the majority of rock friction experiments have been performed on Euclidean surfaces. For Euclidean surfaces, the relationship between wear zone thickness and displacement is nonlinear. More work is needed to incorporate the observed self-similar character of natural rock surfaces into our understanding of rock friction, fault zone mechanics, and the earthquake source.

ACKNOWLEDGMENTS

This study was supported by National Science Foundation grants EAR-8509014 and EAR-8610088. We wish to express special thanks to all those who discussed earlier versions of this work with us.

REFERENCES

Adler, R. J., and Firman, D., *Phil. Trans. Royal Soc. London Ser. A* **303**, 433 (1981).
Aki, K., *J. Geophys. Res.* **89**, 5867 (1984).
Aviles, C. A., Scholz, C. H., and Boatwright, J., *J. Geophys. Res.* **92**, 331 (1987).
Barton, N. R., *Geotechniq.* **36**, 147 (1986).
Barton, N., Bandis, S., and Bakhtar, K., *Int. J. Rock Mech. Min. Sci. Geomech. Abstr.* **22**, 121 (1985).
Bendat, J. S., and Piersol, A. G., *Random Data: Analysis and Measurement Procedures* (Wiley, New York, 1986).

Berry, M. V., and Lewis, Z. V., *Proc. Roy. Soc. London Ser. A* **370**, 459 (1980).
Blanpied, M. L., Tullis, T. E., and Weeks, J. D., *Geophys. Res. Lett.* **14**, 554 (1987).
Blenkinsop, T. G., *J. Struct. Geol.* **11**, 1051 (1989).
Bracewell, R. N., *The Fourier Transform and Its Applications* (McGraw-Hill, New York, 1986).
Brown, S. R., "Fundamental Study of the Closure Property of Joints," Ph.D. diss., Columbia University, 1984.
———, *Geophys. Res. Lett.* **14**, 1095 (1987).
Brown, S. R., and Scholz, C. H., *J. Geophys. Res.* **90**, 12,575 (1985).
Burrough, P. A., *Bull. Inst. Math. Appl.* **20**, 36 (1984).
Byerlee, J. D., *Geophys. Res.* **72**, 3639 (1967).
———, *Int. J. Rock Mech. Min. Sci. Geomech. Abstr.* **7**, 577 (1970).
———, *Pure Appl. Geophys.* **116**, 615 (1978).
Cao, T., and Aki, K., *Pure Appl. Geophys.* **124**, 487 (1986).
Carr, J. R., *Bull. Assoc. Eng. Geol.* **26**, 253 (1989a).
———, in *Proceedings of the 25th Symposium on Engineering Geology and Geotechnical Engineering, Reno, Nevada* (Balkema, Rotterdam, 1989b), pp. 297–303.
———, in *Rock Mechanics as a Guide for Efficient Utilization of Natural Resources*, Proceedings of the 30th US Symposium, West Virginia University, Morgantown (Balkema, Rotterdam, 1989c), pp. 193–200.
Carr, J. R., and Warriner, J. B., in *Proceedings of the 28th US Symposium on Rock Mechanics, Tucson, Arizona* (University of Arizona, 1987), pp. 73–80.
Cox, S. J. D., in *Deformation Mechanics, Rheology, and Tectonics* (R. J. Knipe, and E. H. Rutter, eds.) (Geological Society of London Special Publication, 1990).
Cox, S. J. D., and Scholz, C. H., *J. Struct. Geol.* **10**, 413 (1988).
Dieterich, J. H., *Pure Appl. Geophys.* **116**, 790 (1978).
———, *J. Geophys. Res.* **84**, 2161 (1979).
———, in *Mechanical Behavior of Crustal Rocks* (American Geophysical Union Monograph, 1981), Vol. 24 (N. L. Carter, M. Friedman, J. M. Logan, and D. W. Stearns, eds.) pp. 103–20.
———, in *Earthquake Source Mechanics*, American Geophysical Union Monograph, Vol. 37 (S. Das, J. Boatwright, and C. H. Scholz, eds.), pp. 37–47.
Engelder, J. T., *Geol. Soc. Am. Bull.* **85**, 1515 (1974).
Fox, C. G., *Pure Appl. Geophys.* **131**, 211 (1989).
Gentier, S., Billaux, D., and van Vliet, L., *Journal of Rock Mechanics and Rock Engineering* **22**, 149 (1989).
Goff, J. A., and Jordan, T. H., *J. Geophys. Res.* **93**, 13,589 (1988).
Gu, J. C., Rice, J. R., Ruina, A., and Tse, S., *J. Mech. Phys.* **32**, 167 (1984).
Hancock, P. L., and Barka, A. A., *J. Struct. Geol.* **9**, 573 (1987).
Hough, S.E., *Geophys. Res. Lett.* **16**, 673 (1989).
Hull, J., *J. Struct. Geol.* **10**, 431 (1988).
Jackson, R. E., and Dunn, D. E., *J. Rock Mech. Min. Sci. Geomech. Abstr.* **11**, 235 (1974).
Jaeger, J. C., *Geotechniq.* **21**, 97 (1971).
Keller, K., and Bonner, B. P., *Rev. Sci. Inst.* **56**, 330 (1985).
King, G., and Nabelek, J., *Science* **228**, 984 (1985).
Logan, J. M., Friedman, M., Higgs, N. G., Dengo, C., and Shimamoto, T., "Experimental Studies of Simulated Gouge and Their Application to Studies of Natural Fault Gouge," US Geological Survey Open-File Report 79-1239, 1979, pp. 276–304.
Lorenzetti, E. A., and Tullis, T. E., *J. Geophys. Res.* **94**, 12,343 (1989).
Maerz, N. H., and Franklin, J. A., in *Proceedings of the First International Workshop on Scale Effects in Rock Masses, Leon, Norway, June, 1990* (A. Pinto da Cunha, ed.) (A. A. Balkema, Rotterdam, 1990), pp. 121–26.
Mandelbrot, B. B., *The Fractal Geometry of Nature* (W. H. Freeman, New York, 1983).
Mandelbrot, B. B., Passoja, D. E., and Paullay, A. J., *Nature* **308**, 721 (1984).
Marone, C., Raleigh, C. B., and Scholz, C. H., *J. Geophys. Res.* **95**, 7007 (1989).
Marone, C., and Scholz, C. H., *Geophys. Res. Lett.* **15**, 621 (1988).
Martel, S. J., Pollard, D. D., and Segall, P., *Geol. Soc. Am. Bull.* **100**, 1451 (1988).
Means, W.D., *J. Struct. Geol.* **9**, 585 (1987).
Myers, N. O., *Wear* **5**, 182 (1962).
Nayak, P. R., *J. Lubr. Tech.* **93**, 398 (1971).
Ohnaka, M., *J. Phys. Earth.* **21**, 285 (1973).
———, *J. Phys. Earth* **23**, 87 (1975).

Okubo, P. G., and Dieterich, J. H., *J. Geophys. Res.* **89**, 5817 (1984).
Otsuki, K., *J. Geol. Soc. Jpn.* **84**, 661 (1978).
Power, W. L., and Tullis, T. E., *J. Geophys. Res.* (1991a)
———, "The Contact Between Opposing Fault Surfaces at Dixie Valley, Nevada, and Implications for Fault Mechanics," submitted to *J. Geophys. Res.* (1991b).
Power, W. L., Tullis, T. E., Brown, S. R., Boitnott, G. N., and Scholz, C. H., *Geophys. Res. Lett.* **14**, 29 (1987).
Power, W. L., Tullis, T. E., and Weeks, J. D., *J. Geophys. Res.* **93**, 15,268 (1988).
Priestley, M. B., *Spectral Analysis and Time Series* (Academic Press, London, 1981).
Queener, C. A., Smith, T. C., and Mitchell, W. L., *Wear* **8**, 391 (1965).
Reeves, M. J., *Int. J. Rock Mech. Min. Sci. Geomech. Abstr.* **22**, 429 (1985).
Rengers, N., in *Proceedings of the 2d Congress of the International Society for Rock Mechanics*, (International Society for Rock Mechanics, Belgrade, 1970), Vol. 1, pp. 229–34.
Robertson, E. C., *Mining Eng. (New York)* **35**, 1426 (1983).
———, *Minubg Eng. (New York)* **36**, 1677 (1984).
———, in *Proceedings of the 28th US Symposium on Rock Mechanics, Tucson, Arizona*, US Symposium on Rock Mechanics (28th: 1987: University of Arizona) Rock mechanics: Proceedings of the 28th US Symposium, University of Arizona, Tucson, June 21–1 July 1987/edited by I. W. Farmer et al., Rotterdam; Boston (A. A. Balkema, 1987), pp. 65–72.
Ruina, A., *J. Geophys. Res.* **88**, 10,359 (1983).
Sayles, R. S., and Thomas, T. R., *Nature* **271**, 431 (1978).
Schneider, H. J., in *Proceedings of the 3d Congress of the International Society of Rock Mechanics, Denver, Colorado*, Vol. 2A (National Academy of Sciences, Washington, 1974), pp. 311–15.
Scholz, C. H., *Geology* **15**, 493 (1987).
———, *Nature* **336**, 761 (1988).
Scholz, C. H., and Aviles, C. A., in *Earthquake Source Mechanics: American Geophysical Union Monograph*, Vol. 37 (S. Das, J. Boatwright, and C. H. Scholz, eds., 1986), pp. 145–55.
Segall, P., and Pollard, D. D., *J. Geophys. Res.* **88**, 555 (1983).
Sibson, R. H., *Nature* **316**, 248 (1985).
———, *Phil. Trans. Royal Soc. London Ser. A* **A317**, 63 (1986).
Swan, G., *Rock Mechanics and Rock Engineering* **16**, 19 (1983).
Thomas, T. R., *Rough Surfaces* (Longman, New York, 1982).
Tse, S., and Rice, J. R., *J. Geophys. Res.* **91**, 9452 (1987).
Tullis, T. E., *Pure Appl. Geophys.* **124**, 375 (1986a).
———, *Eos Trans. Am. Geophys. Union* **67**, 1187 (1986b).
———, *Pure Appl. Geophys.* **126**, 555 (1988).
———, "The Relationship between Roughness and Contact between Surfaces and Decay Distances of Friction Constitutive Laws, in preparation (1991).
Tullis, T.E., and Weeks, J. D., *Pure Appl. Geophys.* **124**, 375 (1986).
Turk, N., Grieg, M. J., Dearman, W. R., and Amin, F.F., in *Proceedings of the 28th US Symposium on Rock Mechanics, Tucson, Arizona* (University of Arizona, 1987), pp. 1223–236.
Waterman, G.C., *Min. Eng.* **36**, 1677 (1984).
Weissbach, G., *Int. J. Rock Mech. Min. Sci. Geomech. Abstr.* **15**, 131 (1978).
Wesnousky, S. G., *Nature* **335**, 340 (1989).
Wilder, D. G., *Min. Eng.* **36**, 1677 (1984).
Will, T. M., and Wilson, C. J. L., *J. Struct. Geol.* **11**, 657 (1989).
Yoshioka, N., *J. Phys. Earth* **34**, 335 (1986).
Yund, R. A., Blanpied, M. L., Tullis, T.E., and Weeks, J. D., *J. Geophys. Res.* **95**, 15,589 (1990).

6

Fractals and Ocean Floor Topography: A Review and a Model

A. Malinverno

6.1. INTRODUCTION

The topography of the Earth's surface, as many other geophysical variables, can be separated into two components: a large-scale, long-wavelength component that can be predicted by simple models of physical processes, and that we may call deterministic; and a small-scale, random component that is very difficult to predict in a deterministic fashion (Mandelbrot, 1975). An example is the overall topography of the ocean floor (Bell, 1979; Gilbert and Malinverno, 1988; Goff and Jordan, 1988; Mareschal, 1989). As the oceanic lithosphere moves away from the midocean ridge axis, it thickens by cooling and subsides isostatically. The overall depth of the ocean floor is proportional to the square root of its age, following the predictions of simple thermal models (Davis and Lister, 1974; Langseth and others, 1966; McKenzie, 1967; Oldenburg, 1975; Parsons and Sclater, 1977). While the cooling model explains the overall deepening of the ocean floor away from the ridge axis, it is also obvious that there are a number of apparently random residual features on the ridge flank that are not accounted for (see Fig. 6.1). These ridge-and-trough features or abyssal hills (Heezen and others, 1959), are elongated in a direction parallel to the ridge axis (i.e., perpendicular to the page in Fig. 6.1). These volcanic constructions and/or fault-bounded blocks formed near the ridge axis and were subsequently transported onto the ridge flanks (Dietz, 1961; Kappel and Ryan, 1986; Larson, 1971; Lewis, 1979; Lonsdale, 1977; Macdonald and Atwater, 1978; Macdonald and Luyendyk, 1985; Menard and Mammerickx, 1967; Pockalny and others, 1987; Pockalny and others, 1988; Rea, 1975; Searle and Laughton, 1977).

Admittedly this is a simplified example. Although abyssal hills cover extensive

---

A. Malinverno • Lamont–Doherty Earth Observatory, Columbia University, Palisades, New York; *present address*: Schlumberger–Doll Research, Ridgefield, Connecticut.

Fractals in the Earth Sciences, edited by Christopher C. Barton and Paul R. La Pointe. Plenum Press, New York, 1995.

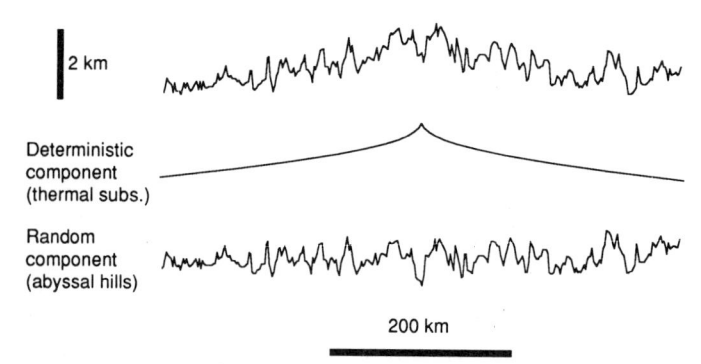

FIGURE 6.1. (a) An east-west profile crossing the Atlantic Ocean at about 27° north culminates in the crest of the mid-Atlantic ridge. This profile can be subdivided in two components: (b) a smooth subsidence curve that follows model predictions of thermal contraction and (c) a rough, seemingly random abyssal hill topography.

portions of the ocean floors and are probably the most common landforms on Earth (Menard, 1964), there are a number of additional topographic features on midocean ridge crests and flanks (e.g., axial valleys or highs, transform faults and fracture zones, seamounts, midplate swells). Also some of these features are progressively buried by sedimentation as they are transported away from the ridge axis by sea floor spreading. Sedimentation, mass wasting, and tectonism control the morphology of the ocean margins. Topographic features in the ocean basins and margins can be classified as deterministic or random, since the boundary between these two domains is often arbitrary. In any case, it is fair to conclude that while we have physical models and sufficient measurements to predict the topography of the ocean floor at scales greater than a few hundreds of kilometers, our predictive ability and knowledge decrease drastically for smaller length scales (Bell, 1979; Fox and Hayes, 1985).

In the recent past, statistical models linked to fractal geometry have been used to model the randomly varying component of topography. While simple geometrical shapes can be found in nature (e.g., spherical celestial bodies, regular crystals), they represent the exception rather than the rule. Most natural objects are irregular and contain a variety of shapes on a wide range of scales. Fractal geometry holds a particular fascination for the student of the Earth's morphology, since its constructions are strikingly similar to the jagged shapes found in nature (Mandelbrot, 1983; Peitgen and Saupe, 1988). However fractals are often viewed with some skepticism; the ability to *generate* a synthetic shape similar to a natural object may just be a mathematical gimmick, and by itself it may not advance in any way our *understanding* of how natural objects are created. In Chapter 6, I review a number of recent studies of the Earth's topography and attempt to show that by emulating the fundamental processes that shape topography, we can indeed gain some insight into the critical parameters of such processes. Emphasis is placed on the overall random topography of ocean basins and margins and particularly on a subset of ocean floor topography, abyssal hills. The purpose of Chapter 6 is to provide a starting point and convince the reader that such studies have the potential to advance significantly our understanding of how the Earth works.

6.2. ELEMENTARY THEORY AND EXAMPLES

Fractals have two basic characteristics that make them suitable for reproducing topography: self-similarity and randomness. Self-similarity refers to the well-known observation that the Earth's morphology looks the same at a variety of scales. A distant mountainous horizon resembles the profile of a fracture in a small stone; the coastline of an island on an aerial photograph is as jagged as the coastline of a rocky tidal pool (Huang and Turcotte, 1989; Mandelbrot, 1967). This loose concept of self-similarity also contains randomness, since the resemblance of the Earth's morphology at different scales is not exact but statistical. The difference between these two cases and the concept of fractal dimension are best illustrated by some simple examples.

To construct a self-similar figure, let us first take a linear segment (see Fig. 6.2, Step 0). Let us then break it into three pieces, replacing the central piece with two pieces of equal length, thus creating a protrusion in the shape of an equilateral triangle (Step 1). Now we have four linear segments, each of which has a length equaling one-third the length of the initial segment. If we repeat this procedure on the linear segments at each step (Steps 2, 3, etc.), the shape becomes more and more complex, and as the number of steps tends to infinity, we obtain a geometrical construct called the von Koch curve (Mandelbrot, 1983; Voss, 1988). The von Koch curve illustrates exact self-similarity: If we magnify any small portion of the curve, we obtain an identical copy to the original.

The jaggedness of fractals is quantified by a characteristic parameter, the fractal dimension D. Geometrical objects are usually classified by an integer Euclidean dimension; for example, a line has dimension 1, a plane figure dimension 2, and a solid figure dimension 3. The concept of dimension can be extended to irregular objects, such as the von Koch curve, as illustrated in Fig. 6.3. If we subdivide a segment of unit length into $N = 4$ equal parts, each part is scaled by a factor $r = \frac{1}{4}$. If we subdivide a unit square into $N = 4$ parts, each part is scaled by a factor $r = \frac{1}{2}$. In general we can define the dimension D of an object as in the relationship (Mandelbrot, 1967; Voss, 1988).

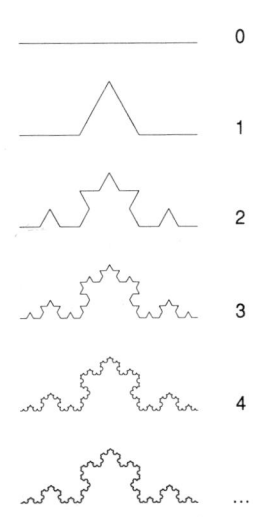

FIGURE 6.2. A jagged self-similar curve (the von Koch curve) can be constructed by starting with a line (*step* 0), replacing the middle-third with two segments of length one-third and repeating this procedure recursively (Steps 1, 2, 3, 4, . . .).

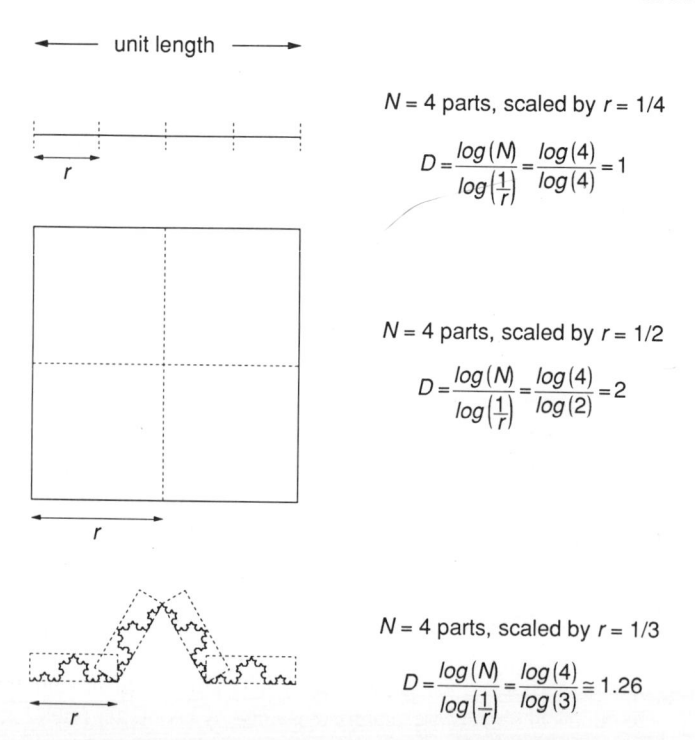

FIGURE 6.3. (a) A unit segment can be subdivided in four equal parts, each scaled by a factor $r = ¼$. (b) A unit square can be subdivided in four equal parts, scaled by a factor $r = ½$. (c) The four equal parts of a von Koch curve are scaled by a factor $r = ⅓$. The von Koch curve has a fractal dimension $D = 1.26$ that is intermediate between the dimension of a line and the dimension of a square.

$$N = \frac{1}{r^D} \tag{1}$$

so that the dimension is

$$D = \frac{\log(N)}{\log(1/r)} \tag{2}$$

With this definition, the dimensions of the segment and the square in Fig. 6.3 are 1 and 2, respectively. For the von Koch curve and $N = 4$, we have $r = ⅓$; therefore, D is not an integer but about 1.26. In general the fractal dimension D of an object varies between its Euclidean dimension and its Euclidean dimension plus 1 and quantifies the extent to which the object fills space. For example, as the fractal dimension of a curve varies between 1 and 2, the jaggedness of the curve increases until the curve completely fills the two-dimensional plane where it lies.

This definition of fractal dimension immediately suggests a method of measuring D, the so-called divider or ruler method. If we measure the length of the von Koch curve by walking a divider of fixed aperture along its perimeter, this length increases as the aperture decreases. If we take the scaling factor r used before as the aperture of the divider, the length of the von Koch curve is simply $L(r) = Nr$ (Figs. 6.2 and 6.3) and from Eq. (1), $L(r) = r^{1-D}$.

6. FRACTALS AND OCEAN FLOOR TOPOGRAPHY: A REVIEW AND A MODEL

If $D = 1$, as for a straight line, the length of the figure is constant for any aperture r of the divider; if $D > 1$, the length increases without bounds as r decreases, and a plot of $L(r)$ versus r estimates the fractal dimension D. Using this method, it has been shown that the western coast of Britain is fractal with a dimension close to that of the von Koch curve (Mandelbrot, 1967). However the self-similarity of an irregular coastline is not exact. As we magnify a segment of coastline, we do not expect to obtain an identical copy of the original but rather a figure that differs from the original in the details yet has the same statistical properties, i.e., the same overall roughness. In general we cannot follow a simple rule, such as that used to construct the von Koch curve, to obtain fractal objects that adequately reproduce the irregularity of natural shapes, but we need to include some degree of randomness.

A fractal construction that includes randomness and has been used to model topographic profiles is fractional Brownian motion (Mandelbrot, 1983; Mandelbrot and Van Ness, 1968; Mandelbrot and Wallis, 1969a; Voss, 1985). The simplest example of fractional Brownian motion is a random walk. Suppose we have a gambler playing a game of heads and tails; suppose also that the coin is fair and the gambler gains a fixed amount each time he or she wins (or alternatively loses the same amount). A graph of the gambler's cumulative gain (and loss) in time is a one-dimensional random walk resembling a profile of the Earth's topography (Mandelbrot, 1983). A less crude model for a random walk, one-dimensional Brownian motion or Brownian noise, has discrete increments that are not simply a positive or negative constant but are drawn from a normal distribution.

To make an analogy with the Earth's topography, imagine a hiker who walks along a straight line in a rough terrain and that the profile followed is Brownian motion. Each step may bring the hiker up or down, and the height differences at each step are normally distributed. An important point is that like the toss of an unbiased coin, in a random walk, the height difference at each step is independent from the height differences nearby; thus the hike is rather rough. Alternatively imagine that the height differences are positively correlated in space. In other words, if a hiker takes a step upward, it is more likely that the next step will be upward again, rather than down, and the hike is smoother. In the opposite case, height differences are negatively correlated, and the hike is extremely rough. This simple concept of spatial correlation is quantitatively expressed by fractional Brownian motion, which is characterized by a parameter H varying between 0 (very rough) and 1 (very smooth), with the case of Brownian motion having an intermediate value of $H = 0.5$.

Fractional Brownian motion is not self-similar but rather self-affine. The von Koch curve is an example of a self-similar fractal: An enlargement rescaled by a constant factor in all coordinates resembles the original. Self-affine fractals, such as fractional Brownian motion, require rescaling by different factors in different coordinates for an enlargement to look like the original (Mandelbrot, 1983; Mandelbrot and Van Ness, 1968; Mandelbrot and Wallis, 1969a; Voss, 1985). This characteristic is inherent in the definition of fractional Brownian motion. If we denote as $\sigma^2(L)$ the variance of a series of length L, fractional Brownian motion is defined to have a variance that increases with length

$$\sigma^2(L) \propto L^{2H} \tag{3}$$

so that its standard deviation, or rms amplitude $\sigma(L)$ is

$$\sigma(L) \propto L^H \tag{4}$$

where the symbol $\propto$ denotes proportionality. Equation (4) shows that if rescaled in the horizontal coordinate by a factor r, a sample of fractional Brownian motion must be rescaled in the vertical coordinate by a factor r^H; these two factors are the same only for $H = 1$. The different status of the spatial coordinates in self-similar and self-affine fractals can be physically interpreted as follows. In constructing the von Koch curve, we let features grow in all directions, or in other words, we assigned equal status to the two spatial coordinates in the plane. In the von Koch curve in Fig. 6.2 however, there are overhangs that are never found in a profile of topography. In fact, horizontal and vertical coordinates in a profile cannot have the same status because of gravity, which acts in the vertical direction only and effectively limits maximum slopes.

The fractal dimension of a self-affine figure can be defined in terms of its box dimension (Voss, 1985). Consider a profile of fractional Brownian motion encompassed in a unit square (Fig. 6.4). Analogous to what was done with the von Koch curve, D can be defined in terms of the number N of square boxes of linear size r needed to cover the profile. Intuitively we need $1/r^2$ boxes if the profile is rough enough to fill the unit square and $1/r$ boxes if the profile is smooth. Because of self-affinity, a segment of length r of the unit-range fractional Brownian motion in Fig. 6.4 covers on average a vertical span r^{1-H}, so that the number of boxes is

$$N = \frac{1}{r} \frac{1}{r^{1-H}} = \frac{1}{r^{2-H}} \tag{5}$$

and by comparison with Eq. (1), we obtain

$$D = 2 - H \tag{6}$$

FIGURE 6.4. The fractal dimension D of a profile can be defined in terms of its box dimension, i.e., the number of square boxes N of size $r < 1$ necessary to cover a profile enclosed in a unit square.

$$n = \frac{1}{r} \frac{1}{r^{1-H}} = \frac{1}{r^{2-H}} = \frac{1}{r^D}$$

Using the definition of dimension in Fig. 6.3, it can be shown that a profile of fractional Brownian motion with a rescaling parameter H ($0 \leq H \leq 1$) has a fractal dimension $D =$ to $2 - H$.

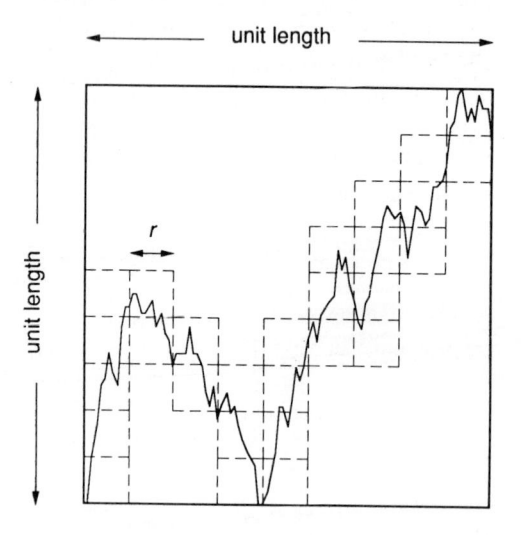

Number of square boxes of size r needed to cover the trace of fractional Brownian motion:

$$N = \frac{1}{r} \frac{1}{r^{1-H}} = \frac{1}{r^{2-H}} = \frac{1}{r^D}$$

6. FRACTALS AND OCEAN FLOOR TOPOGRAPHY: A REVIEW AND A MODEL

so that for $0 \leq H \leq 1$, $2 \geq D \geq 1$. For $D = H = 1$, rescaling factors in the vertical and horizontal directions are the same, and in this case fractional Brownian motion becomes truly self-similar.

Because of the different status of vertical and horizontal coordinates, the fractal dimension of a profile should not in general be estimated by the ruler method; it can be shown that the ruler method gives reliable results only when the length of the ruler is smaller than a crossover length, which is a function of the variance of the profile (Brown, 1987; Mandelbrot, 1985; Wong, 1987). A method widely used to estimate the fractal dimension of a profile is based on the power spectrum. The power spectrum (or power spectral density) of a series of measurements of elevation quantifies how its overall variance is partitioned in a corresponding series of harmonic components whose wavelength varies between the length of the profile and twice the sampling interval (e.g., Priestley, 1981). If a profile is self-affine, the power spectrum follows a power law (Mandelbrot and Van Ness, 1968; Saupe, 1988; Voss, 1985)

$$P(f) = Af^{-\beta} \tag{7}$$

where $P(f)$ is the power spectrum, f is spatial frequency (the reciprocal of the wavelength λ), and A and β are positive constants. Intuitively a self-affine series must have a spectrum that follows a power law, because a power function is invariant under rescaling, i.e., operations of multiplication or division. The relationship between the spectral exponent β and the fractal dimension D is best illustrated by considering the rescaling properties of fractional Brownian motion; it can be shown to be (Saupe, 1988)

$$D = \frac{5 - \beta}{2} \tag{8}$$

For $3 \geq \beta \geq 1$, the fractal dimension D varies between 1 and 2 (Fig. 6.5).

It should be stressed that the power spectrum does not contain information about the phase of each harmonic component. Two profiles may have the same power spectrum despite being quite different—having different phases for the component sinusoids. For example, consider a series of independent samples drawn from a normal distribution (i.e., height differences that when integrated constitute Brownian motion). Such a series is typically called white noise, since it can be regarded as the sum of a series of harmonic components with a constant amplitude and random phases (Blackman and Tukey, 1959). Fractional Brownian motions are red noises; component sinusoids have random phases, but long wavelength components have amplitudes much larger than the short wavelength components.

FIGURE 6.5. Synthetic profiles of fractional Brownian motion with different values of the fractal dimension D and the spectral exponent β. All profiles have the same overall variance, but roughness increases as the fractal dimension D increases from 1 to 2.

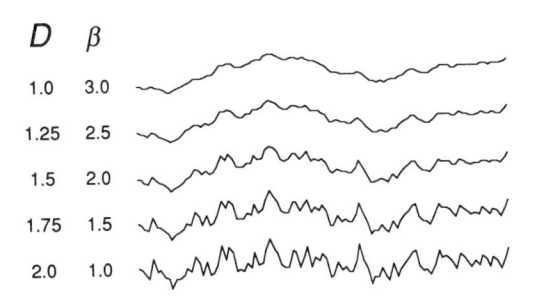

Other methods used to estimate the fractal dimension of profiles essentially consider the increase in variance with the length of the series. For example, it can be shown that the variance of the increments or mean square increments of a profile z_x of fractional Brownian motion is related to the fractal dimension (Mandelbrot and Wallis, 1969a)

$$E[(z_x - z_{x+s})^2] \propto s^{2H} = s^{4-2D} \tag{9}$$

where the operator $E\{\}$ denotes the expected value and the coordinate x and the lag s are integers (i.e., $x = 0, 1, 2 \ldots$; the result can be generalized to any sampling interval Δx). The variance of the increments can be estimated for a series of measurements from the variogram, which is the average squared difference at a given spatial lag (Journel and Huijbregts, 1978). Alternatively we can consider the variation of the rms amplitude $\sigma(L)$ with the length of the sample L, which should be proportional to L^{2-D} [Eqs. (4) and (6)]. This characteristic is used in the rescaled range method of Mandelbrot and Wallis (1969a).

The concepts and methods illustrated for profiles can be immediately extended to two-dimensional surfaces. The fractal dimension of a surface varies between 2 and 3, and it is 1 plus the fractal dimension of a profile crossing the surface. The appropriate relationship between the spectral exponent β and the fractal dimension D for a two-dimensional spectrum is discussed by Goff (1990), and is

$$D = \frac{8 - \beta}{2} \tag{10}$$

Finally it is worth mentioning the concept of an upper and lower fractal cutoff or limit (Mandelbrot, 1983). These limits quantify the possibility that a given object can be fractal on a finite range of scales, being essentially smooth at scales greater than the upper limit and/or smaller than the lower limit. For example, Eq. (3) shows that for the general case of $D < 2$ ($H > 0$), the variance of a profile increases without bounds as its length increases. In reality the variance of topographic profiles may reach an upper bound for lengths larger than a limit size, resulting in a topography that is flat at large scales. The significance of an upper fractal limit is discussed in depth later.

6.3. USING FRACTALS TO PARAMETERIZE TOPOGRAPHY

Fractal theory thus provides an attractive framework for studying and classifying topographic data. In this section, I review available data analyses to test how well a self-affine model fits the observations. A simple test of self-affinity is provided by spectral analysis: If a profile is self-similar, its power spectrum should follow a power law as in Eq. (7). At first order, this prediction is usually fulfilled. For example, in an early study, Neidell (1966) published a power spectrum for a topographic profile in the Carlsberg Ridge, northeastern Indian Ocean) that follows a power law with an exponent β of about 2 ($D = 1.5$). Also the phases of the harmonic components of topographic profiles have been observed to be randomly distributed (Fox and Hayes, 1985). This is an important requirement for a profile to be comparable to a random, self-affine figure (Hough, 1989). Other studies however show that topography in many cases does not conform exactly to a simple fractal model, and a review is presented in the rest of this section.

Some authors advanced the hypothesis that the spectral exponent β is a universal constant, i.e., the Earth's topography has a unique fractal dimension. Nye (1970) observed that units of power spectral density of a topographic profile have a dimension of length

6. FRACTALS AND OCEAN FLOOR TOPOGRAPHY: A REVIEW AND A MODEL

cubed, and he concluded that a hypothetical self-similar profile must have a power law spectrum as in Eq. (7), with an exponent $\beta = 3$ ($D = 1$). Later Nye (1973) noted that power spectra of land (Jaeger and Schuring, 1966) and sea ice topography (Hibler and LeSchack, 1972) indeed followed his prediction. Another preferred value for the spectral exponent β is 2, as in Brownian motion ($D = 1.5$). Bell (1975) showed a compilation of spectra of ocean floor and land topography (see Fig. 6.6) that follow a power law with an exponent $\beta = 2$ for wavelengths between about $0.5-10^4$ km (Balmino and others, 1973; Bretherton, 1969; Cox and Sandstrom, 1962; Warren, 1973). Turcotte (1987) showed that the same relationship holds for the large-scale topography of the Moon, Venus, and Mars. Sayles and Thomas (1978) noted that the variance of topographic profiles seems to be proportional to the length of the profiles and that this implies a power law spectrum with an exponent $\beta = 2$; an analogous result was obtained by Bell (1979) for abyssal hill topography. Sayles and Thomas (1978) coined the term topothesy for the constant A in Eq. (7). They argued that, since β seems to be a universal constant of value 2, the topothesy (with dimensions of length) uniquely defines the statistical properties of a topographic surface. Note that some authors, assuming the units of power spectral density to be length squared, suggest that profiles with $\beta = 2$ and $D = 1.5$ are self-similar (Fox and Hayes, 1985; Huang and Turcotte, 1989; Turcotte, 1987); this is obviously not the case, since the parameter H to be used for rescaling is 0.5 (see Fig. 6.7). Finally Mandelbrot (1975; 1983) argued from a comparison of synthetic fractal landscapes to actual topography that a fractal dimension D of 1.5 (i.e., a spectral exponent β of 2) is too high, and he proposed that the Earth's topography has a fractal dimension around 1.2.

These discrepancies contradict the concept that the spectral exponent β and the fractal dimension D are universal constants for topography. For example, Berry and Hannay (1978) point out that exponents in the spectra examined by Sayles and Thomas (1978) are not all equal to 2, but rather their value ranges between 1–3 (i.e., D varied between 2–1). Mark and Aronson (1984) examine digitized topographic maps of the United States to find that the fractal dimension, as calculated from the variogram, varies in different areas and in the same area, varies at different scales. Similarly Brown and Scholz (1985) examine the power

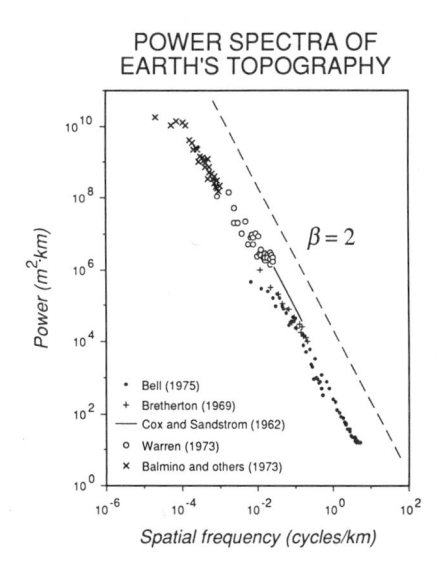

FIGURE 6.6. The power spectra of profiles of the Earth's surface when plotted in log-log space seem to follow a straight line with slope of -2. This corresponds to a power law with an exponent $\beta = 2$, i.e., a fractal dimension $D = 1.5$ (Bell, 1975).

SELF-AFFINITY OF 1-D BROWNIAN MOTION

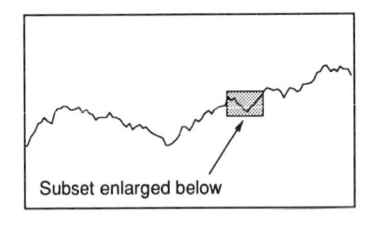

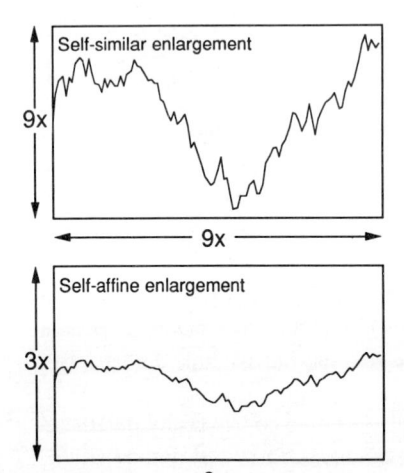

FIGURE 6.7. A profile of Brownian motion (*top*; fractal dimension $D = 1.5$, spectral exponent $\beta = 2$) is not self-similar but self-affine; an enlargement has statistical characteristics similar to the original not when both coordinates are rescaled by the same factor (*middle*), but when the vertical coordinate is rescaled by a factor equal to the square root of the horizontal coordinate (*bottom*).

spectra of natural rock surfaces at wavelengths of 20 μm–1 m, and note that the spectral exponent β is not constant over the whole frequency band, varying between 2–3. As far as ocean floor topography is concerned, spectral studies based on data collected by sonar for wavelengths ranging from few hundreds of meters to hundreds of kilometers (Berkson and Matthews, 1983; Fox and Hayes, 1985; Urick, 1975), and on data obtained by bottom photography or direct measurement at scales of a few meters to a few centimeters (Briggs, 1989; Fox and Hayes, 1985) agree in showing a substantial variation of the exponent β on different profiles. A possible explanation for the discrepancy with the results of Bell (1975) is that when a spectral analysis is carried out on a composite profile containing segments with different statistical characteristics, the resulting spectrum has an exponent that is an average of the spectral exponents of each segment. In many cases, this average β is about 2, as for a random walk (Fox and Hayes, 1985; Hough, 1989). Alternatively, as suggested by Berry and Hannay (1978), the spectral exponent β may vary and have an average value of 2; plotting a variety of spectra measured on a wide frequency band may obscure a variation of β with frequency, such as that observed by Brown and Scholz (1985). This point is illustrated in Fig. 6.8, where power spectra of abyssal hills from the central Pacific (Bell, 1975) and south Atlantic (Gilbert and Malinverno, 1988) are compared. While the spectrum number 1 of Bell (1975) follows a power law with $\beta = 2$, spectra 2 and 3 have a β close to that of the south Atlantic, i.e., about 2.6.

FIGURE 6.8. Power spectra of profiles of abyssal hill topography in the Pacific (Spectra 1, 2, and 3 in Bell, 1975) and the south Atlantic (Spectra E and W in Gilbert and Malinverno, 1988) differ in the overall power at a given wavelength, but they have a similar shape: a high-frequency segment that follows a power law with an exponent β between 2–2.6 and a low-frequency segment with a much lower exponent. The corner frequency separating these two segments is on the order of 0.1 cycles/km, corresponding to a wavelength on the order of 10 km.

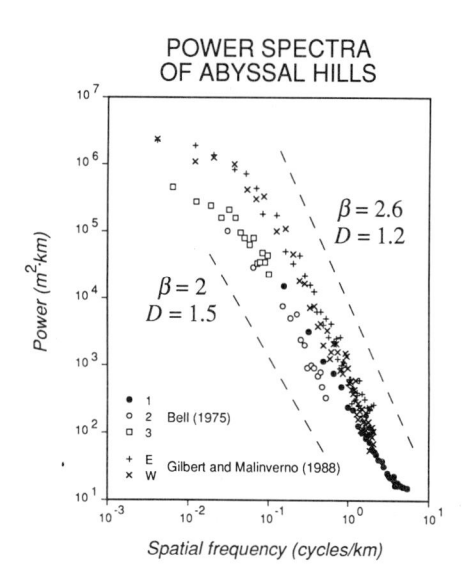

A number of studies also noted a limit in the increase of variance with increasing wavelength predicted by theory, i.e., an upper fractal limit. In the spectral domain, the upper limit translates into a corner frequency (or corner wavelength). The spectra follow a power law for frequencies higher (wavelengths shorter) than the corner frequency, and these flatten at lower frequencies (longer wavelengths). For example, Nye (1973) noted that the exponent β decreases significantly in sea ice spectra for wavelengths longer than about 100 m, reflecting the fact that sea ice topography is flat on the large scale. Bell (1975) noted that the power spectra of abyssal hills seem to flatten for wavelengths longer than about 40 km (see Fig. 6.8), and he attributed this long-wavelength flattening to the absence of large hills. In a later study, Bell (1979) suggested that the observed increase in variance with increasing profile length [Eq. (3)] must stop at lengths of a few tens of kilometers, since abyssal hills are generally less than 10 km wide. Malinverno (1989b) examined spectra of abyssal hills profiles between wavelengths of about 300 m–50 km that follow a power law with an exponent $\beta = 2.28$ ($D = 1.36$); yet extrapolating the power law at long wavelengths gave unrealistically large amplitudes (4 km for a wavelength of 1000 km); therefore he concluded that the spectra had to flatten at long wavelengths. Gilbert (1989) and Gilbert and Malinverno (1988) studied the power spectrum of a 3000 km long profile of abyssal hill topography in the south Atlantic and observed that the spectrum followed a power law with an exponent β of about 2.6 ($D = 1.2$) for wavelengths shorter than about 10 km, while it flattened at longer wavelengths (see Fig. 6.8). Similar results were obtained by Gilbert and Courtillot (1987), who observed that the spectra of topography in the south Atlantic followed a power law but with different exponents at long wavelengths (where β is about 1) and short wavelengths (β about 3). It should be noted that a data set comprising both continents and oceans is bound to contain large features, and in fact power spectra of the global Earth's topography show a flattening only for wavelengths longer than 10^4 km (see Fig. 6.6).

To summarize these results, we conclude that the power spectra of profiles of ocean floor topography tend to follow a power law, as predicted for self-affine series. The characteristic exponents β of the spectra vary substantially, while their average may be around 2. However for length scales grater than a few tens of kilometers, the power of the long-wavelength components is not so large as predicted from extrapolating the power law trend observed at shorter wavelengths. In the space domain, the corresponding observation is that the variance of a profile does not increase without bounds as the profile length increases. This is most likely due to the fact that the features making up the topography of the ocean floor, i.e., the abyssal hills, have a size limit.

An approach based on the spectral analysis of profiles however has some limitations and presents some difficulties. A first set of problems is related to the intrinsic nature of topography. A topography characteristic not directly addressed in a one-dimensional approach is anisotropy, i.e., the variation of statistical properties as measured on profiles with different orientations (Bell, 1975; Fox and Hayes, 1985; Malinverno, 1989a). Since abyssal hills are elongated in a direction parallel to the midocean ridge axis, the ocean floor fabric is inherently anisotropic. Typically investigations of abyssal hill topography are conducted on profiles perpendicular to the direction of elongation of the abyssal hills (Bell, 1975; Gilbert and Malinverno, 1988). Results obtained from these investigations can be extended to a direction parallel to the orientation of abyssal hills considering that their typical length-to-width ratio is around 5 (Bell, 1979). Because of anisotropy, it is clearly desirable to investigate the properties of topography from a full two-dimensional data set. For example, Huang and Turcotte (1989) analyze the topography of Arizona obtained from a two-dimensional gridded database, but they averaged the spectra in all directions, which is equivalent to assuming isotropy.

An important limitation of a two-dimensional approach to investigations of ocean floor topography however is that most measurements are available in the form of profiles; systems that collect bathymetry along a swath (e.g., Sea Beam; Farr, 1980) and the availability of two-dimensional gridded data sets are only recent developments. Spectral analyses of two-dimensional topographic data obtained from bottom photography (Akal and Hovem, 1978) and multibeam sonar surveys (Czarnecki and Bergin, 1986) have been published, and a method of statistically characterizing swath bathymetry data based on the histogram of local slopes was devised by Shaw and Smith (1987; 1990). Goff and Jordan (1988) formulated a general model to parameterize statistical characteristics of an anisotropic topography as sampled by swath bathymetry. Their model includes five parameters: a rms amplitude, a lineation direction, two characteristic scale parameters λ_n and λ_s (respectively orthogonal and parallel to the lineation direction), and a fractal dimension D. The scale parameters are essentially upper fractal limits, i.e., the corner wavelengths at which the power spectrum flattens; D applies to wavelengths shorter than λ_n and λ_s. In abyssal hills terrain, the lineation direction is the average strike of the hills, λ_n their average width, and λ_s their average length. An anisotropy parameter a is defined as λ_s/λ_n, i.e., as the average length-to-width ratio of abyssal hills. Goff and Jordan (1989a) estimated characteristic parameters of abyssal hill topography from swath bathymetry data in the Pacific and Atlantic oceans and obtained widths of 1.2–15 km, lengths of 7–41 km, and anisotropy parameters of 2–10, in agreement with other results (Bell, 1975; Bell, 1979; Gilbert and Malinverno, 1988).

An additional problem is that statistical characteristics of topography are generally not uniform in space but vary due to the effects of different geological processes (e.g., volcanism, faulting, sedimentation). This characteristic of topography is commonly referred to as nonstationarity or inhomogeneity (Bell, 1979; Fox and Hayes, 1985; Goff and

Jordan, 1988; Malinverno, 1989a). At the ocean basin scale, Bell (1979) showed that in agreement with qualitative observations (Macdonald, 1982; Menard, 1967; Van Andel and Bowin, 1968), there is an inverse relationship between the rms amplitude of abyssal hill topography and spreading rate: For a given observation scale, the rms amplitude in the slow-spreading Atlantic is at least twice as large as the amplitude in the fast-spreading Pacific Ocean. This relationship is also illustrated in Fig. 6.8: The power at any given wavelength is consistently greater in the spectra from the south Atlantic (Gilbert and Malinverno, 1988) than in spectra from the central Pacific (Bell, 1975). At smaller observation scales, Fox and Hayes (1985) note that profiles of seafloor topography typically contain segments with different statistical characteristics (e.g., different spectral exponents β) and statistical characteristics measured on these nonstationary profiles as a whole are a rather unpredictable mixture. For these measurements to be meaningful, profiles must be separated into segments that are at first order statistically stationary or homogeneous. Nonstationary profiles can be separated into homogeneous segments by placing boundaries where local statistical characteristics of the profiles vary most rapidly (Fox and Hayes, 1985; Malinverno, 1989a).

The statistical models illustrated here rest on the assumption that the topography to be modeled is Gaussian; in this case, statistical properties of topography are completely specified by its first- and second-order moments, i.e., mean and covariance (or equivalently mean and power spectrum; Goff and Jordan, 1988). We can infer that topography is Gaussian from the central limit theorem, which states that a variable is normally distributed if it is the sum of many independent random causes. For example, Bell (1975) found that the distribution of abyssal hill topography is not significantly different from normal if large seamounts are ignored. However the distribution has a positive skew (third moment), i.e., an asymmetry toward high elevations, and a kurtosis (fourth moment) greater than that of a normal distribution. Specifically Goff and Jordan (1988) note that to treat systematic slope asymmetries and the presence of smooth, sedimented valleys in ocean floor topography properly, it is necessary to consider moments up to fourth order. Alternatively the succession of rough and smooth areas can be treated as spatial nonstationarity; however Malinverno (1989a) found that in many cases, quasi-stationary segments do not conform to a Gaussian model. While statistical approaches based on some assumption of normality are commonly used, it is important to bear in mind their limitations. Methods to deal with non-Gaussian seafloor statistics have been described by Goff (1989b), Malinverno (1989a), and Shaw and Smith (1990).

Other problems are related to methods used to analyze data (the reader uninterested in technical details may wish to skip the rest of this section). One difficulty is that because topography spectrum has very powerful long-wavelength components, spectral analyses are not straightforward (Bell, 1979). For example, profiles should be prewhitened, i.e., filtered to resemble white noise before a spectral estimate is taken. The final spectrum should be the spectrum of the prewhitened series multiplied by the frequency response of the prewhitening filter (Blackman and Tukey, 1959). If prewhitening is not carried out, power in the long wavelengths can leak into the short wavelengths, and while the resulting spectrum may still follow an approximate power law, the exponent will be quite different (Fox and Hayes, 1985). Some investigators also subtract the average trend from input data before taking a spectral estimate (Brown, 1987; Czarnecki and Bergin, 1986; Goff and Jordan, 1988); this operation decreases the amplitude of long-wavelength components, and it may thus affect the final value of the spectral exponent. Another procedure involves removing a trend that corresponds to a definite process; for example, Gilbert and Malinverno (1988) subtracted the trend due to the thermal subsidence of the oceanic lithosphere from their profiles of abyssal hill topography before computing spectral estimates. In this case, it should be kept in mind that results are applicable only to the residual topography rather than to the topography as a whole. Many apparently conflicting values of the spectral exponent β are probably due to different procedures used for spectral estimation.

Another difficulty of a fractal interpretation of power spectra is that while profiles should have an exponent β between 1–3 to have a well-defined fractal dimension [see Eq. (8)], spectral exponents larger than 3 have been measured (Berkson and Matthews, 1983; Fox and Hayes, 1985; Malinverno, 1989a). Because the second moment of the power spectrum (the variance of the slopes) is finite for β ⩾ 3, Brown and Scholz (1985) argue that profiles with a spectral exponent β ⩾ 3 are differentiable and should have a fractal dimension $D = 1$; a similar argument has been presented by Hough (1989). However profiles with different values of β ⩾ 3 have different roughnesses, and a classification based on fractal dimension does not distinguish between them (Malinverno, 1989a). An important question then is whether the fractal dimension D based on the spectral exponent β [Eq. (8)] is the same as the dimension obtained using an alternative method. This does not seem to be the case: Numerical experiments show that fractal dimensions estimated by the ruler method (Aviles and others, 1987; Fox, 1989) and using the sample variogram (Malinverno, 1989a) are not 1 when β = 3 but rather tend asymptotically to 1 as β increases beyond 3. Mandelbrot (1985) concludes that different *methods* of estimating the fractal dimension D of self-affine figures in fact correspond to different *definitions* of D. Fractal dimensions estimated using the ruler method on topographic profiles (Barenblatt and others, 1984) and map contours (Turcotte, 1989) or the sample variogram (Burrough, 1981; Malinverno, 1989b; Mark and Aronson, 1984) may not be directly comparable to fractal dimensions estimated from the spectral exponent β.

An apparently similar problem was noted by Huang and Turcotte (1989), who found that fractal dimensions estimated by using the power spectrum of profiles did not differ by a factor of 1 from fractal dimensions estimated from two-dimensional spectra, as expected from theory. Goff (1990) showed however that fractal dimensions agreed with the theoretical prediction when the appropriate relationship between the spectral exponent β and the fractal dimension was used [Eq. (10)].

6.4. FRACTALS AND RELIEF-FORMING PROCESSES

An analysis of topographic data based on a fractal model can provide useful information. For example, rather than having to deal with a high-resolution description of an irregular surface, we summarize all its characteristics in a few parameters. If a profile has a power law spectrum as in Eq. (7), it can be characterized by only two parameters: the spectral exponent β (related to the fractal dimension D) and the constant A (related to the overall variance of the profile). A more complete model includes upper fractal limits and anisotropy (Bell, 1979; Gilbert and Malinverno, 1988; Goff and Jordan, 1988). These parameters are quantitative estimates rather than subjective, poorly constrained measures. As previously discussed, a fractal model by no means accounts for all the observed features of ocean floor topography; it is however a good first approximation over a finite band of spatial frequencies.

A first-order model for topographic roughness can be used to map areas with different statistical characteristics (Fox and Hayes, 1985; Goff and Jordan, 1989a; Huang and Turcotte, 1989; Malinverno, 1989a), and it has a number of applications. For example, Bechtel and others (1987) obtained a value for the average crustal thickness in East Africa from a downward continuation of surface gravity anomalies, assuming that the topography of the crust-mantle boundary has a power law spectrum. Mareschal (1989) used fractal interpolation to construct seafloor topographic profiles with a realistic roughness from a few data points. Goff and Jordan (1989a) generated synthetic ocean floor topographies at scales smaller than those sampled in reality, assuming that self-similarity persists at these scales. Fractals can also be viewed as useful noise models or fundamental null hypotheses to compare with real data (Burrough, 1984). In many cases, useful information can be extracted when observations indeed do not conform to an idealized fractal model. For example, upper fractal limits provide the maximum size of topographic features (Bell, 1975; Bell, 1979), and variations in fractal dimension in different frequency bands may be related to the effects of different physical processes (Brown and Scholz, 1985).

Another important consequence of self-affinity is the presence of apparent periodicities. These prominent cycles may simply be the consequence of the power law form of the spectrum, where long-wavelength components have an amplitude much larger than short-wavelength components. In fact finite samples of fractional noises appear to contain cycles with a wavelength roughly equal to one-third the total sample length (Mandelbrot and Wallis, 1969a). Dominant wavelengths for abyssal hills have been identified by some authors [e.g., 30 km in the north Atlantic (Heirtzler and Le Pichon, 1965)]. The preceding results suggest that there is indeed no dominant periodicity in abyssal hills; these wavelengths appear to dominate the topographic signal because the power spectra tend to flatten at wavelengths longer than a few tens of kilometers. Fractal theory shows that care must be taken when dividing a geophysical variable into separate wavelength domains of variation (e.g., small scale, mesoscale, large scale, etc.). If the variable has a continuous power law spectrum and it is self-affine, such subdivision may have no physical basis and thus are essentially arbitrary (Lorenz, 1969; Mandelbrot and Wallis, 1969b).

There is however an unaddressed question: Given that topography is fractal, why is that so? In other words, what characteristics of the basic geological processes are necessary to produce a fractal topography? A way of investigating this question is to consider the energy necessary to construct topography. For example, Bell (1975) shows that if the energy of formation were distributed uniformly over the size fo the features (i.e., there were a large number of small features and a small number of large features), then the resulting topography would have a power law spectrum with an exponent $\beta = 2$, in agreement with his observations.

Another way of addressing this problem is to attempt to generate fractal topography by geologically realistic models. In fact, there are a number of methods available for creating synthetic fractal profiles and surfaces (Fournier and others, 1982; Fox, 1987; Goff and Jordan, 1988; Huang and Turcotte, 1989; Malinverno, 1989a; Saupe, 1988; Voss, 1985, 1988). These methods however are designed with computational simplicity and efficiency as a constraint, and they do not have a relationship to geomorphic processes (Mark and Aronson, 1984). An exception is the method used by Mandelbrot (1975) to create a fractal surface by cutting an initially flat topographic surface with a large number of vertical faults with random offsets and orientations. This approach has a fundamental flaw at large scales, since it assumes that the effect of a fault is a constant vertical offset of the surface topography to an infinite distance from the fault. Applying this method on the planetary scale, as done by Voss (1988), is obviously not a simulation of a realistic geological process.

Accounting for the fact that the effect of faulting events is limited to some finite horizontal distance, a faulting model can be applied to generating abyssal hill topography. A number of authors suggest that in slow-spreading oceans, the floor of the depression typically found at the ridge axis is faulted and uplifted, thus forming abyssal hills (Atwater and Mudie, 1968; Laughton and Searle, 1979; Macdonald and Atwater, 1978). Malinverno and Gilbert (1989) formulated a model in which a profile of abyssal hill topography results from the sum of the effects of a number of uplift events of random amplitude. To obtain a model topography with a power spectrum similar to that of abyssal hills in the south Atlantic, the width of the response of topography to a faulting event must be about 3.5 km; significantly different widths resulted in spectra with corner wavelengths different from that observed. An important implication of this model is that, since amounts of uplift must vary in time to produce abyssal hills, the shape of the axial valley cannot be strictly steady-state; this notion is supported by other studies (Kappel and Ryan, 1986; Lewis, 1979; Malinverno, 1990; Vogt and others, 1969).

To complement the faulting model and demonstrate that this approach can provide valuable insight, I present in the rest of this section a simple alternative model for generating abyssal hills by volcanic construction. In the interest of simplicity, the model is one-dimensional, i.e., it creates a profile of abyssal hills rather than a two-dimensional surface. In this approach, there is no attempt to model the along strike variation in abyssal hill topography. Since abyssal hills are clearly elongated, the model should provide an acceptable first approximation.

Similarly to the method used to create the von Koch curve, we can construct a rough topographic profile by superpositioning a number of self-similar features of varying size, where the number of features is inversely proportional to their size (i.e., the smallest are the most numerous). Since the model is formulated in terms of realistic geological processes, there are a number of differences from the method used for the von Koch curve. In the model, abyssal hill topography results from superpositioning a number of volcanic constructions, each having a height h (defined with respect to the preexisting topography) and a constant slope of the flanks α (defined with respect to the horizontal; see Fig. 6.9). While features in the von Koch curve are located to obtain an exactly self-similar figure, the location of volcanic constructions in the model is random. Also while the size distribution of features in the von Koch curve is also self-similar, following a power law, heights of volcanic constructions are samples drawn from an exponential distribution. The cumulative distribution function of an exponential distribution is

$$\text{Prob}\,(h \leq \eta) = 1 - e^{-\eta/\bar{h}} \tag{11}$$

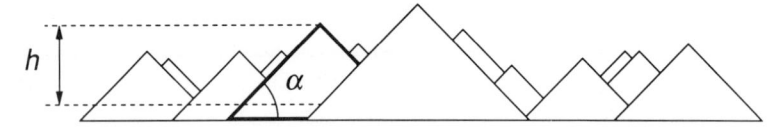

FIGURE 6.9. The process of creating abyssal hills by volcanism can be simulated by superpositing a number of constructions characterized by a random height h and a flank slope α.

where Prob $(h \leq \eta)$ is the probability that the height h is smaller than or equal to η and $\bar{h}$ is the average height of the constructions. An exponential distribution is chosen because it follows the principle of having many small features and few large features; it is mathematically convenient, being completely defined by only one parameter (the average height $\bar{h}$); and the rapid increase of Prob $(h \leq \eta)$ with increasing η imposes an effective upper limit on feature size. Also while the size distribution of some geological objects, such as rock fragments, follows a power law and therefore is self-similar (Turcotte, 1986), heights of seamounts in the Pacific are exponentially distributed (Smith and Jordan, 1988). Finally while features are added on the von Koch curve ad infinitum, volcanic constructions are added in the model only until the average thickness of the volcanic layer is 1 km, in agreement with estimates obtained from seismic experiments in the oceanic crust (e.g., Shor and Raitt, 1969).

Note that locating the volcanic constructions at random is not exactly reproducing abyssal hill generation, since abyssal hills are created near the midocean ridge axis and later transported on the ridge flanks by seafloor spreading. To simulate seafloor spreading, it is necessary to add features in a location that moves along the profile at a rate equal to the half-spreading rate. Nevertheless, since the height h of the features is an uncorrelated random variable, adding the same number of features at random or in a location that steadily moves along the profile results in a topography with identical statistical characteristics.

The only two free parameters in this model are the average height $\bar{h}$ of the volcanic constructions and the slope of the flanks α. Intuitively the overall roughness of a profile increases as $\bar{h}$ and α increase; this is shown for a few synthetic profiles in Fig. 6.10. It is now interesting to determine whether this model can reproduce the basic characteristics of actual abyssal hills and what are the characteristic parameters $\bar{h}$ and α. As an example, the south Atlantic profile studied by Gilbert and Malinverno (1988) is compared to model results. Because average half-spreading rates in the last 80 Ma as computed from the magnetic anomaly map of Cande and others (1988) differ on the two flanks (about 26 mm/yr on the west flank and 20 mm/yr on the east flank), each flank is considered separately. The fit between data and model is assessed by constructing synthetic profiles with various values of

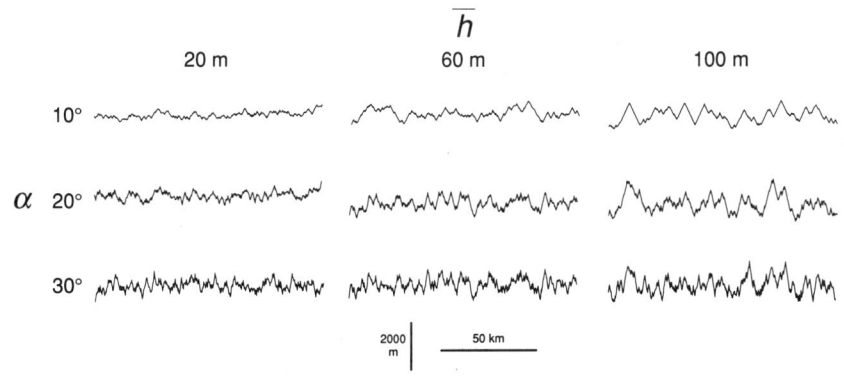

FIGURE 6.10. Synthetic abyssal hills created by the method illustrated in Fig. 6.10 are markedly different for different values of the average height $\bar{h}$ and the flank slope α. As $\bar{h}$ increases, the amplitude of the oscillations increases; as α increases, profiles become more and more jagged; i.e., their fractal dimension increases.

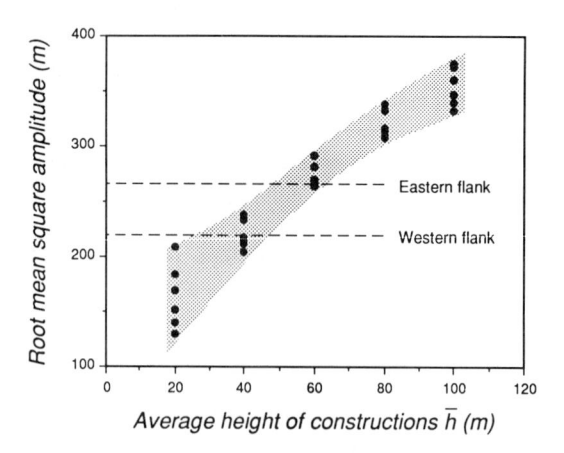

FIGURE 6.11. The rms amplitude of synthetic abyssal hill profiles is mainly a function of the average height $\bar{h}$. The shaded area highlights the overall trend. For each value of $\bar{h}$, six synthetic profiles with α ranging from 5°–30° have been examined. To obtain synthetic profiles with the same rms amplitude as that measured on the eastern and western flank of the south Atlantic midocean ridge, average heights of about 55 m and 40 m, respectively, should be used.

characteristic parameters $\bar{h}$ and α and by comparing statistical characteristics of the synthetics with those of the actual profiles.

The profiles in Fig. 6.10 suggest that while the average height ($\bar{h}$) controls the overall amplitude, i.e., the variance of the profile, the slope of the hill flanks (α) controls the degree of correlation between neighboring points, i.e., the fractal dimension. The relationship between $\bar{h}$ and the rms amplitude of the profiles is illustrated in Fig. 6.11. The rms amplitude of the eastern flank profile is 266 m and that of the western flank profile 219 m, corresponding to values of the average height $\bar{h}$ of about 55 m and 40 m, respectively. Note that the rms amplitude is estimated both for data and synthetics on very long intervals (about 1500 km); as previously discussed, the variance of these profiles reaches a limit if measured on intervals longer than a few tens of kilometers. A best-fit value for the slope of the constructions α can be estimated by comparing power spectra of the data with power spectra of the synthetics and by determining the average misfit; results are illustrated in Fig. 6.12. The minimum misfits for the eastern and the western flank profiles correspond to a value of

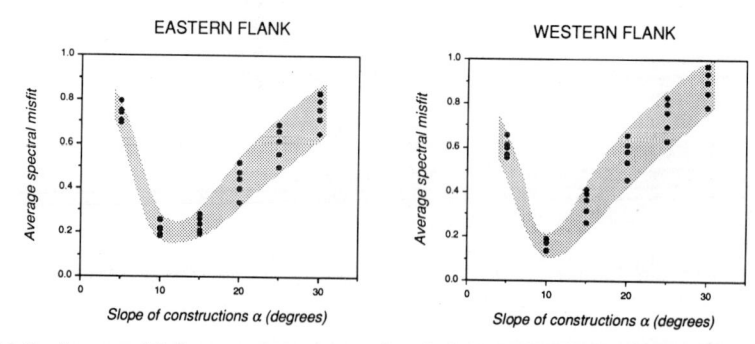

FIGURE 6.12. Average misfit between power spectra of synthetic abyssal hills and spectra of actual topography from the eastern and western flank of the south Atlantic midocean ridge calculated for wavelengths from 1–300 km. For each value of α, five synthetic profiles with $\bar{h}$ ranging from 20–100 m have been examined. The misfit is mainly a function of the slope of the volcanic constructions α. It reaches a minimum for values of α about 12° (eastern flank) and 10° (western flank). Shaded areas highlight the overall trend.

6. FRACTALS AND OCEAN FLOOR TOPOGRAPHY: A REVIEW AND A MODEL

α of about 12° and 10°, respectively. Spectra of synthetic profiles constructed with these values of the parameters $\bar{h}$ and α indeed have a corner frequency and a short-wavelength spectral exponent close to those of the actual data (see Fig. 6.13).

Because the assumption of a constant slope for the volcanic constructions is somewhat questionable, the best fit values of 10°–12° have to be considered as only averages over the length of the profiles. Slopes obtained from the model are comparable to estimates of typical slopes of abyssal hills (Pockalny and others, 1987; Pockalny and others, in press) and submarine volcanoes (Fornari and others, 1984). In agreement with the general rule that slower spreading rates correspond to greater ridge flank roughness the rms amplitude and the average height $\bar{h}$ of abyssal hills on the slower spreading eastern flank are greater than those measured on the faster spreading western flank. The model presented here also provides some insight into how the volcanic layer in the oceanic crust may be generated: Small average heights of the population of volcanic constructions suggest that the volcanic layer of the oceanic crust is composed of a number of small units and large abyssal hills result from the superposition of several volcanic constructions. In fact the cumulative distribution function in Eq. (14) implies that only one in a hundred constructions has a height greater than 4.6 times the mean (i.e., greater than 180–250 m) and only one in a thousand has a height greater than 6.9 times the mean (280–380 m).

The resemblance between the observed data and the synthetics is not, however, completely satisfactory (see Fig. 6.14). While actual abyssal hill topography contains sharp peaks and flat depressions, synthetic topography is essentially symmetrical with respect to the horizontal. The reader may verify this statement by turning Figs. 6.10 and 6.14 upside down; while the real topography looks quite different when turned upside down, the synthetic topography does not. This may be viewed as a consequence of the central limit theorem: By adding a number of independent, small features, we eventually obtain a topography that is Gaussian, i.e., symmetrically distributed about its mean. This observa-

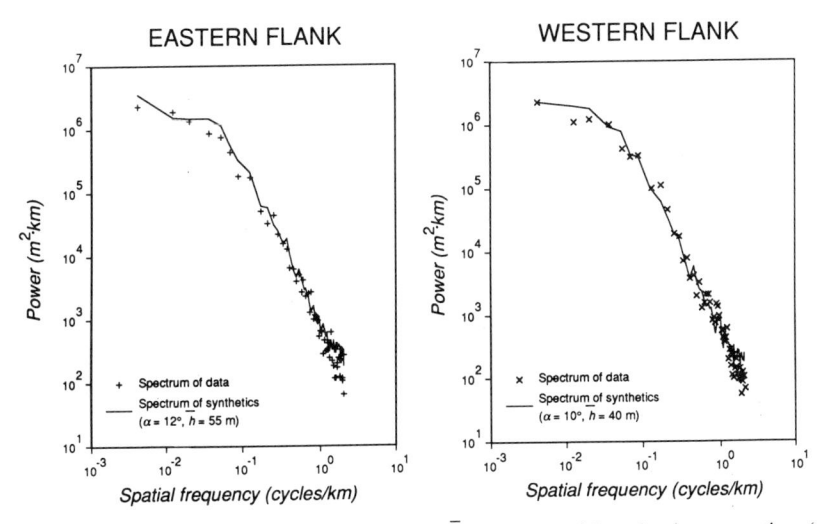

FIGURE 6.13. For the best fit values of the average height $\bar{h}$ and slope α of the volcanic constructions (see Figs. 6.11 and 6.12), the power spectra of the synthetics closely match the spectra of actual abyssal hills in the south Atlantic.

EASTERN FLANK PROFILE

SYNTHETIC PROFILE ($\alpha = 12°$, $\overline{h} = 55$ m)

WESTERN FLANK PROFILE

SYNTHETIC PROFILE ($\alpha = 10°$, $\overline{h} = 40$ m)

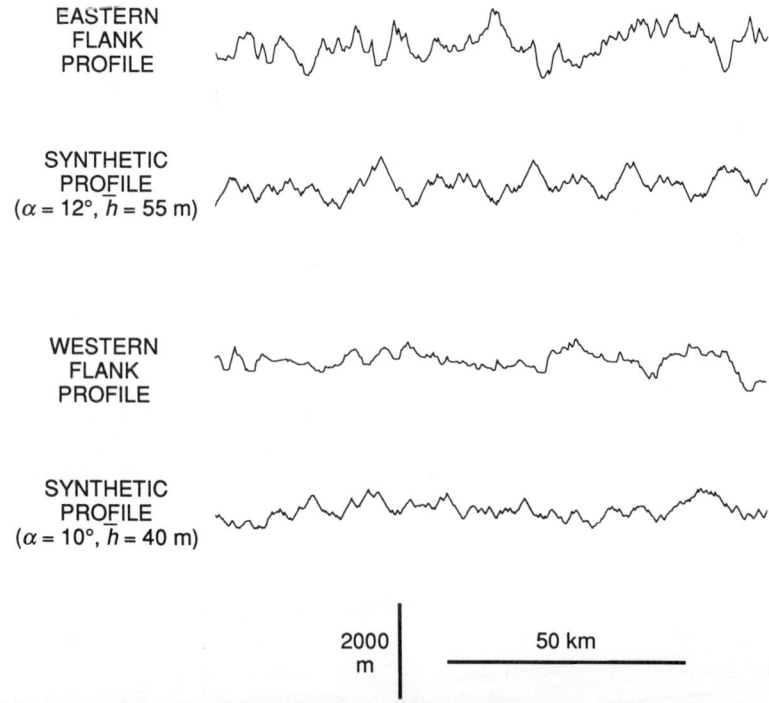

2000 m | 50 km

FIGURE 6.14. While synthetics resemble actual abyssal hills in the space domain, the similarity is not wholly satisfactory. For example, synthetic abyssal hills are symmetric with respect to the horizontal (the reader may check this by turning Fig. 6.14 upside down), while actual abyssal hills are not, since they contain sharp peaks and flat valleys. This is particularly evident in the profile from the eastern flank of the south Atlantic midocean ridge.

tion underscores the importance of high-order statistical moments in characterizing ocean floor topography (Goff and Jordan, 1988). If only moments up to second order (e.g., the power spectrum) are used, the resemblance between synthetics and actual data is not complete.

These discrepancies between the model and observations have some interesting geological implications. For example, volcanism is unlikely to produce topographic features with a constant slope of the flanks. The slope is expected to vary according to such variables as eruption rate and lava viscosity, generating a whole range of features, from steep-sided constructions to subdued lava flows (Larson, 1971). Also the cross-section of a volcanic construction may have a slope that is not constant but varies with distance from the summit. Incorporating these characteristics into a more refined model is likely to produce a better fit with the actual topography; this is not done here to keep the discussion as simple as possible.

6.5. CONCLUSIONS AND SPECULATIONS

From the statistical analyses presented here, we conclude that profiles of abyssal hills can be effectively described as self-affine, with an upper fractal limit related to the upper size limit of hills. Also, since abyssal hills are clearly elongated, a full description has to incorporate the variation of statistical properties with direction, i.e., anisotropy. Starting from this first-order description, it is interesting to devise models simulating actual geological processes and resulting in a self-affine topography similar to that observed. In the example presented here, abyssal hill topography is created by superpositioning a number of volcanic constructions, and characteristic parameters of the population of constructions (the average height $\bar{h}$ and the slope of the flanks α) can be estimated from a comparison with actual topographic data.

However, using fractals and statistical models has in general deeper consequences on how we approach the study of nature. In some sense, the physical scientist abhors using statistical models, since it seems that we need to resort to statistics when we do not have a very clear idea of underlying processes. Also a statistical approach by itself merely provides a succinct description of observations rather than physical insight into underlying phenomena. While these are all valid objections, the pertinent question to ask is whether we can afford to ignore a statistical approach when studying some geological phenomena. The argument offered here is that we cannot: Since faulting and volcanism near mid ocean ridge crests intrinsically result in a chaotic, jumbled morphology, it seems more appropriate to study statistical characteristics of abyssal hills rather than the details of a few features. While there is no substitute for extensive measurements, statistical models presented here offer an opportunity to gain some insight even from a limited data set, since statistical characteristics of a large population can be estimated from a relatively small sample. It should also be pointed out that if topography is essentially self-affine, increasing measurement resolution does not necessarily improve our understanding.

Using statistical models may not really be a surrender to the notion that we will never be able to account for all causative factors in the natural processes examined. It has been realized recently that even simple, completely deterministic physical systems may behave in an unpredictable fashion if governing equations contain nonlinear terms, and this phenomenon has been referred to as chaos.* An important feature of these systems is their sensitivity to initial conditions: If we compare the behavior of two identical systems with a slightly different initial state, we observe very large differences at subsequent times (Lorenz, 1969). In these cases, it becomes a semantic question to qualify the observed behavior of the system as random or as chaotic, albeit deterministic. Indeed we may qualify games of chance (e.g., the throw of a dice) as chaotic systems, where small differences in the initial state (position and velocity of the dice when thrown) result in widely different behavior (orientations of the dice when it comes to rest).

All of this is reasonable, but what does it have to do with abyssal hills? The key point is that processes of faulting and volcanism are also likely to be very sensitive to small changes in external conditions. Two similar parcels of rock break in different ways, although placed in the same stress field; earthquakes and volcanic eruptions are extremely difficult to predict, since large events seem to be triggered by vanishingly small changes in stress and/or magma supply. It has been shown that there are physical systems that attain a critical state where small perturbations trigger events whose size distribution is self-similar (Bak and others, 1987). It is then to be expected that faulting and volcanism result in an intrinsically complex topography, so that using a statistical approach may then be the first step toward a

---

*General introductions on this topic are provided by Crutchfield and others (1986), Gleick (1987), and Jensen (1987).

more complete physical understanding. In contrast, a search for simple periodicities in topography may in fact have no physical justification.

Interesting results have already been obtained by studying ocean floor topography in terms of fractal processes, and the application of concepts from chaotic dynamics to the study of topography has just started (Smith and Shaw, 1989). Future work in these directions seems to promise a more complete understanding of the fundamental processes active at midocean ridges. It is the author's hope that the evidence presented here convinced the reader of the usefulness of a statistical approach and will encourage similar studies.

ACKNOWLEDGMENTS

Discussions with Bernie Coakley, Lewis Gilbert, John Goff, Jill Karsten, Rob Pockalny, Chris Scholz, and comments by Denny Hayes, Sue Hough, and two anonymous reviewers provided fruitful inspiration. This work was supported by the Office of Naval Research under contract numbers N00014-87-K-0204 and N00014-89-J-1159. Lamont–Doherty Earth Observatory contribution number 5253.

REFERENCES

Akal, T., and Hovem, J., *Marine Geotech.* **3**, 171 (1978).
Atwater, T. M., and Mudie, J. D., *Science* **159**, 729 (1968).
Aviles, C. A., Scholz, C. H., and Boatwright, J., *J. Geophys. Res.* **92**, 331 (1987).
Bak, P., Tang, C., and Wiesenfeld, K., *Phys. Rev. Lett.* **59**, 381 (1987).
Balmino, G., Lambeck, K., and Kaula, W. M., *J. Geophys. Res.* **78**, 478 (1973).
Barenblatt, G. I., Zhivago, A. V., Neprochnov, Y. P., and Ostrovskiy, A. A., *Oceanol.* **24**, 695 (1984).
Bechtel, T. D., Forsyth, D. W., and Swain, C. J., *Geophys. J. R.* **90**, 445 (1987).
Bell, T. H., *Deep Sea Res.* **22**, 883 (1975).
———, *Deep Sea Res.* **26A**, 65 (1979).
Berkson, J. M., and Matthews, J. E., in *Acoustic in the Sea Bed* (N. G. Pace, ed.) (Bath University Press, Bath, England, 1983), pp. 215–23.
Berry, M. V., and Hannay, J. H., *Nature* **273**, 573 (1978).
Blackman, R. B., and Tukey, J. W., *The Measurement of Power Spectra from the Point of View of Communications Engineering* (Dover, New York, 1959).
Bretherton, F. P., *Q. JRMA* **95**, 213 (1969).
Briggs, K. B., *IEEE J. Ocean. Eng.* **14**, 360 (1989).
Brown, S. R., *Geophys. Res. Lett.* **14**, 1095 (1987).
Brown, S. R., and Scholz, C. H., *J. Geophys. Res.* **90**, 12,575 (1985).
Durrough, P. A., *Nature* **294**, 240 (1981).
———, *J. Inst. Math. and its Applications* **20**, 36 (1984).
Cande, S. C., LaBrecque, J. L., and Haxby, W. F., *J. Geophys. Res.* **93**, 13,479 (1988).
Cox, C., and Sandstrom, H., *J. Oceanogr. Soc. Jpn.* **20**, 499 (1962).
Crutchfield, J. P., Farmer, J. D., Packard, H. D., and Shaw, R. S., *Sci. Am.* **255**, 46 (1986).
Czarnecki, M., and Bergin, J. Statistical characterization of small-scale bottom photography as derived from multibeam sonar data: Proceedings of the Fourth Working Symposium on Oceanographic Data Systems, La Jolla, IEEE Computer Society, p. 15–24.
Davis, E. E., and Lister, C. R. B., *Earth Plan.* **21**, 405 (1974).
Dietz, R. S., *Nature* **190**, 854 (1961).
Farr, H. K., *Marine Geodesy* **4**, 77 (1980).
Fornari, D. J., Ryan, W. B. F., and Fox, P. J., *J. Geophys. Res.* **89**, 11,069 (1984).
Fournier, A., Fussell, D., and Carpenter, L., *Commun. ACM* **25**, 371 (1982).

Fox, C. G., *Comput. Geosci.* **13**, 1 (1987).
——, *Pure Appl. Geophys.* **131**, 211 (1989).
Fox, C. G., and Hayes, D. E., *Rev. Geophys.* **23**, 1 (1985).
Gibert, D., and Courtillot, V., *J. Geophys. Res.* **92**, 6235 (1987).
Gilbert, L. E., *Pure Appl. Geophys.* **131**, 241 (1989).
Gilbert, L. E., and Malinverno, A., *Geophys. Res. Lett.* **15**, 1401 (1988).
Gleick, J., *Chaos: Making a New Science*, (Viking, New York, 1987).
Goff, J. A., *J. Geophys. Res.* **95**, 5159 (1990).
Goff, J. A., and Jordan, T. H., *J. Geophys. Res.* **93**, 13,589 (1988).
——, *Geophys. Res. Lett.* **16**, 45 (1989a).
——, Use of the 3-point correlation function in the characterization of abyssal hill asymmetries: *Eos (Trans. Am. Geophys. Union)* **70**, 1306 (1989b).
Heezen, B. C., Tharp, M., and Ewing, M., The Floors of the Oceans: 1. The North Atlantic, Geological Society of America Special Paper, v. 65, 1959, p. 122.
Heirtzler, J. R., and Le Pichon, X., *J. Geophys. Res.* **70**, 4013 (1965).
Hibler, W. D. I., and LeSchack, L. A., *J. Glaciol.* **63**, 345 (1972).
Hough, S. E., *Geophys. Res. Lett.* **16**, 673 (1989).
Huang, J., and Turcotte, D. L., *J. Geophys. Res.* **94**, 7491 (1989).
Jaeger, R. M., and Schuring, D. J., *J. Geophys. Res.* **71**, 2023 (1966).
Jensen, R. V., *Am. Sci.* **75**, 168 (1987).
Journel, A. G., and Huijbregts, C. J., *Mining Geostatistics* (Academic Press, New York, 1978).
Kappel, E. S., and Ryan, W. B. F., *J. Geophys. Res.* **91**, 13,925 (1986).
Langseth, M. G., Pichon, X. L., and Ewing, M., *J. Geophys. Res.* **71**, 5321 (1966).
Larson, R. L., *Geol. Soc. Am. Bull.* **82**, 823 (1971).
Laughton, A. S., and Searle, R. C., in *Deep Drilling Results in the Atlantic Ocean: Ocean Crust* (M. Talwani, C. G. Harrison, and D. E. Hayes, eds.) (American Geophysical Union, Washington, DC, 1979), pp. 15–32.
Lewis, B. T. R., *Geophs. Res. Lett.* **6**, 753 (1979).
Lonsdale, P., *Mar. Geophys. Res.* **3**, 251 (1977).
Lorenz, E. N., *Tellus* **21**, 289 (1969).
Macdonald, K. C., *Annual Review of Earth and Planetary Sciences* **10**, 155 (1982).
Macdonald, K. C., and Atwater, T. M., *Earth Plan.* **39**, 319 (1978).
Macdonald, K. C., and Luyendyk, B. P., *Mar. Geophys. Res.* **7**, 515 (1985).
Malinverno, A., *IEEE J. Ocean. Eng.* **14**, 348 (1989a).
——, *Pure Appl. Geophys.* **319**, 139 (1989b).
——, *J. Geophys. Res.* **95**, 2645 (1990).
Malinverno, A., and Gilbert, L. E., *J. Geophys. Res.* **94**, 1665 (1989).
Mandelbrot, B. B., *Science* **156**, 636 (1967).
——, *Proceedings of the Nat. Academy of Sciences, USA* **72**, 3825 (1975).
——, *The Fractal Geometry of Nature* (W. H. Freeman, New York, 1983).
——, **32**, 257 (1985).
Mandelbrot, B. B., and Van Ness, J. W., *SIAM Rev.* **10**, 422 (1968).
Mandelbrot, B. B., and Wallis, J. R., *Water Resour. Res.* **5**, 228 (1969a).
——, *Water Resour. Res.* **5**, 321 (1969b).
Mareschal, J.-C., *Pure Appl. Geophys.* **131**, 197 (1989).
Mark, D. M., and Aronson, P. B., *Math. Geol.* **16**, 671 (1984).
McKenzie, D. P., *J. Geophys. Res.* **72**, 6261 (1967).
Menard, H. W., *Marine Geology of the Pacific* (McGraw-Hill, New York, 1964).
——, *Science* **157**, 923 (1967).
Menard, H. W., and Mammerickx, J., *Earth Plan.* **2**, 465 (1967).
Neidell, N. S., *Geophysics* **31**, 122 (1966).
Nye, J. F., *Proc. R. Soc. Lond. A* **315**, 381 (1970).
——, *AIDJEX Bulletin* **21**, 20 (1973).
Oldenburg, D. W., *Geophys. J. R.* **43**, 425 (1975).
Parsons, B., and Sclater, J. G., *J. Geophys. Res.* **82**, 803 (1977).
Peitgen, H. O., and Saupe, D., eds., *The Science of Fractal Images* (Springer-Verlag, New York, 1988).
Pockalny, R. A., Detrick, R. S., and Fox, P. J., *Eos (Trans. Am. Geophys. Union)* **68**, 1491 (1987).

———, *J. Geophys. Res.* **93**, 3179 (1988).
Pockalny, R. A., Fox, P. J., and Detrick, R. S., "A Morphological Comparison of Abyssal Hill Topography Using High-Resolution, Multibeam Bathymetry Data," *J. Geophys. Res.*, in press.
Priestley, M. B., *Spectral Analysis and Time Series* (Academic Press, London, 1981).
Rea, D. K., *Geology* **3**, 77 (1975).
Saupe, D., in *The Science of Fractal Images* (H. O. Peitgen and D. Saupe, eds.) (Springer-Verlag, New York, 1988), pp. 71–136.
Sayles, R. S., and Thomas, T. R., *Nature* **271**, 431 (1978).
Searle, R. C., and Laughton, A. S., *J. Geophys. Res.* **82**, 5313 (1977).
Shaw, P. R., and Smith, D. K., *Geophys. Res. Lett.* **14**, 1061 (1987).
———, *J. Geophys. Res.* **95**, 8705 (1990).
Shor, G. G., Jr., and Raitt, R. W., in *The Earth's Crust and Upper Mantle*, Geophysical Monograph, Vol. 13 (J. P. Hart, ed.) (American Geophysical Union, Washington, DC, 1969), pp. 225–30.
Smith, D. K., and Jordan, T. H., *J. Geophys. Res.* **93**, 2899 (1988).
Smith, D. K., and Shaw, P. R., *Eos (Trans. Am. Geophys. Union)* **70**, 1306 (1989).
Turcotte, D. L., *J. Geophys. Res.* **91**, 1921 (1986).
———, *J. Geophys.. Res.* **92**, E597 (1987).
———, *Pure Appl. Geophys.* **319**, 171 (1989).
Urick, R. J., *Principles of Underwater Sound* (McGraw-Hill, New York, 1975).
Van Andel, T. H., and Bowin, C. O., *J. Geophys. Res.* **73**, 1279 (1968).
Vogt, P. R., Schneider, E. D., and Johnson, G. L., in *The Earth's Crust and Upper Mantle* (P. J. Hart, ed.) (American Geophysical Union, Washington, DC, 1969), pp. 556–617.
Voss, R. F., in *Fundamental Algorithms for Computer Graphics* (R. A. Earnshaw, ed.) (Springer-Verlag, Berlin, 1985), pp. 805–35.
———, in *The Science of Fractal Images* (H. O. Peitgen and D. Saupe, eds.) (Springer-Verlag, New York, 1988), pp. 21–70.
Warren, B. A., *Deep Sea Res.* **20**, 9 (1973).
Wong, P., in *Physics and Chemistry of Porous Media II*, AIP Conf. Proc., Vol. 154 (J. Banavar, J. Koplik, and K. Winkler, eds.) (American Institute of Physics, New York, 1987), pp. 304–18.

7

Fractal Transitions on Geological Surfaces

C. H. Scholz

7.1 INTRODUCTION

In recent years, many papers have demonstrated that a wide variety of geological objects are fractal; examples are coastlines (Mandelbrot, 1967), seafloor topography (Fox and Hayes, 1985; Malinverno and Gilbert, 1989), mountain topography (Gilbert, 1989), and joints and bedding planes (Brown and Scholz, 1985). In reporting these findings, the emphasis has usually been on the success of fractal geometry in describing these forms. However Burrough (1981), who noted the similarity of the fractal description to earlier geostatistical measures, such as the variogram, also pointed out that many objects may exhibit self-similarity over only a limited range of scales and different characteristics at widely separated scales. It is this latter theme that I take up here.

One limitation of a simple fractal picture of the world that has been pointed out many times is that natural objects are often fractal over only a limited bandwidth, whereas fractal geometry is not band-limited. The Earth, for example, may have fractal topography at short wavelengths, whereas at long wavelengths, it is very well-described by a Euclidean shape: an ellipsoid of revolution. The ends of fractal bands are often referred to as the upper and lower fractal limits. In addition there are changes in the fractal character of surfaces that occur within the fractal band. These fractal transitions appear to be of two types: gradual fractal transitions, where the fractal dimension changes continuously as a function of spatial wavelength, and abrupt transitions, where there is a sharp change at a particular wavelength. Gradual transitions seem to be the rule rather than the exception: It appears to be the case that the fractal dimension of natural geological surfaces cannot be considered to be constant over a bandwidth greater than about two decades in spatial wavelength (Brown and

---

C. H. Scholz • Lamont–Doherty Earth Observatory and Department of Geological Sciences, Columbia University, Palisades, New York.

Fractals in the Earth Sciences, edited by Christopher C. Barton and Paul R. La Pointe. Plenum Press, New York, 1995.

Scholz, 1985). Note that emphasis is on abrupt transitions because the evidence so far indicates that these occur at wavelengths corresponding to characteristic lengths in the system that generates the surface and these lengths may provide clues to the underlying physics.

7.2. OBSERVATIONS

The fractal nature of objects can be measured with a variety of techniques. The one used herein, the spectral method, consists of analyzing the spectrum of a surface profile. The surface profile is considered to be composed of an infinite series of sinusoids of different spatial frequencies or wavelengths. The square of the amplitude of each sinusoid is the power at that frequency, which is a measure of roughness at that scale. This quantity plotted versus the frequency (or wavelength) is the power spectrum of the profile. If the surface is fractal, its power spectral density (power normalized by the frequency) decays as $\omega^{-\xi}$, where ω is the spatial frequency. Thus a log-log plot of the power spectrum of a fractal surface yields a straight line of slope $-\xi$, which may be related to the fractal dimension **D** by $(5 - \xi)/2 = \mathbf{D}$. Schematic examples are shown in Fig. 7.1.

If $\xi < 3$, then $\mathbf{D} = 1$ and the surface is differentiable and Euclidean. Such surfaces consist of rounded topography at any point at which the tangent can be defined. In the surfaces I describe, $2 < \xi < 3$; hence $1 < \mathbf{D} < 1.5$, and these are nondifferentiable outward-facing fractals of a type called fractional Brownian. As **D** increases, the jaggedness of the surface increases. Thus the Brownian trace, with $\mathbf{D} = 1.5$, is the most jagged surface that does not contain overhangs. Surfaces where $\mathbf{D} > 1.5$ contain interior space, and these become increasingly spongelike as **D** becomes greater.

7.2.1. Joints

The spectrum of a joint in diabase over the band 1m–10^{-5}m is shown in Fig. 7.2. Although the spectrum has the general power law form characteristic of fractals over this entire bandwidth, the slope of the spectrum is not constant. It gradually steepens over the

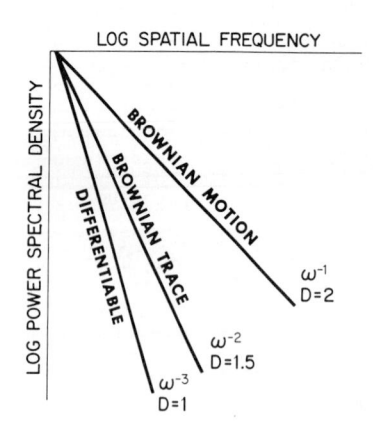

FIGURE 7.1. Schematic diagram showing the power spectra of profiles of surfaces of different fractal dimension. Profiles with spectra that fall off faster than ω^{-3} are differentiable and not fractal. The Brownian trace, $D = 1.5$ is the roughest surface that does not contain overhangs, and Brownian motion, $D = 2$ fills the plane.

7. FRACTAL TRANSITIONS ON GEOLOGICAL SURFACES

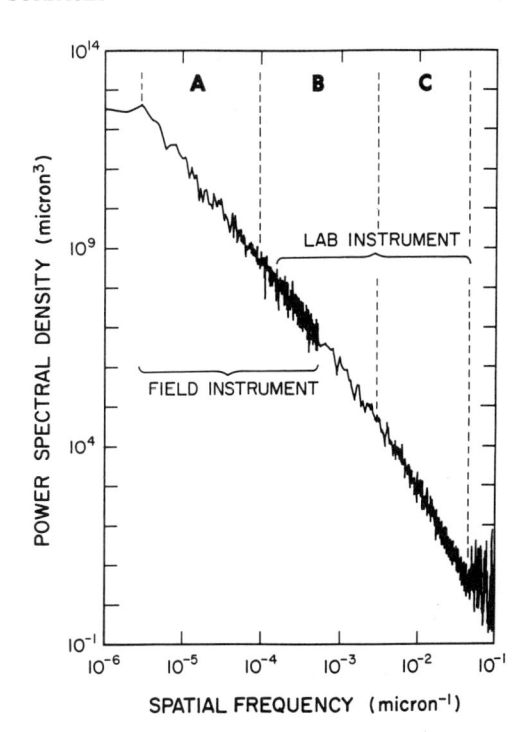

FIGURE 7.2. Power spectrum of the surface topography of a joint in the Palisades diabase, Palisades, New York.

bands labeled A and B, indicating a continuous decrease in fractal dimension with spatial frequency over that range. The average slope for each range yields the value $D = 1.24$ in band A and 1.15 in band B. There is a sharp change in slope between regions B and C, however. within region C, D is constant and approximately equal to one, indicating that in this band, the surface is Euclidean and differentiable. The BC transition occurs at the grain size of this rock. In this application, the fractal dimension can be considered as a measure of the degree of surface jaggedness: The joint surface becomes progressively more jagged as the scale increases from that of the hand specimen (B) to the outcrop (A). In the C band below the grain size, when we look at the surface of weathered single crystals, the surface is no longer fractal but though rugged has rounded contours. So in this case, an abrupt fractal transition occurs at a characteristic dimension of the system, here, the grain size of the rock.

Surfaces of course do not always have this random character over all wavelengths. One case measured by Brown and Scholz (1985) was a joint perpendicular to bedding in a thinly bedded (1–5 cm) siltstone. The profiles parallel to bedding on that joint surface showed a general fractal character similar to Fig. 7.2, but the profiles perpendicular to bedding showed enhanced power in the period range of 1–10 cm as a result of the bedding. In that case, we could treat the fractal spectrum as noise and subtracting the bedding parallel spectrum from the bedding perpendicular spectrum obtain the spectrum of the bedding alone. In another case, measurements parallel and perpendicular to the striae of a glacially polished surface showed that the surface was anisotropic over a wide bandwidth. Although no formalism has yet been developed to treat anisotropic fractals, they are rather common in geology. Another case, described later, is a slickensided fault surfaces.

7.2.2. Seafloor Topography

Gilbert (1989) measured seafloor topography on a 1500-km line measured parallel to the spreading direction adjacent to the mid-Atlantic Ridge in the south Atlantic. After correcting for the square root of age-depth dependence caused by the thermal contraction of the oceanic lithosphere, spectra show that the seafloor has a fractal character over the wavelength band 1–10 km, but it departs markedly from a fractal at wavelengths above several tens of kilometers (see Fig. 7.3). This constitutes an abrupt fractal transition at about 10–40 km, with a strong reduction of topography at longer wavelengths in comparison with what we would expect if the surface continued to be fractal. Two plausible explanations for this transition have been put forward by Malinverno and Gilbert (1989). The first is based on the assumption that all seafloor topography is created within the neovolcanic center of the median valley, which for the mid-Atlantic Ridge ranges in width between 10–40 km, and this produces an upper fractal limit on the topography that is introduced by the generating mechanism. The alternative model shows that regardless of the topography produced, isostasy relaxes the long-wavelength components by flexure.

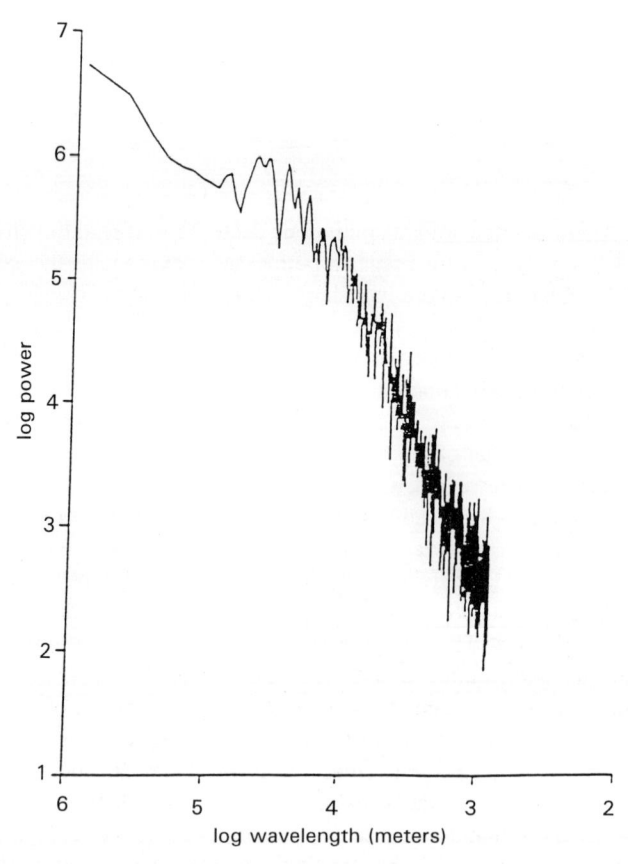

FIGURE 7.3. Power spectrum of a 1800 km long profile taken parallel to the spreading direction on one side of the mid-Atlantic ridge in the south Atlantic (Gilbert, 1989). The spectrum is plotted against wavelength rather than frequency, but the interpretation is otherwise the same as in Fig. 7.2.

7. FRACTAL TRANSITIONS ON GEOLOGICAL SURFACES

Either of these models can fit data in Fig. 7.3 equally well, although the isostasy model requires a flexural rigidity of about 5×10^{17} N-m, which is considerably less than what we obtain from independent determinations. In either case however, this abrupt fractal transition corresponds to a characteristic length: on the one hand to the width of the median valley and on the other to the relaxation dimension of the lithosphere. In contrast data for the Sierra Nevada (Gilbert, 1989), while showing curvature in their spectrum at long wavelengths, do not exhibit the strong flattening seen in the seafloor spectra, suggesting a lack of the same characteristic length scale in the continental case. That is either the topography-generating mechanism is not so band-limited or the flexural rigidity of the continent produces a relaxation length greater than the longest wavelengths studied (100 km).

7.2.3. Faults and Earthquakes

The topography of fault surfaces has been measured over a particularly broad spectral bandwidth, as shown in Fig. 7.4. We note first that these data exhibit a typical fractal power law spectrum over an enormous scale range, 10^5–10^{-5}m. Several fractal transitions can be

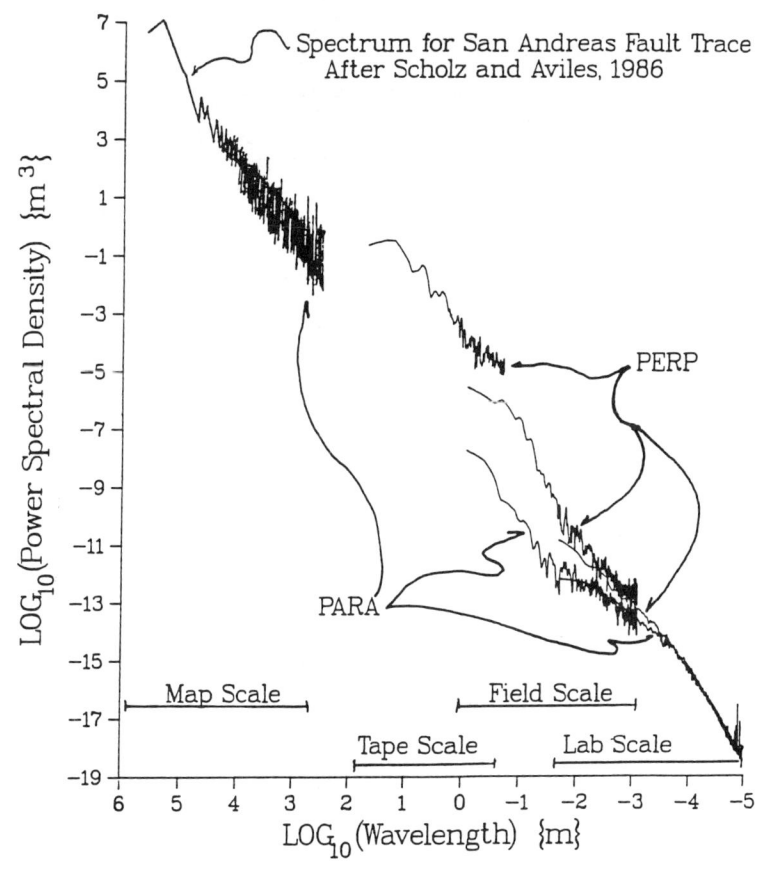

FIGURE 7.4. Power spectra of fault surfaces measured over a wide range of spatial bands. Para and perp refer to profiles parallel and perpendicular to slickensides (Power et al., 1987).

observed. An abrupt transition is marked by a sudden change in slope at about 0.5 mm—this again corresponds to the grain size transition (see Power and others, 1988, for more detailed examples). The surface becomes highly anisotropic at wavelengths greater than the grain size, with profiles parallel to the slip direction (Para) showing much less roughness than in the perpendicular direction (Perp). These profiles are thus parallel and perpendicular to the direction of slickensides, and just as in the case of glacia-striated surfaces, this anisotropy reflects the anisotropy of the wear process. Notice however that the anisotropy extends over a much broader band than individual slickenside groove lengths—indeed it is likely to extend to the map scale, and it can account for the common observation that surface traces of strike-slip faults are much straighter than dip-slip faults (over and above the effect of dipping fault interference with topography; see Scholz, 1990).

The jump in the spectrum shown in Fig. 7.4 at the tape scale is not due to a fractal transition; it is an artifact of a change to a much less precise measuring technique. On the map scale however, an important fractal transition is observed at about 10 km. In the longer wavelength band, 10–100 km, the fault has dimension $\mathbf{D} \sim 1$, so it is a differentiable, Euclidean surface in that range. However in the range of 10–0.5 km, the spectrum indicates a very jagged fault with $\mathbf{D} = 1.45$. As we see however, this transition is even more profound than this measurement indicates, and in fact the spectral technique is not even suitable for measuring it.

Figure 7.5 shows a sequence of maps at different scales, of the surface rupture of the Dasht-e-Bayez (Iran) earthquake of 1968 (Tchalenko and Berberian, 1975). The scale bar in map 5a is 10 km, and in Map 5d, it is 100 m, so from top to bottom, the maps progressively increase in magnification by a factor of about 100, and the transition previously referred to can be seen. At a scale larger than 10 km, we can approximate the fault as an irregular surface with rounded contours; at higher magnification, this description progressively breaks down. The fault appears more irregular and to consist of many strands, which at the finest scale are seen to be discontinuous. This segmentation of faults into strands has been described by Wallace (1973, 1989). Thus in Maps 5c and 5d, the fault is a zone of discontinuous strands that occupies a volume. This is why the spectral measurement of Fig. 7.4 which is strictly applicable only to continuous surfaces, is not suitable for characterizing this scale range and why measurements of the San Andreas fault using the box method, which takes the volume-filling nature of the fault into account (Okubo and Aki, 1987), yields a higher fractal dimension than the ruler method, which does not (Aviles and others, 1987). To summarize, at the longest wavelengths, the fault is a smoothly curved (differentiable) surface, but abruptly at about 10 km, it changes to a zone of discontinuous fractures that occupy a volume. Surfaces of individual strands are found to be anisotropic fractals down to the grain size, where another fractal transition occurs.

Another consequence of the scale independence of fractals is that a set of fractal objects exhibits a power law size distribution. The size distribution of fault strands, shown in Fig. 7.6, is such a power law, indicating that strands themselves constitute a fractal set. It is further known that the distribution of faults within a tectonic region is fractal (Hirata, 1989; Scholz and Cowie, 1990), and when we look at the gouge within a fault, we find that it can also be fractal (Sammis and Biegel, 1989). Thus the entire faulting structure can be considered a nested hierarchy of fractals.

We now consider what characteristic length may control the abrupt fractal transition at 10 km. The most likely possibility is that it reflects the seismogenic thickness of the crust. For most crustal faults, the brittle plastic transition occurs at about 10–15 km, and this

7. FRACTAL TRANSITIONS ON GEOLOGICAL SURFACES

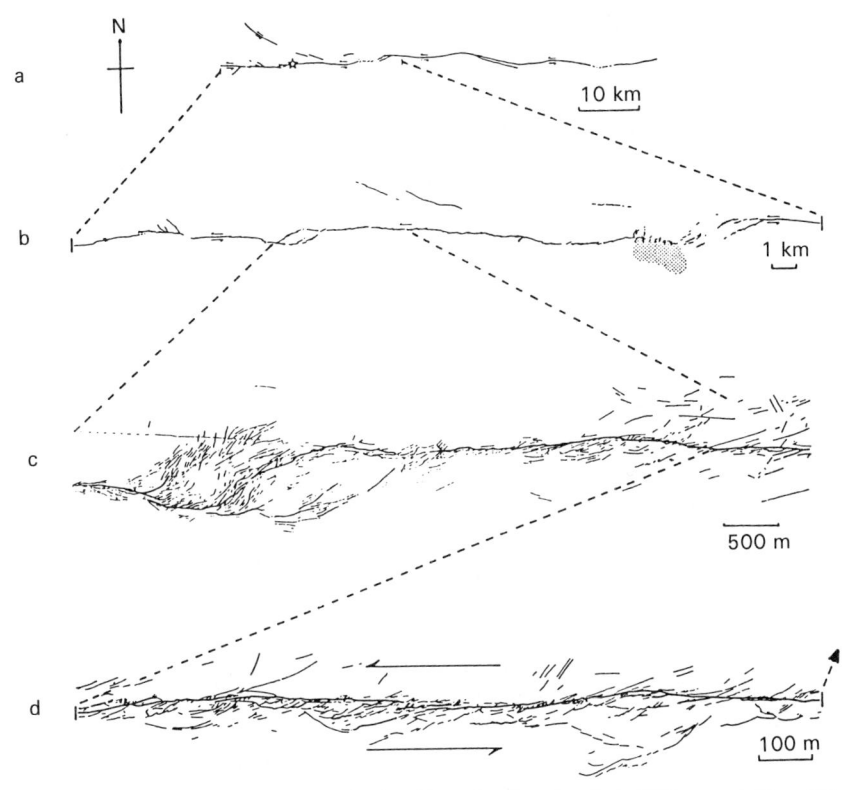

FIGURE 7.5. Maps of the surface rupture of the Dasht-e-Bayaz (Iran) earthquake in 1968 mapped at four different scales. This earthquake was a right-lateral strike–slip event (modified from Tchalenko and Berberian, 1975).

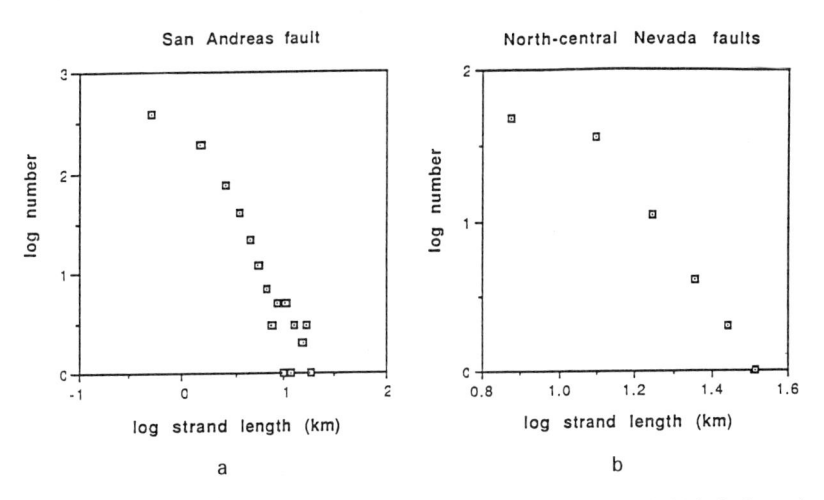

FIGURE 7.6. Length distribution of strand lengths for (a) the San Andreas fault and (b) faults in north central Nevada. These distributions indicate that strand lengths belong to fractal sets, with upper fractal limits of approximately 10 km (data from Wallace, 1989).

controls the maximum depth of brittle faulting. Hence this is the first characteristic dimension that affects rupture above the grain scale of the rock. In support of this, we also note that this appears to constitute an upper fractal limit for strand lengths (see Fig. 7.6). Thus the fault seems to be segmented into discrete surfaces up to a scale set by seismogenic thickness.

This view is reenforced by earthquake observations, which exhibit the same type of segmentation and a fractal transition at the same scale length. The linear measure of earthquake size is the seismic moment $M_0 = \mu u A$, where u is the mean slip in the earthquake, A is the rupture area, and μ is the shear modulus. If W is the thickness of the seismogenic layer and L is a linear measure of the rupture dimension, we find that a profound change occurs in earthquake-scaling relations when L = W. For small earthquakes, when L < W, then $M_0 \propto L^3$; whereas for large earthquakes, when L > W, $M_0 \propto L^2$ (Scholz, 1982; Shimazaki, 1986). This is because when the rupture is smaller than W, it can grow in all directions in the rupture plane; but when it reaches the dimension of W, it can grow further only in the along strike direction; that is, the dimensionality of the earthquake changes fundamentally.

This means that small and large earthquakes, so defined, belong to different fractal sets. Thus, although earthquakes are fractal and have a power law size distribution similar to that shown in Fig. 7.6 (the Gutenberg–Richter frequency-magnitude relation is a measure of this), small and large earthquakes belong to separate size distributions (Davison and Scholz, 1985; Scholz and Aviles, 1986).

This has important practical implications for earthquake hazard analysis. If we try to predict the recurrence time of the single large earthquake that ruptures a fault segment by extrapolating the frequency-magnitude relation of the small earthquakes for the same segment, we find that the size of the large earthquake is always underestimated by about 1.5 orders of magnitude.

The internal structure of earthquakes bears some marked resemblances to that of faults, as described. The high-frequency, strong ground motions generated by earthquakes are just shown to be a type of random noise typical of a fractal (Hanks and McGuire, 1981), and it is suggested that this arises because of a fractal distribution of stress on the fault, which presumably is due to the fractal nature of fault topography (Andrews, 1980). At a larger scale, earthquakes are found to be composed of a large number of subevents that evidently represent coherent rupture on individual patches of a fault. In a survey of western US earthquakes, Doser and Smith (1989) found that the size of subevents is sharply reduced at about 15 km, i.e., at the same upper limit as found for strand lengths. Although there is not yet enough data to show that subevents obey a fractal size distribution, what is available suggests that subevents represent a rupture of individual fault strands and their size distribution is therefore similar to that shown in Fig. 7.6. Again then the upper limit of subevent sizes is controlled by the seismogenic thickness of the crust.

We are now beginning to understand the physics behind this abrupt transition. As discussed by Bak and Chen (see Chapter 11), earthquakes can be interpreted as a form of self-organized critical phenomena. The researchers' cellular automata model produces the fractal size distribution (the Gutenberg–Richter relation) without a priori assumptions. However since they were careful to study the phenomena far from the boundary of their model, they observed only what we have called small earthquakes. To examine the effect of a characteristic dimension on this behavior, Brown and others (1990) constructed a similar model in the form of a narrow, infinite strip. The characteristic length, the width of the strip, was found to affect the behavior of the model in a profound way. Now both small and large earthquakes occurred, and large earthquakes did not belong to the size distribution of the small events in just the way observed in nature. Furthermore large earthquakes were found to have lognormally distributed recurrence times, just as observed in nature (Nishenko and Buland, 1987). The introduction of a characteristic length therefore introduces a characteristic time in the problem. The effect of the characteristic dimension on the behavior of the model was profound because almost all energy or moment release in the model was then dominated by large earthquakes and small earthquakes became a minor process. This effect was qualitatively different than the introduction of a scale length to a linear problem, and it could not have been predicted prior to running the model.

7.3. DISCUSSION

The review of observations to date indicates that although many types of geological surfaces and surface phenomena are fractal over very wide scale ranges, it is the rule rather than the exception that their fractal dimension changes gradually as a function of spatial wavelength. Abrupt fractal transitions are also encountered; these seem to occur at a wavelength corresponding to the presence of a characteristic length scale in the system. Fractals themselves have no characteristic length, so it appears that when a characteristic length occurs in the physical system, it creates a boundary between two distinct fractal domains. The most important length scale for crustal faulting is the seismogenic thickness of the crust, below which the fault is a three-dimensional zone composed of a fractal set of strands with an upper fractal limit set by that length scale. Similarly although earthquakes can grow much larger than that scale, they are composed of many subevents, which are also limited in size to that dimension. At scales above that dimension, both faults and earthquakes can be treated as irregular surfaces and ruptures. However earthquakes smaller than that dimension belong to a different fractal set than larger ones.

The greatest attraction of fractals in geological applications lies in their scale invariance, because that allows observations made at one scale to serve as the basis for constructing models at another scale, perhaps one not readily accessible to observation. My observations indicate that this should be done with caution. The ubiquitous presence of gradual fractal transitions means that extrapolation over greater than two orders of magnitude in scale length begins to produce erroneous results and errors grow as the length of extrapolation is increased. The occurrence of abrupt fractal transitions at characteristic length scales means that any problem should be carefully examined for the presence of such characteristic lengths and the fractal nature of the object in question should be *expected* to be different at scales above and below that length. My experience with earthquakes of dimension above and below the seismogenic depth shows that comparing them is similar to comparing apples and oranges: It is not useful unless we wish to prove only that they are both fruits!

ACKNOWLEDGMENTS

The basic message of Chapter 7 was first presented at a United States–Japan seminar on Fracture, Form, and Fractals, September 24–27, 1989, which was organized by C.

Sammis and M. Sato. C. C. Barton encouraged me to prepare this material for the present publication. The work was supported by NSF (EAR 88-03835) and the USGS (14-08-0001-G1668). D. Simpson, S. Martel, and A. Malinverno are thanked for their critical reviews. Lamont–Doherty Geological Observatory contribution number 4828.

REFERENCES

Andrews, D., *J. Geophys. Res.* **85**, 3867 (1980).
Aviles, C. A., Scholz, C. H., and Boatwright, J., *J. Geophys. Res.* **92**, 331 (1987).
Bak, P., and Chen, K., in this volume, Chap. 11.
Brown, S. R., and Scholz, C. H., *J. Geophys. Res.* **90**, 12,575 (1985).
Brown, S. R., Rundle, J. B., and Scholz, C. H., *Geophys. Res. Lett.* **18**, 215 (1991).
Burrough, P. A., *Nature* **294**, 240 (1981).
Davison, F., and Scholz, C. H., *Bull. Seism. Soc. Am.* **75**, 1349 (1985).
Doser, D. I., and Smith, R. B., *Bull. Seism. Soc. Am.* **79**, 1383 (1989).
Fox, C. G., and Hayes, D. E., *Rev. Geophys.* **23**, 1 (1985).
Gilbert, L. E., in *Fractals in Geophysics* (C. H. Scholz and B. B. Mandelbrot, eds.), *Pure and Applied Geophysics* **131**, 24 (1989).
Hanks, T. C., and McGuire, R., *Bull. Seism. Soc. Am.* **71**, 207 (1981).
Hirata, T., in *Fractals in Geophysics* (C. H. Scholz and B. B. Mandelbrot, eds.), *Pure and Applied Geophysics* **131**, 157 (1989).
Malinverno, A., and Gilbert, L. E., *J. Geophys. Res.* **94**, 1665 (1989).
Mandelbrot, B. B., *Science* **155**, 636 (1967).
Nishenko, S., and Buland, R., *Bull. Seism. Soc. Am.* **77**, 1382 (1987).
Okubo, P., and Aki, K., *J. Geophys. Res.* **92**, 345 (1987).
Power, W., Tullis, T., Brown, S., Boitnott, G., and Scholz, C. H., *Geophys. Res. Lett.* **14**, 29 (1987).
Power, W., Tullis, T., and Weeks, J., *Geophys. Res.* **93**, 15,268 (1988).
Sammis, C., and Biegel, R., in *Fractals in Geophysics* (C. H. Scholz and B. B. Mandelbrot, eds.), *Pure and Applied Geophysics* **131**, 255 (1989).
Scholz, C. H., *Bull. Seism. Soc. Am.* **72**, 1 (1982).
———, *The Mechanics of Earthquakes and Faulting* (Cambridge University Press, Cambridge, 1990).
Scholz, C. H., and Aviles, C. A., *Earthquake Source Mechanics*, AGU Monograph 37 (S. Das, J. Boatwright, and C. Scholz, eds.) (American Geophysical Union, Washington DC, 1986), pp. 747–55.
Scholz, C. H., and Cowie, P. A., *Nature* **346**, 837 (1990).
Shimazaki, K., in *Earthquake Source Mechanics*, AGU Monograph 37 (S. Das, J. Boatwright, and C. Scholz, eds.) (American Geophysical Union, Washington DC, 1986), pp. 209–16.
Tchalenko, J., and Berberian, M., *Geol. Soc. Am. Bull.* **86**, 703 (1975).
Wallace, R. E., in *Proc. Conf. Tectonic Problems of the San Andreas fault System*, *Spec. Publ. Geol. Sci. 13* (R. Kovach and A. Nur, eds.) (Stanford University Press, Palo Alto, CA, 1973), pp. 248–50.
———, in *Fault Segmentation and Controls of Rupture Initiation and Termination* (D. Schwartz and R. Sibson, eds.) (US Geological Survey Open File Rept. 89–315, 1989), pp. 400–408.

8

Fractal Analysis of Scaling and Spatial Clustering of Fractures

C. C. Barton

8.1. INTRODUCTION

Fractures exist over a wide range of scales, from the largest faults to microfractures, and this range is primarily responsible for scaling effects observed in fractured-rock hydrology and bulk mechanical properties of fractured rock (Witherspoon and others, 1979; Thorp and others, 1983; DeMarsily, 1985).

Fractal geometry is a branch of mathematics that can identify and quantify how the geometry of patterns repeats from one size to another. The repetition of fracture patterns over a wide range of scales is qualitatively demonstrated by the need to place an object of known size, such as a coin, hammer, or person, into photographs or a scale bar on photomicrographs and maps to establish scale. This is illustrated in Fig. 8.1, which shows a series of photographs of fracture patterns whose scales, rock types, ages, and deformation histories are different. Figure 8.1 also illustrates that fracture patterns can range from ordered (8.1d) to disordered (8.1a–c). Fractal geometry provides a method for quantifying the size scaling and spatial clustering of the full range of complexity found in networks of fractures. Fractal geometry also provides a means for extrapolating fracture properties from topologically limited samples, such as boreholes, which are one-dimensional samples, to three-dimensional fracture networks. Finally fractal geometry can be used to determine or constrain size scaling and spatial clustering of synthetic computer-generated fracture networks.

Quantitative understanding of size scaling and spatial clustering is fundamental to a quantitative understanding of fractured-rock hydrology and the bulk mechanical properties of fractured rock. This is because fluid flow and mechanical deformation do not use fractures at any one size scale but integrate the contribution of fractures at all scales, from

---

C. C. Barton • US Geological Survey, MS 940, Denver, Colorado.

Fractals in the Earth Sciences, edited by Christopher C. Barton and Paul R. La Pointe. Plenum Press, New York, 1995.

FIGURE 8.1. Photographs of fracture patterns at different scales. (a) Volcanic tuff, Calico Hills, NV; lens cap is 5 cm in diameter. (b) Sandstone, Morrison, CO; scale bar is 10 cm long. (c) Volcanic tuff, Yucca Mountain, NV; arrow is 2 m long. (d) Dolomite, Sturgeon Bay, WI; trees are approximately 12 m high. Fracture traces in (d) are outlined by alfalfa plants drawing water and nutrients from bedrock fractures underlying 0.3 m of soil. (Photo courtesy of Kenneth Bradbury.)

8. FRACTAL ANALYSIS OF SCALING AND SPATIAL CLUSTERING OF FRACTURES

microfractures to the largest scale of interest. While there is a power law increase in the abundance of fractures as we move to smaller sizes (Barton and Hsieh, 1989), the contribution of smaller and smaller fractures to fluid flow and bulk mechanical deformation depends on the geometry of fracture connections within the network (Barton and Scholz, in press). For fluid flow, we must also consider the parameter of fracture volume (aperture times length times width), which must be finite for any given volume of rock. When the geometry of connectivity is convergent, even though there is a power law increase in the number of smaller fractures, the contribution of fractures of smaller sizes is correspondingly less important. Fractal behavior of trace length tells us how to scale the fractures of smaller sizes, but not the relative contribution of the fracture-trace length to flow. For flow-through fractures interconnected in parallel networks (in contrast to series networks), the largest fractures contribute most, and the effect of fractures at sizes less than one to two orders of magnitude less than the scale of the problem are minimal. In natural systems, the lower size limit may be reached even before the convergence limit. For fluid flow through rock, there is a transition at the size scale of the pores below which scale the fracture flow problem reduces to that of flow through a porous media whose pattern of flow has been shown to be fractal (see Oxaal and others, 1987). Fractal behavior of fractures also provides a basis for extrapolation in size from the scale of fracture data collection.

The concept of a representative elementary volume (REV) of rock was introduced by Bear (1972) as a means of characterizing and extrapolating hydrologic properties of porous media. Because the largest pore size is usually limited to a few millimeters in most rock types, a representative elementary volume need be only some small multiple of the pore size. Extending the REV concept to fractured rocks has been suggested by Long and others (1985) but it is problematic because there is no characteristic size limit to fractures. Moreover the REV concept assumes linear scaling, while fracture networks exhibit fractal (power law) scaling. Because fracture networks are fractal, the concept of characterizing a small part and extrapolating to the scale(s) of interest is possible, but not in the linear manner of the REV concept.

Connectedness within a fracture network is particularly important to the fluid flow properties of the network. Scaling and spatial-clustering distributions of connectedness lead to patches of high and low conductivity. Dead-end fractures contribute to the fluid storage capacity of the network, but not to flow across the network. Crossing and abutting intersections permit fracture segments in between to participate in flow across the network. In an attempt to quantify scaling and spatial-clustering patterns of interconnected segments in fracture networks, scaling and spatial-clustering distributions of crossing and abutting intersections should be analyzed.

Scales for the study of intermediate-sized fractures occurring as natural outcrops and roadcuts range from approximately 0.5–200 m in length. Until recently the study of rock fractures over this range has focused primarily on measuring the orientation of the fracture planes (for a summary, see Barton and Hsieh, 1989). Orientation frequency is normally plotted on rose diagrams (strike azimuth only) or stereographic projections (strike and dip), and it often reveals higher frequencies in one or more orientations that define what are termed fracture sets. Fractures not included in the set(s) are usually dropped from further analysis and interpretation. Alternatively some advocate a biased sampling procedure whereby the geologist visually judges what fracture set(s) are present in an outcrop, then records the orientation of only those fractures.

Interpreting fracture history; the relation of fractures to folds, large faults, and other major tectonic features and fabrics; *in situ-* and paleo-principal stresses; the direction of fluid flow through fracture networks; and mechanical properties of fractured rock are defined traditionally in terms of fracture sets. Grouping fractures in sets defined by orientation frequency suppresses the heterogeneity of a complex system. The reduction of complex fracture patterns to highly ordered patterns has been practiced because the existing mathematics could best deal with highly ordered patterns. Ordered fracture patterns are normally found where only one or two generations of fractures are present except where mineralization has healed early generations. Fractal geometry is a mathematics especially well-suited to quantifying and modeling highly complex as well as ordered patterns.

A complete sampling of all fractures within some designated range of length or aperture is necessary to provide a representative sample for fractal analysis. Unfortunately most published geologic maps do not provide a representative sample of faults and other fractures either because of incomplete exposure and/or because no consistent criteria were used to show, not show, or interpolate fault traces. Often unstated criteria established during mapping are inconsistently applied to different rock types on the same map. Thus most published fault maps are too highly censored to permit a meaningful fractal analysis of spatial clustering and scaling of faults.

Chapter 8 touches a broad range of issues inherent in the study of scaling and spatial clustering of fractures, including one-, two-, and three-dimensional sampling of fracture networks, a review of fractal and nonfractal approaches for mathematically analyzing scaling and spatial clustering of fracture data sets, an introduction to the fractal box methods of measuring fracture data sets, and my own approaches for generating synthetic one- and two-dimensional models of fracture spacing and networks. A method of dissecting a fracture network into age generations (based on abutting relations) is presented, and this is the basis for explaining the transition from ordered to disordered fracture patterns; it is also the basis for generating synthetic fractal fracture networks. Comments on the implications of fractal behavior for the evolution of fracture networks are integrated throughout Chapter 8.

8.2. SAMPLING SPATIAL CLUSTERING AND SCALING OF FRACTURES IN ROCK

The methods of sampling spatial clustering and scaling of fractures in rock can be one-, two-, or three-dimensional. One-dimensional sampling is based on measuring the spacing between fractures along traverses across surface exposures or in boreholes in the subsurface. Two-dimensional sampling is based on mapping fracture traces exposed on subplanar exposures. Three-dimensional sampling requires geophysical imaging and mapping of fracture surfaces in a volume of rock. To my knowledge, there are no published studies of three-dimensional fracture networks detailed enough to permit quantitative analysis of their spatial and scaling properties. At present geophysical-imaging methods do not have the resolution necessary to image fractures adequately over the range of scales (one order of magnitude or more) required for fractal analysis, although the future is promising (Ramirez and Daily, 1987; Majer and others, 1988). A three-dimensional sample could be constructed by interpolating between closely spaced one-dimensional samples or between a sequence of parallel, closely spaced two-dimensional fracture trace maps. Such interpolation is commonly done for large faults, but usually sampling distance is too large to allow interpolation of the smaller fractures discussed in Chapter 8. Measuring the scaling of

8. FRACTAL ANALYSIS OF SCALING AND SPATIAL CLUSTERING OF FRACTURES

fracture networks requires mapping fractures over a wide range of size scales. Optimally all sizes sampled can be shown on a single map; usually however as in Chapter 8, a series of maps is used, with each map sampling a range of size scale.

8.3. FRACTAL MEASURE OF SPATIAL AND SCALING PROPERTIES

Fractal geometry is a branch of mathematics that provides methods for quantifying the spatial and scaling properties of geometric data sets that are uncorrelated, positively correlated (persistent), or negatively correlated (antipersistent) as a function of scale or spatial distribution. Feder (1988) describes persistence and antipersistence. Persistence means that an increasing trend in preceding increments implies an increase in the next increment; conversely a decreasing trend in preceding increments implies a decrease in the next increment. Antipersistence means that an increasing trend in preceding increments implies a decreasing trend in the next and vice versa. In terms of fracture spacing, increasing persistence means increased clumping, and antipersistence leads to an even spacing. Fractal geometry is particularly well-suited to both positively and negatively correlated data sets. By comparison geostatistics (Journel and Huijbregts, 1978; Hohn, 1988) are applicable only to positively correlated data sets. Fractal patterns can be highly ordered or disordered, and this can be understood in terms of the procedures for generating fractal patterns, as described in the discussion of Cantor dusts. Fractal geometry also provides methods for creating synthetic analogs of natural geometries.

The geometry of a fractal pattern is represented by a fractional number, termed the fractal dimension (D). The size scale over which a fractal dimension applies is bounded by an upper and lower fractal limit. Fractal methods for analyzing spatial and scaling properties of objects are applicable to one-, two-, three-, or n-dimensional data sets.

8.4. METHODS OF MEASURING THE FRACTAL DIMENSION OF FRACTURE NETWORKS

8.4.1. Box Method

The box method is used in Chapter 8 to measure the fractal dimension of the spatial and scaling distribution of fractures. The method is robust in that it is applicable to self-similar and with certain restrictions, to self-affine data sets (see Chap. 4). The method is applicable to one- and two-dimensional data sets. In principle the method is simple: A sequence of grids, each with a different cell size, is placed over maps of fracture traces, then the number of cells intersected by fracture traces is counted. The fractal distribution is

$$Nr^D = 1 \tag{1a}$$

or equivalently

$$D = \frac{\log N}{\log (1/r)} \tag{1b}$$

where N is the number of cells containing portions of one or more fracture traces, r is the length of the side of the cell, and the fractal dimension D is the slope of straight-line segments fitted to the N, $1/r$ points plotted on logarithmic axes. A derivation of Eq. (1) is given by Feder (1988) and by Pruess (Chap. 3). A schematic diagram of the counting procedure and the fractal plot are shown in Fig. 8.2.

Theoretically the fractal dimension is taken at the limit where the cell size goes to zero (Hausdorff, 1919) as discussed by Pruess (Chap. 3). This is not possible when analyzing real data sets with finite lower size limits. However, in Chapter 8 and my previous papers, the distribution of cell sizes is logarithmic, so there are progressively more smaller sized cells.

Fitting straight lines to points on the fractal plot is done using a least squares linear regression. Performing the regression in log-log space leads to a logarithmic weighting that favors smaller cell sizes. Breaks between lines of different slopes are located by visual inspection. Usually upper and lower fractal limits of data are exceeded by cell sizes that are too large and too small. These extra points are sequentially stripped off the ends of the plot until the slope of the line fit to the data stabilizes. This is a nonrigorous method that could be accomplished by such rigorous statistical method as jackknifing or bootstrapping, although the improvement would probably be minimal. The slope of the line is the fractal dimension, and the end points of the line are the upper and lower fractal limits. The least squares method of fitting a straight line to data permits calculating a goodness of fit by means of a correlation coefficient ranging from 0, for no correlation, to 1 for a perfect fit. I have tested the box method on fractal figures of known dimension (Koch curves) and found the error in dimension to be as great as 0.05. To improve the accuracy and reproducibility of the box method, I explored variations of the box method, as described in the following section.

8.4.2. Box-Rotate Method

In practice the procedure of overlaying grids is complicated by the need to overlay the grid so that for each cell size the minimum number of cells is occupied, a boundary condition stated by Hausdorff (1919) and incorporated into the derivation of Eq. 1. One way of accomplishing this is to rotate the grid relative to the data until for each cell size, the minimum number of cells is occupied. The effect of grid rotation on the fractal dimension

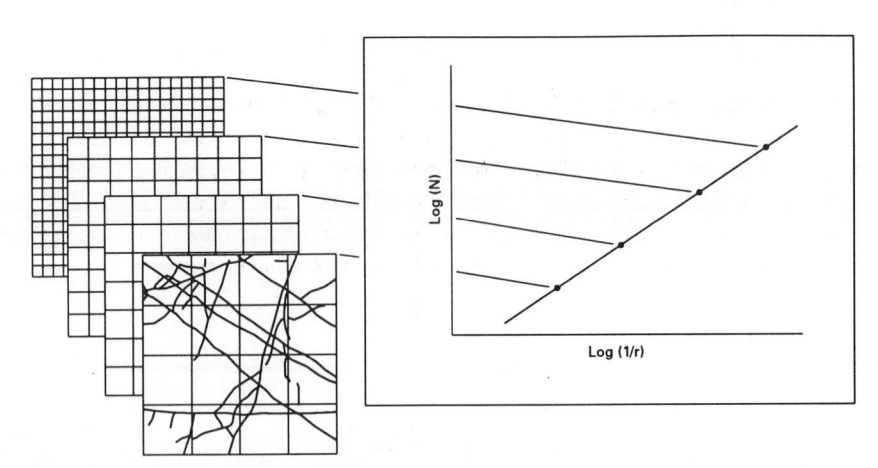

FIGURE 8.2. Illustration of the box method of measuring fractal dimension by overlaying a sequence of grids, each with a different cell size. For fractal data sets, a plot of the log of the number of occupied cells (N) versus the log of inverse cell size ($1/r$) yields data points that can be fit by a straight line whose slope is the fractal dimension (D), as stated in Eq. (1).

8. FRACTAL ANALYSIS OF SCALING AND SPATIAL CLUSTERING OF FRACTURES

of a single straight line (whose true dimension is 1.000 . . . .) is shown in Fig. 8.3. Here the grid was rotated to the angle θ, and then the sequence of cells was counted. In Fig. 8.3, the increment of rotation angle θ is 2.5 degrees, and we can show analytically that for a single straight line, the minimum number of cells is occupied when θ equals 0, 45, or 90 degrees. At other orientations, additional cells are crossed; the number of such cells is given by the following equation:

$$N(r) = N_o(r) + \left[\frac{a}{r(\sin\theta)}\right] \quad (2)$$

Where $N(r)$ is the number of boxes crossed of size r, $N_o(r)$ is the number of boxes crossed of size r when $\theta = 0$, a is the length of the line (equal to 1 for unit length), and brackets [ ] indicate the integer part of the function within the brackets.

The consequence of not properly orienting the grid in the analysis of a straight line for each cell size is the introduction of an error of as much as 0.06 in the fractal dimension (Fig. 8.3). A single straight line is a worst case test because it is so highly anisotropic. Note that for complex shapes (Koch curves, for example), the equation is much more complicated than Eq. (2). Tests on Koch curves show that the error introduced when the grid is not oriented so that N is a minimum is as much as 0.05. When N is a minimum for each cell size, the error can still be as much as 0.02, because cells are squares that are not geometrically compatible with the Koch curve, whose angles are 60 and 120 degrees. Therefore for complex patterns, the box-rotate method is expected to have an error of approximately 0.02.

8.4.3. Box-Flex Method

Pruess (see Chap. 3) found that decreasing the increment between cell sizes so that tens, hundreds, or even thousands of cell sizes are used eliminates the need to orient grids properly. I tested Pruess's approach on shapes of known fractal dimension (Koch curves)

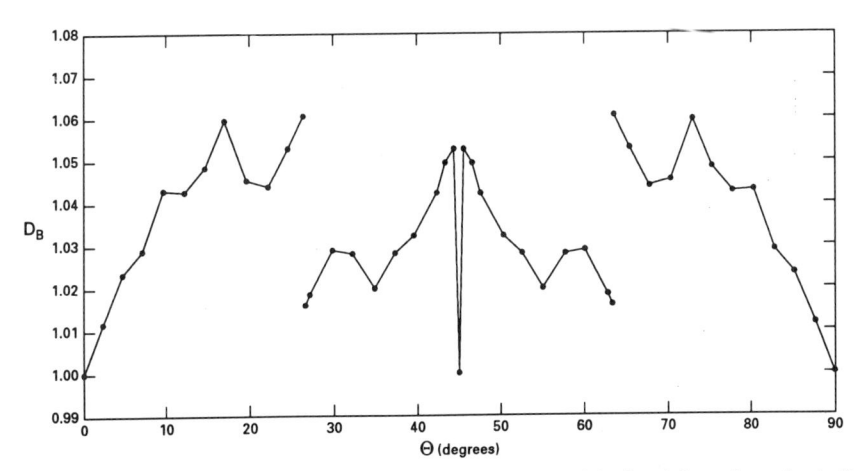

FIGURE 8.3. A plot of the variation in fractal dimension (D) for a straight line (whose dimension is 1) as a function of the orientation θ of the line measured in increments of 2.5 degrees. Note that the minimum number of cells, and therefore the true D, is occupied where θ = 0, 45, and 90 degrees.

and found that the error was only 0.005 when using approximately 50 different cell sizes. To achieve small increments in cell size, it is necessary to allow the outer boundaries of the grid to expand and contract slightly because there must always be an integer number of boxes in the grid; there can be no fractional boxes. I term this the box-flex method and in Chapter 8, fractal dimensions are rounded up to the hundredth's decimal place.

8.4.4. Box-Density Method

In the box-counting methods just described, the size-scaling properties of a data set are measured by simply counting the number of occupied cells. La Pointe (1988) used a variation of the box method where the number of blocks bounded by fractures contained in each cell are counted. I call this the box-density method; it measures the spatial-clustering variation in a data set.

The procedure is to overlay a sequence of grids, then count the number of data points in each occupied cell. For each cell size, the maximum count is divided by the number of cells on one side of the grid. This value (Z) is then used to normalize the data such that for each cell in the grid, the sum of the normalized count is stored as N. The fractal dimension is calculated using Eq. (1) where N is this normalized value. A fractal dimension measured in this way ranges between 0–3.

8.5. PREVIOUS FRACTAL STUDIES OF THE SCALING AND SPATIAL DISTRIBUTION OF FRACTURES

Barton (1990), Velde and others (1990), Barton and Zoback (1992), and Manning (1994) reported fractal analyses of fracture spacing along a line sample. Barton (1990) reported on the spacing pattern of quartz/gold veins in cores from the Perseverance mine, which is described in detail later. Velde and others (1990) measured the fractal dimension in terms of a probability of finding a fracture-free zone in the following length unit to be measured. To study the effect of anisotropy in fracture patterns, researchers measured line samples at various orientations on two-dimensional fracture trace maps. The fractal dimension was observed to vary by as much as 0.33 for the most anisotropic of their fracture trace maps. The range of fractal dimension reported was 0.10–0.68. Visual observation of their fracture trace maps suggest that the very low fractal dimensions they obtained (all but four values less than 0.50) may be due to the sparseness of their data sets. Barton and Zoback (1992) fit the frequency of fracture-spacing intervals versus spacing interval from the Cajon Pass well, California, with a power law with a scaling exponent of 1.03. Note that their analysis of fracture spacing is different from that presented in Chapter 8 and other studies cited because they did not analyze a pattern of fractures with a box-counting procedure; rather they fit a spacing interval versus frequency plot with a power law. In such an analysis, the spatial sequence of fractures is lost, and results can not be compared with those reported elsewhere in Chapter 8. Manning (1994) reports on spacing metamorphic veins along linear transects in a variety of geologic settings. He finds the fractal dimension to range between 0.25–0.46 for sites in continental crust and to be 0.81 fo hydrothermally altered oceanic crust.

There have been several published fractal analyses of fracture trace maps. The first study was by Barton and Larsen (1985), who reported fractal dimensions for three fracture patterns mapped at nearby locations in the same Miocene volcanic tuff unit at Yucca Mountain, Nevada. They used the box method and reported fractal dimensions ranging from 1.12–1.16. These values were calculated for cell sizes at and below the shortest fracture length and thus resulted in an improper analysis, unrepresentative of the fracture network. These same three patterns were reanalyzed using the box method over a wider range of cell sizes, which in part exceeded the size at which the fracture pattern was completely covered, by Barton and others (1986) along with four other fracture patterns and a fault map of the southern half of Yucca Mountain. Fractal dimensions of 1.5–1.9 were reported. The same three maps were reanalyzed by Barton and Hsieh (1989) using the box-rotate method in which again cell sizes in part exceeded the size at which the fracture pattern was completely covered, and fractal dimensions ranging from 1.6–1.7 were reported. I analyzed them yet again for this chapter, using the box-flex method over an appropriate range of box sizes, and I found the dimension ranges from 1.38–1.52. For the reasons stated in Section 8.4.3, the box-flex method provides the most accurate results.

Chiles (1988) used the box method to analyze fracture patterns mapped on drift walls in a granite mine. The drift walls were 2 m high and 50–100 m long. The minimum fracture trace length mapped was 0.2 m. Because of the limited height of the walls and a lower trace length cutoff at 0.2 m, the range of scales sampled vertically is considerably less than one order of magnitude. Examining the fractal plots in Chiles (1988) reveals that he permitted the cell size to be as small as 0.01 m, smaller than the smallest fracture trace (0.2 m) and thus improperly analyzed his maps in the same manner as Barton and Larsen (1985). A better range of box sizes for his maps would be 0.2–0.5 m rather than the 0.01–10 m he used. A fractal analysis of the spacing pattern of fractures along a line sample 50–100 m long (with the smallest cell size two times the shortest distance between two fractures and the largest cell size one-half the length of the longest distance between two fractures) would be a more appropriate way of analyzing data such sets as Chiles's.

La Pointe (1988) introduced the box-density method to the analysis of fracture trace maps. He counted the number of blocks bounded by fractures per cell rather than the number of fracture traces. He analyzed the three maps (reproduced in Figs. 8.7a–c) at Yucca Mountain published in Barton and Larsen (1985), a map (reproduced in Fig. 8.7i) from the Lannon area, Wisconsin, published in La Pointe and Hudson (1985), and several computer-generated synthetic maps. Instead of analyzing entire maps, he analyzed a strip taken as a representative subset of the map. His fractal dimensions for strips across the maps in Fig. 8.7a–c are 2.52, 2.37, and 2.69, respectively.

Hirata (1989) reported fractal dimensions for the pattern of seismogenic faults at various locations in Japan. He used the box method and reported fractal dimensions ranging from 0.72–1.60. Visual inspection of his maps reveals that those maps with fractal dimensions less than 1.5 contain very few fractures, and these are probably censored data sets when compared to the spatial density of fractures reported by King (1983) for seismogenic fault patterns. Also Hirata's fractal analyses include only five cell sizes. Based on my experience, the error in determining a fractal dimension using the box method with only five box sizes is large, as much as 0.15.

Korvin (1989) investigated the size distribution of fault-bounded blocks at the bottom of the southern end of the Gulf of Suez rather than scaling or spatial distributions of fractures traces. His plots of the cumulative frequency of block size can be interpreted as fractal. However because of the rollover at small block sizes and other changes in slope on some of his log-log plots, he interprets his plots as demonstrating nonfractal behavior.

Nonlinear behavior on a log-log plot does not necessarily mean that the data is nonfractal; the data may be scale variant or multifractal. I interpret the rollover he shows as due to data omission at smaller block sizes. His data can be fit with one or more straight-line segments whose slopes are the fractal dimension(s), although I have not done this.

8.6. ONE-DIMENSIONAL SAMPLING AND ANALYSIS OF FRACTURE NETWORKS

8.6.1. Sampling

Although fracture networks are three-dimensional, it is difficult or impossible to obtain a complete three-dimensional sample, as previously discussed. Boreholes, which provide one-dimensional samples of fracture networks, are the most frequently used method of sampling the spatial distribution of fractures in the subsurface. Straight-line, or scan-line traverses along surface outcrops are also one-dimensional samples. Both provide spacing from one fracture to the next. If orientation data are provided, then spacing between fractures of the same set can be studied (Barton, 1983), but this is not normally done. Methods of one-dimensional fracture sampling at the surface can be found in La Pointe and Hudson (1985) and Barton (1983). Methods of sampling subsurface fractures by direct observation in drill core are described in Kulander and others (1979) and by geophysical methods in drill holes in Paillet (1991).

8.6.2. Nonfractal Analysis Methods

There are a number of precedents to any study of scaling and spatial distributions of fractures in rock. Prior to the advent of fractal geometry, scaling and spatial properties were sampled, quantified, and modeled. However data collection was limited primarily to one-dimensional (linear) sampling of spacing between fractures intersected along a traverse, and mathematical treatments were for the most part limited to such linear samples. Few studies have measured the spacing-frequency for individual sets (Barton, 1983); most studies simply include all fractures encountered along the sample without regard for, or knowledge of, orientation, size, or other discriminator (for example, Priest and Hudson, 1976). In studies were fractures are grouped into sets based on orientation, the sampling interval consists of a traverse oriented perpendicular to fracture planes (Barton, 1983).

The mathematical analysis of spatial distribution in early papers was limited to calculating an arithmetic mean, and this approach persists to the present (summarized by Barton, 1983). In a series of papers, Priest and Hudson (1976, 1981), Baecher and others (1977), and Hudson and Priest (1979, 1983) treated frequency plots of fracture spacing as lognormal or negative exponential distributions. The cumulative probability plot for fracture spacing was introduced by Baecher and others (1977) and explored by Barton (1983) for two fracture sets in the same bed. Cumulative probability plots generate a mean and standard deviation, and like all other previous approaches, these simply plot the frequency of various spacing intervals without considering the sequence from one fracture to the next. La Pointe (1980) introduced and explored the use of semivariograms as a method for analyzing fracture spacing along a linear traverse (scan line). Semivariograms plot the second-order moment of the number of fractures per unit length of the scan line versus the length of the sampling

8. FRACTAL ANALYSIS OF SCALING AND SPATIAL CLUSTERING OF FRACTURES 151

increment over some range of increment size analogous to cell size in the box method. Some semivariograms of spacing frequency reveal power law distributions where the power is a fraction rather than an integer. Such semivariogram plots can be recast into fractal plots by plotting the semivariance against sample increment size in log-log space and calculating the fractal dimension from the slope of straight line(s) fit to the points. The approaches just described are limited to spacing data collected along a line sample.

8.6.3. Fractal Analysis of the Spatial Distribution of Quartz/Gold Veins in Exploratory Cores from the Perseverance Mine, Juneau, Alaska

The spacing between gold-bearing quartz-filled fractures (veins) intersected by exploratory drilling from tunnels in the Perseverence mine provides a data set for analysis. Figure 8.4 shows a vertical cross section of the mine, the drilling pattern, and the location of quartz-filled fractures above a specified assay value along bore holes. Qualitatively the spacing has no discernible pattern or structure.

A fractal analysis was performed on the spacing distribution between veins on 23 cores, each approximately 90 m long and intersecting approximately 40–60 veins. The box-rotate method was used, with each drill hole rotated horizontally to minimize the number of occupied cells (see Section 8.4.2). A typical fractal plot of spacing is shown in Fig. 8.5, and the range of fractal dimensions is given in Table 8.1. The range in fractal dimension is 0.41–0.62, with goodness-of-fit coefficients greater than 0.98.

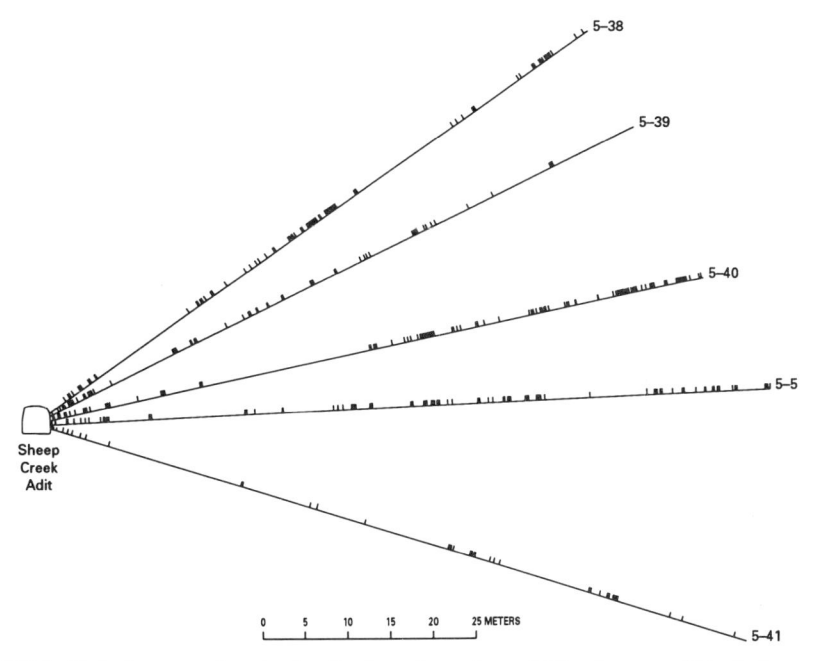

FIGURE 8.4. Vertical cross-section showing the fan pattern of core drilling from an adit in the Perseverence mine, Juneau, Alaska. The position and thickness of gold-bearing quartz veins filling fractures along each drill hole are shown by tick marks of appropriate width.

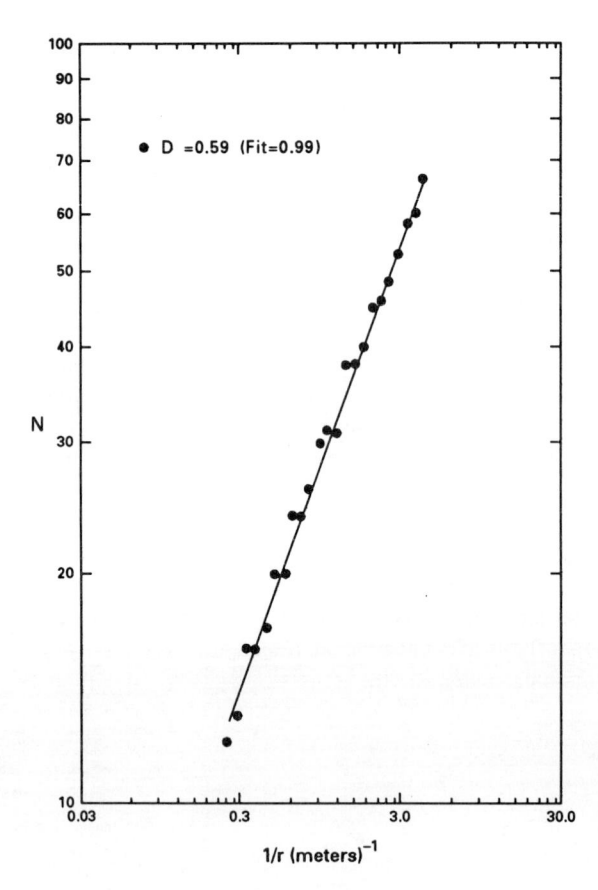

FIGURE 8.5. Fractal plot of vein spacing in Core Hole 7-18 (see Table 8.1) Perseverence mine, Juneau, Alaska; N = number of occupied cells, r = cell size, and D = fractal dimension.

TABLE 8.1. Fractal Spacing Distributions of Quartz/Gold Veins in Exploratory Cores from Perseverance Mine, Juneau, Alaska

Drill Hole Number	Fractal Dimension	Drill Hole Number	Fractal Dimension
1-26	0.52	4-14	0.48
1-27	0.45	4-202	0.51
1-28	0.52	5-5	0.45
1-29	0.41	5-38	0.58
2-2	0.43	5-39	0.51
2-23	0.58	5-40	0.62
3-3	0.55	5-41	0.42
3-20	0.54	7-7	0.47
3-21	0.49	7-18	0.59
4-4	0.51	7-19	0.55
4-12	0.46	9-31	0.47
4-13	0.48		

8. FRACTAL ANALYSIS OF SCALING AND SPATIAL CLUSTERING OF FRACTURES

The significance of the fractal behavior of spacing between veins follows. If the spacing of veins is uncorrelated, then the dimension is 0.5. If the veins is evenly spaced, then the fractal dimension is 1.0. As the dimension decreases from 0.5 to 0.0, spacings become increasingly clumpy, so that at $D = 0$, data are clumped to one point. By analogy to fractional Brownian trails, this behavior is described as having a positive correlation or as being persistent (Mandelbrot, 1983). As the dimension increases from 0.5 to 1, spacings become less clumpy, more evenly spaced. Again by analogy to fractional Brownian trails, this behavior is described as having a negative correlation or as being antipersistent (Mandelbrot, 1983). The range of fractal dimension reported in Table 8.1 indicates that vein spacing can be uncorrelated, persistent, or antipersistent. The average value for all samples listed in Table 8.1 is 0.50, which indicates that on average, the vein spacing is uncorrelated. The concept of a pattern within randomness can be appreciated by constructing a fractal Cantor dust model for fracture spacing.

8.6.4. Fractal Cantor Dust Model for Fracture Spacing

A Cantor dust is a fractal set whose spacing properties are an appropriate model for the spacing properties of fractures. Generating a Cantor set is described in Mandelbrot (1983), and it is illustrated in Fig. 8.6. Begin with a solid straight line (or for illustrative purposes, a bar) and iteratively remove one or more pieces following a prescribed procedure. The procedure in Fig. 8.6a is to remove the middle one-third of the remaining pieces iteratively. This produces an ordered triadic Cantor dust whose dimension is defined by Eq. 1, where N is the number of remaining pieces, and r is the length of the pieces relative to the unit length. In this case, N equals 2 and r equals ⅓, and so D equals $\log(2)/\log(3)$ or 0.6309. The dimension of a Cantor dust is determined from the first generation, and this is the dimension for all generations. The lower size limit of the range over which the dimension is valid is the length between the two closest pieces; thus with each additional generation, the lower limit is extended. Ordered fractals are alternatively referred to as regular, symmetric, or deterministic fractals.

This Cantor dust is not a very good analog because it is too regular and its fractal dimension is slightly greater than the values we observe for spacing gold veins. Now we introduce randomness to the iterative process of forming a Cantor dust by randomly selecting which of the pieces to remove, N and r remain the same, and therefore so does the fractal dimension 0.6309. This disordered triadic Cantor dust is a much better model for the spacing of gold veins, as can be seen in Fig. 8.6b. Disordered fractals are also referred to as stochastic or random fractals. This example illustrates how and why fractals can have patterns within randomness. The fractal dimension of Cantor dusts falls between 0–1. For a given value of r, D increases from zero for $N = 1$ to 1 for $N = r$.

In strictest mathematical terms, the iterative procedure is repeated an infinite number of times to the limit as r approaches zero. Natural data sets always have some lower cutoff—a perceptibility limit—and so the number of objects is finite. Therefore for generating Cantor dusts to simulate or model spatial statistical properties of natural systems, normally only 5–10 iterations are needed. The fifth generation of the random triadic Cantor dust shown in Fig. 8.6b qualitatively evokes the pattern of fracture spacing, including variation in vein width, and quantitatively matches the spatial statistical properties of other dusts with the same fractal dimension.

The pattern of fracture spacing is controlled by unknown boundary conditions of a

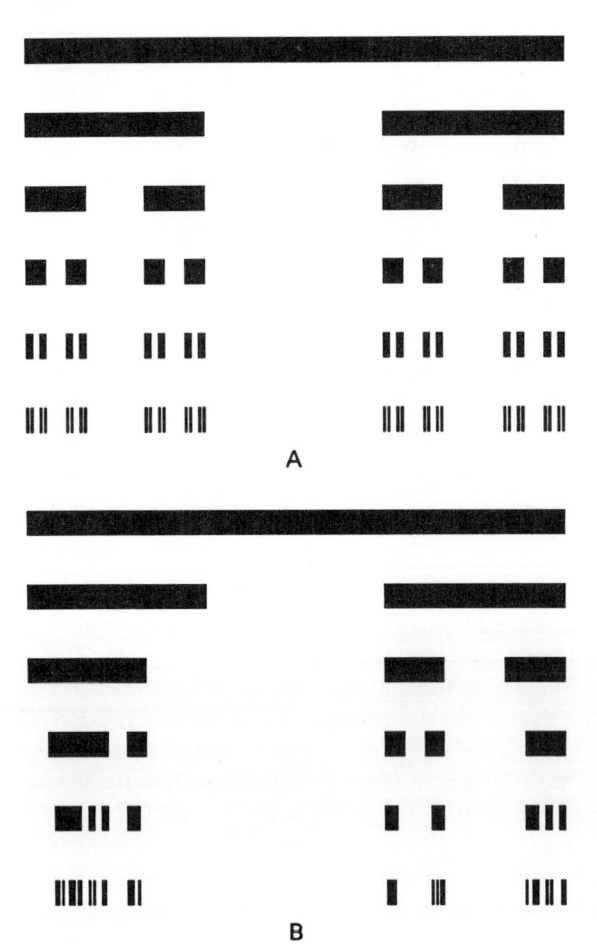

FIGURE 8.6. Triadic Cantor dust, first five generations; $D = 0.6309$. (a) Ordered; (b) disordered. (Modified from Smalley and others, 1987, Figs. 1 and 2).

complex and unknown physics that underlies the generation of fracture arrays in rock. Nevertheless it is likely that veins formed by a process akin to that of Cantor dust formation, whereby initially large blocks are broken into smaller blocks by sequential fracturing, as demonstrated for one of the two-dimensional maps that follow. The narrow range of fractal dimension (0.41–0.62) for vein spacing suggests that one physical process operated within the limits of scale sampled. Like other branches of mathematics, fractal geometry does not provide a physical or mechanistic understanding of the fracture process. Yet is provides a mathematical model and hence some insight as well as a quantitative description of the spatial properties of vein spacing. For fracture spacing, it suggests a mechanism by which larger blocks are reduced to smaller blocks by sequential fracturing, whereby survivability of large blocks (large spacings) is small. This is the mechanism revealed by dissecting fracture network maps, as described in Section 8.7.

Any number of Cantor dust models can be constructed by varying N and r to match the fractal dimension of the spacing and variation in width of veins in each drill hole. The random Cantor dust in Fig. 8.6b is a reasonable model for vein spacing in drill hole 5-40 (Table 8.1) because fractal dimensions are very close.

8.6.5. Two-Dimensional Sampling and Analysis of Fracture Networks

Maps of fracture traces exposed on planar surfaces are two-dimensional samples of fracture networks. Such subhorizontal exposures, ranging in area from less than 1 m^2 to more than 5000 m^2, are called pavements (Barton and Larsen, 1985; Barton and Hsieh, 1989); here the usage is extended to include subvertical and other inclined planar exposures. There are few published maps of fracture trace patterns at the scale of pavements; the only such published maps that I know of are contained in the following seven papers: Kolb and others (1970) mapped fractures in a quartz monzonite near Cedar City, Utah; Segall and Pollard (1983a, 1983b) mapped fracture traces on glacial pavements in the Givens Granodiorite in the Sierra Nevada, California; La Pointe and Hudson (1985) mapped fractures on a quarry floor in the Niagara Dolomite at Lannon, Wisconsin; Olson and Pollard (1989) mapped fractures in the Rico Limestone near Mexican Hat, Utah; Barton and Hsieh (1989) mapped fractures in the Tiva Canyon member of the Paint Brush Tuff at Yucca Mountain, Nevada; and Hill (1990) mapped fractures in the Aztec Sandstone in southern Nevada. All other published maps that I am aware of do not adequately sample the fracture network because one dimension of the map is too small [for example, mine wall maps (Chiles, 1988)] or because the range in the size of blocks bounded by fractures was considerably less than an order of magnitude. An optimal map covers an area large enough to include both ends of most of the largest traces exposed and includes at least two orders of magnitude in the length of fracture traces mapped. The maps analyzed in Chapter 8 only approach this optimal size.

Eight maps (Figs. 8.7d–g, j, m, n, and q) were prepared as a part of this study, and nine (Figs. 8.7a–c, h, i, k, l, o, and p) are taken from the literature.

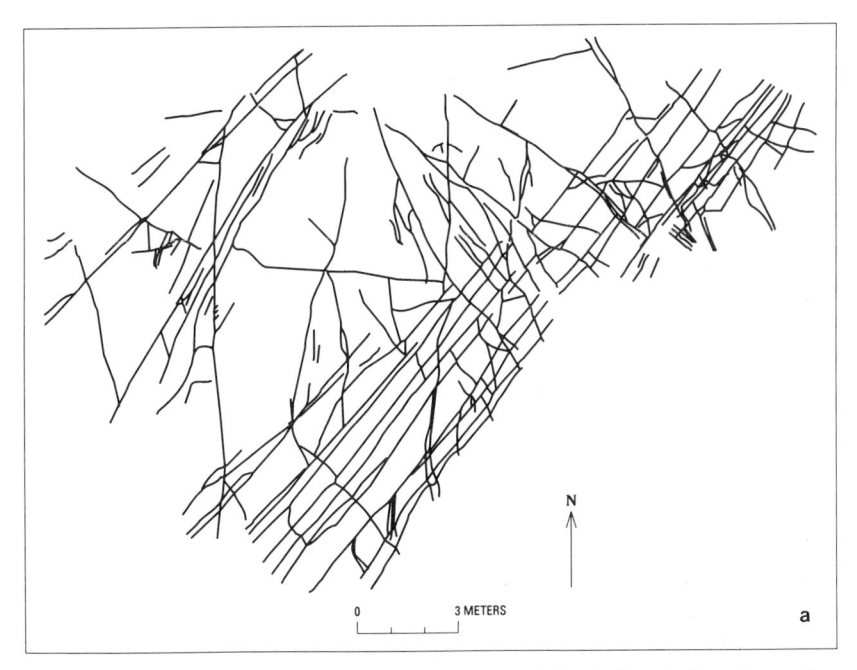

FIGURE 8.7. Fracture trace maps at various scales for 17 sites (8.7a–q). Fractal dimensions and summary descriptions of each site are given in Table 8.2.

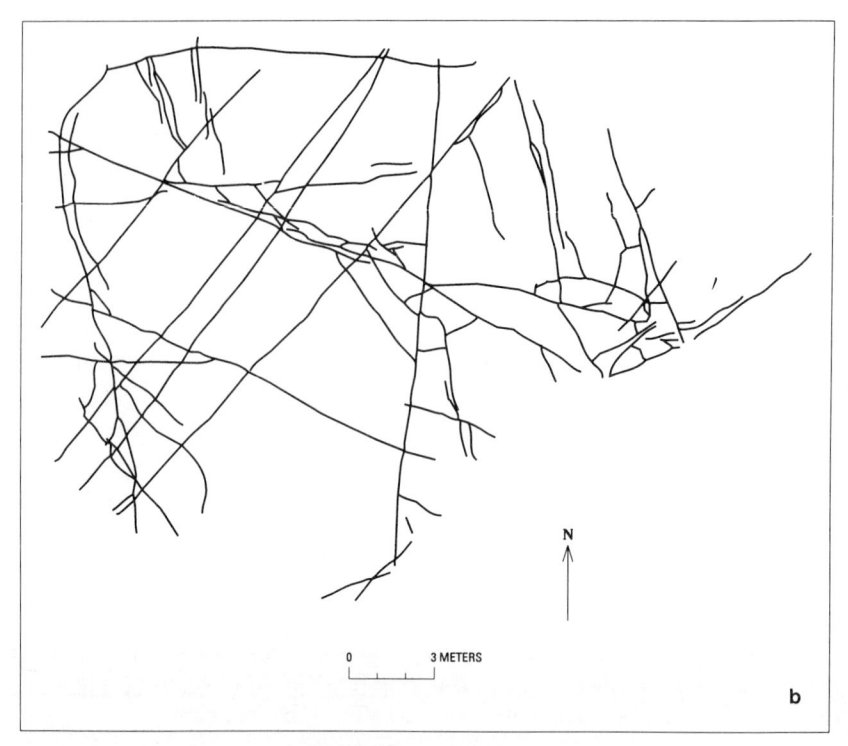

FIGURE 8.7. (*Continued*)

8.6.6. Fracture Trace Maps

Fracture trace maps referred to by figure number are shown in Fig. 8.7. A summary of parameters and sources for each map is given in Table 8.2. Columns in Table 8.2, from left to right, are as follows: map number, fractal dimension (box-flex method), location, rock unit designation and rock type, age of rock, scale at which fractures were mapped in the field, length of shortest fracture, length of longest fracture, and publication references. All maps are planar or subplanar, subhorizontal slices through networks of steeply dipping fractures except the map in Fig. 8.7n, which is subvertical, and maps in Figs. 8.7j and q, whose original orientations are unknown because the rocks were not in place at the time of the mapping. Several types of rock fracture are represented in this collection. Fractures on maps in Figs. 8.7a–i, l, and n formed primarily as joints, based on the absence of shear offset across them. Fractures on the map in Fig. 8.7j formed as deformation bands of reduced grain size due to shear. Fractures on the map in Fig. 8.7k formed as bands of closely packed grains across which there is no demonstrable shear; their mode of origin is unknown at this time (Hill, 1990). Fractures on maps in Figs. 8.7m, o, and p formed as faults. Faults on the map in Fig. 8.7p are transform faults. Fractures on the map in Fig. 8.7q are traces of fluid-inclusion planes as viewed in a thin section, and these exhibit no shear offset across them.

8. FRACTAL ANALYSIS OF SCALING AND SPATIAL CLUSTERING OF FRACTURES

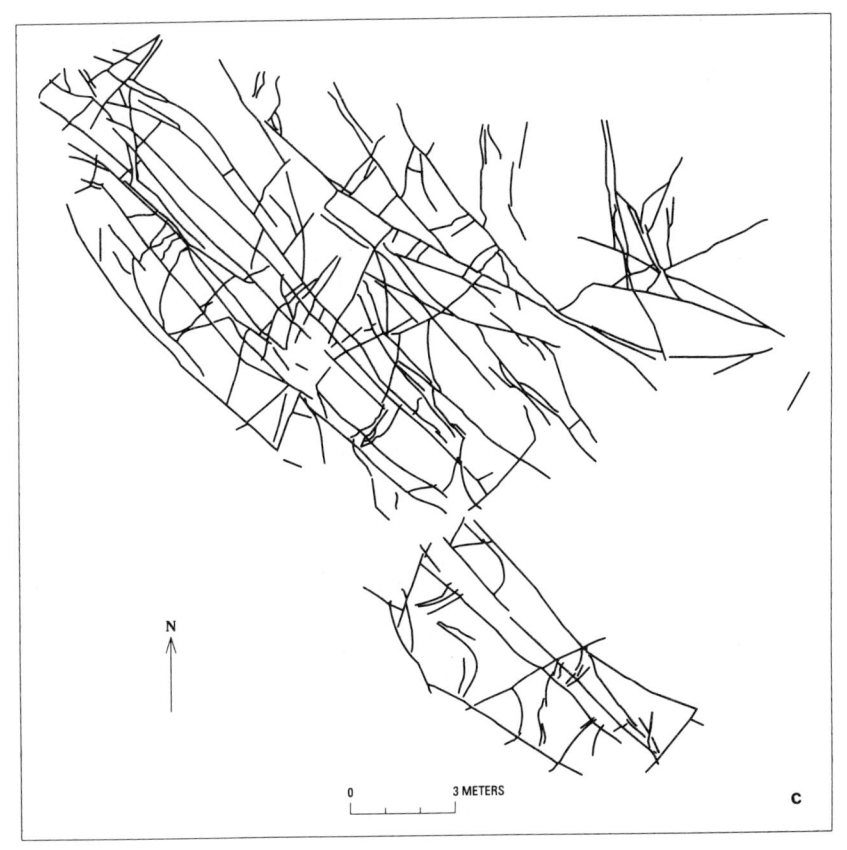

FIGURE 8.7. (*Continued*)

8.6.7. Non Fractal Analysis Methods

I am not aware of areal or volumetric analyses of scaling or spatial characteristics of two- or three-dimensional fracture maps. Previous studies analyzed line samples across two-dimensional maps (Hudson and Priest, 1979, 1983; and La Pointe and Hudson, 1985), a method that reduces a two-dimensional sample to a one-dimensional sample. An advantage of the fractal approach is that the data sample can be analyzed linearly, areally, or volumetrically, as appropriate to the sample.

Based on line samples of two-dimensional data sets, a number of conceptual models and synthetic fracture network generators have been proposed, all of which assume that centers of fractures are distributed in a Poisson manner; that is, there is no spatial correlation (Conrad and Jacquin, 1973; Baecher and others, 1977; Schwartz and others, 1983; Robinson, 1984; Dershowitz, 1984; La Pointe and Hudson, 1985; Long and others, 1985; Watanabe, 1986). None of these models or generators incorporates fractal scaling and spatial clustering observed on fracture network maps. Scaling is either omitted, that is, all fractures are treated as the same length, or a log-normal trace-length frequency distribution is used instead of a fractal power law trace-length frequency distribution. Barton and Hsieh (1989) present the case for fractal power law trace-length frequency distributions. The range

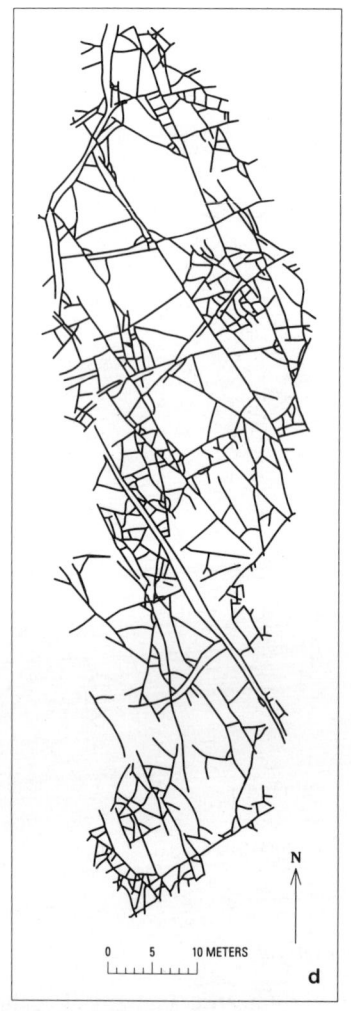

FIGURE 8.7. (*Continued*)

of scales incorporated into published models and generators is usually less than one order of magnitude, which is too limited to provide a realistic model of nature. Spatial clustering if allowed for is either Poisson or parent–daughter (Chiles, 1988), neither of which are good analogs to fractal spatial clustering. Madden (1973) studied the effect of scale and applied the renormalization group approach for modeling the spatial distribution of natural and induced microfractures in rock. This approach is closest to a fractal approach.

8.6.8. Fractal Analysis of Scaling and Spatial-Clustering Distributions of Mapped Fracture Traces

The fracture trace maps (Figs. 8.7a–q) were analyzed using the box-flex method, which measures both scaling and spatial clustering. The range of cell sizes used is no

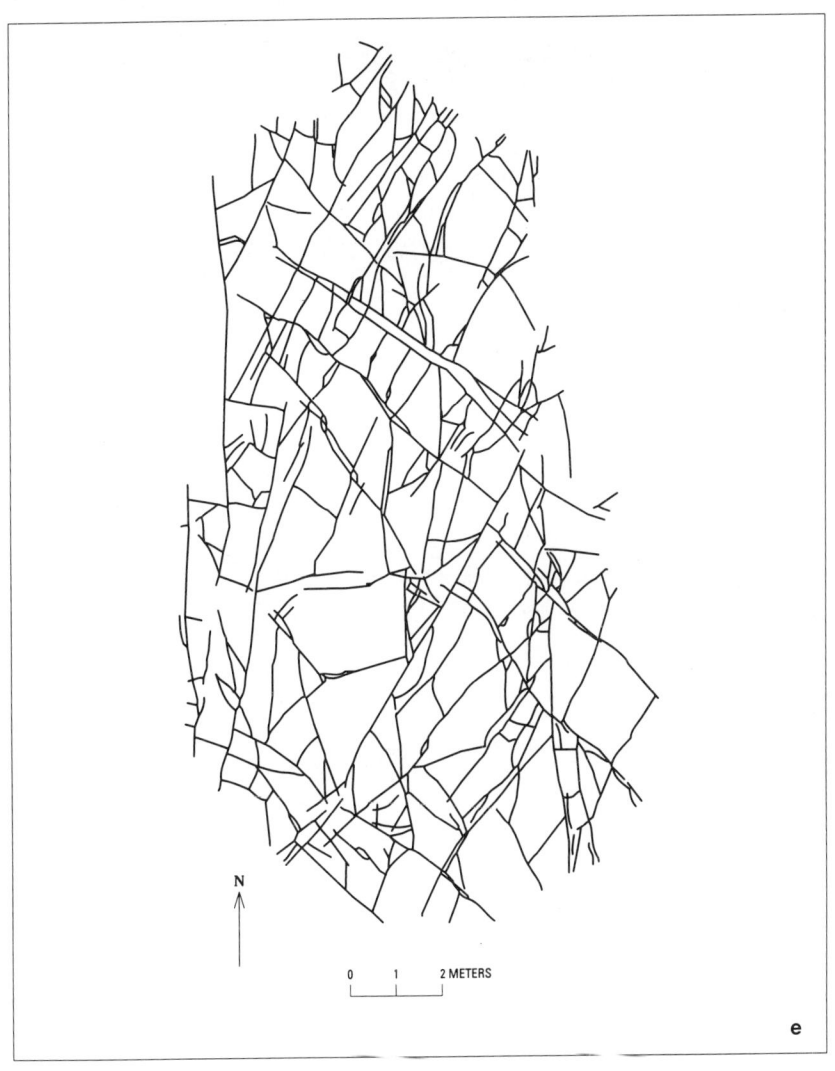

FIGURE 8.7. (*Continued*)

smaller than the shortest trace and no larger than the size at which the number of occupied cells is equal to the number of cells. Even with these limits on cell sizes, rollovers are common at both ends of the fractal plots, which I interpret to be a boundary phenomenon as we approach the upper and lower size limits of a data set. Rollovers were removed from fractal plots by the procedure of the least squares fit to the data as previously described, and the fractal dimension was calculated. Fractal plots of the number of occupied cells versus the inverse of the cell size is shown for the box-flex method in Fig. 8.8. Data points are shown for a few of the spaced best-fit lines to provide visual confirmation of the goodness of fit. Measured by a correlation coefficient, straight-line fits to the data are better than 0.99, where 1.0 is a perfect fit. The range of scales sampled on any one map is between 1–2

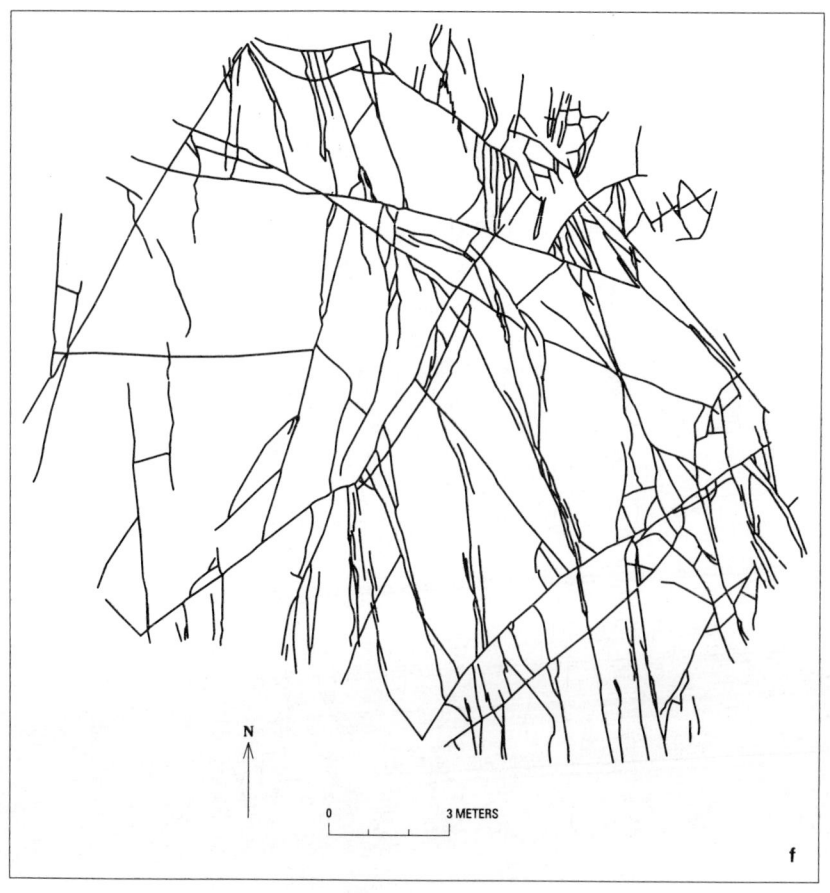

FIGURE 8.7. (*Continued*)

orders of magnitude. The total range of scales sampled in terms of cell size is approximately 0.0002–142,900 m, nearly 10 orders of magnitude.

The fractal dimension for each of the fracture trace maps is shown in Table 8.2. Results of the box-flex analysis show that the fractal dimension ranges from 1.32–1.70. Results for the Yucca Mountain maps show that there can be some change in fractal dimension within a stratigraphic unit over short lateral distances between map locations, on the order of a few tens of meters (Figs. 8.7a–c) or over longer lateral distances between map locations, on the order of a few hundred meters (Figs. 8.7d–f). There is also a difference between different stratigraphic unit—see Fig. 8.7g. The difference from one stratigraphic unit to the next arises because fractures at Yucca Mountain are stratabound at the pavement scales, with individual stratigraphic units or packets of units having different fracture patterns (Barton and Hsieh, 1989). Fractures at the scale of Fig. 8.7m are on the scale of hundreds to thousands of meters, and these cut through many stratigraphic units, including those in Figs. 8.7a–g; the fractal dimension of faults in Fig. 8.7m is 1.50, which is nearly the average (1.55) of values found for fractures in individual units at pavement scales through which they cut.

8. FRACTAL ANALYSIS OF SCALING AND SPATIAL CLUSTERING OF FRACTURES   161

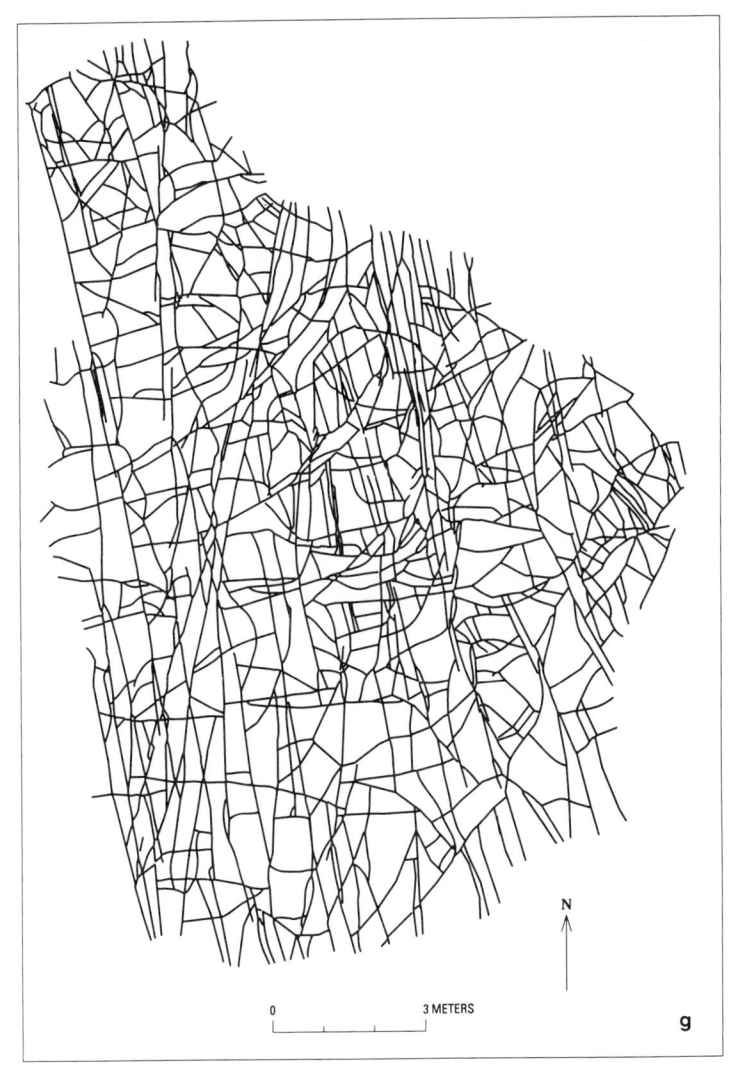

FIGURE 8.7. (*Continued*)

There are at least three fracture mechanisms in the 17 maps analyzed. Fractures on maps in Figs. 8.7a–i, l, n, and q formed primarily as joints; fractures on maps in Figs. 8.7m, o, and p formed as faults. Those on the map in Fig. 8.7j formed as deformation bands by grain size reduction in shear; and those on the map in Fig. 8.7k formed as deformation bands by more closely packing grains without grain size reduction or demonstrable shear. Table 8.2 reveals that there is apparently no correlation between mechanisms of fracture generation and the fractal dimension. This suggests that geometrical constraints on scaling and spatial clustering are independent of the mechanics that divide volumes of rock by fracturing. The narrow range in the fractal dimensions suggests an underlying physics acting over the entire range of scales investigated. The narrow range of fractal dimension

FIGURE 8.7. (*Continued*)

also suggests that geometrical scaling constraints and a common physics apply to a wide range of rock types, ages, and deformation histories represented by maps analyzed in this study.

8.7. DEVELOPMENT PATTERN OF FRACTURE NETWORKS

Fracture networks evolve from initially ordered to increasingly disordered patterns as discussed later. Fracture networks become more complex with time as new fracture generations are added to those that already exist. Fractures generations form during discrete episodes, each of which records a discrete chapter of the tectonic history. Most fracturing episodes are not accompanied by major tectonic deformations, such as folding and faulting (for example, see Barton, 1983). For fractures in a network formed as joints, the network can be disarticulated into generations of joints on the basis of abutting relations—younger fractures abut older ones. Barton and others (1986) first reported this approach for analyzing maps of joints patterns. As a typical example, the map in Fig. 8.7h was disarticulated in this way, and fractures of the same generation are given the same color (see Fig. 8.9). Note that

8. FRACTAL ANALYSIS OF SCALING AND SPATIAL CLUSTERING OF FRACTURES 163

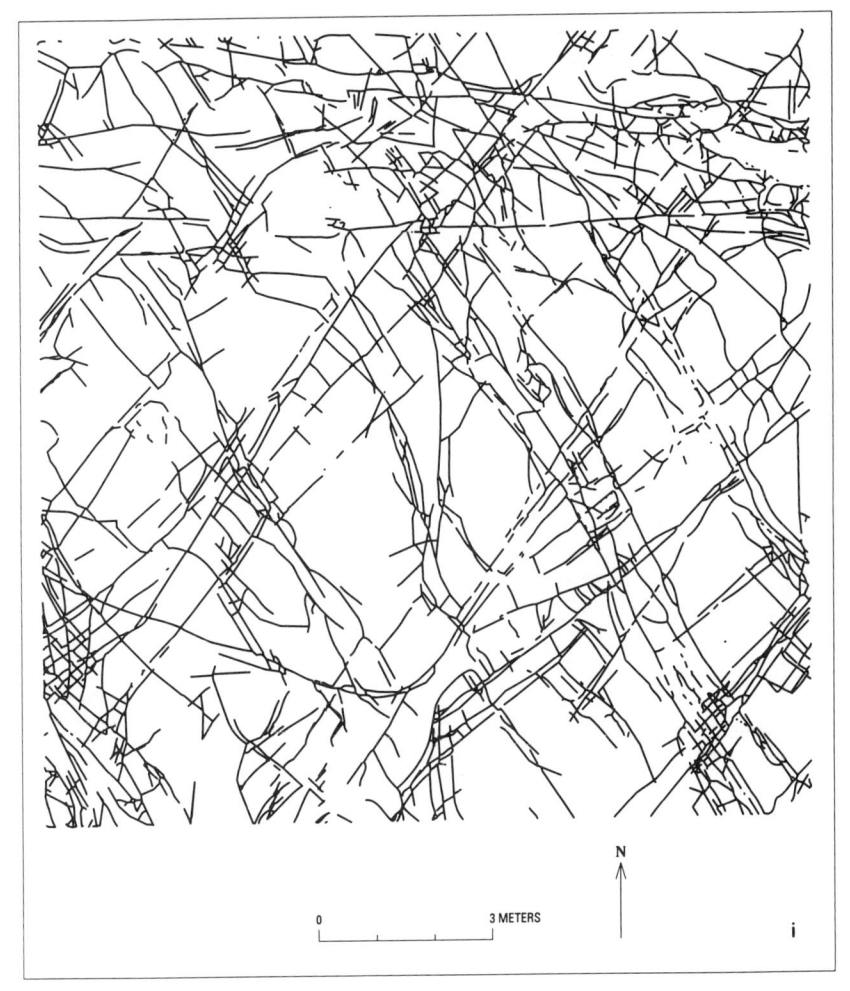

FIGURE 8.7. (*Continued*)

this approach assigns each fracture to the oldest possible generation. This interpretation of abutting relations is based on detailed observations of fracture intersections during pavement mapping. Alternative interpretations, such as the origin of multiple fractures from a shared origin or the chance passing of a younger fracture across the end of an older fracture, are not supported by my field observations.

Analysis of fracture characteristics from one generation to the next (see Figs. 8.9a–f) reveals the following general pattern of fracture network development. The first-generation fractures (see Fig. 8.9a) are long, subparallel, and network connectivity is poor. Second-generation fractures are shorter and abut first-generation fractures, generally at high angles, to form mostly large polygonal blocks (see Fig. 8.9b); network connectivity is improved. Fractures of subsequently younger generations (see Figs. 8.9b–f) are generally shorter, more diversely oriented, and increase network connectivity greatly. Younger fractures generally define small, irregular polygonal blocks bounded by older fractures. Analyzing

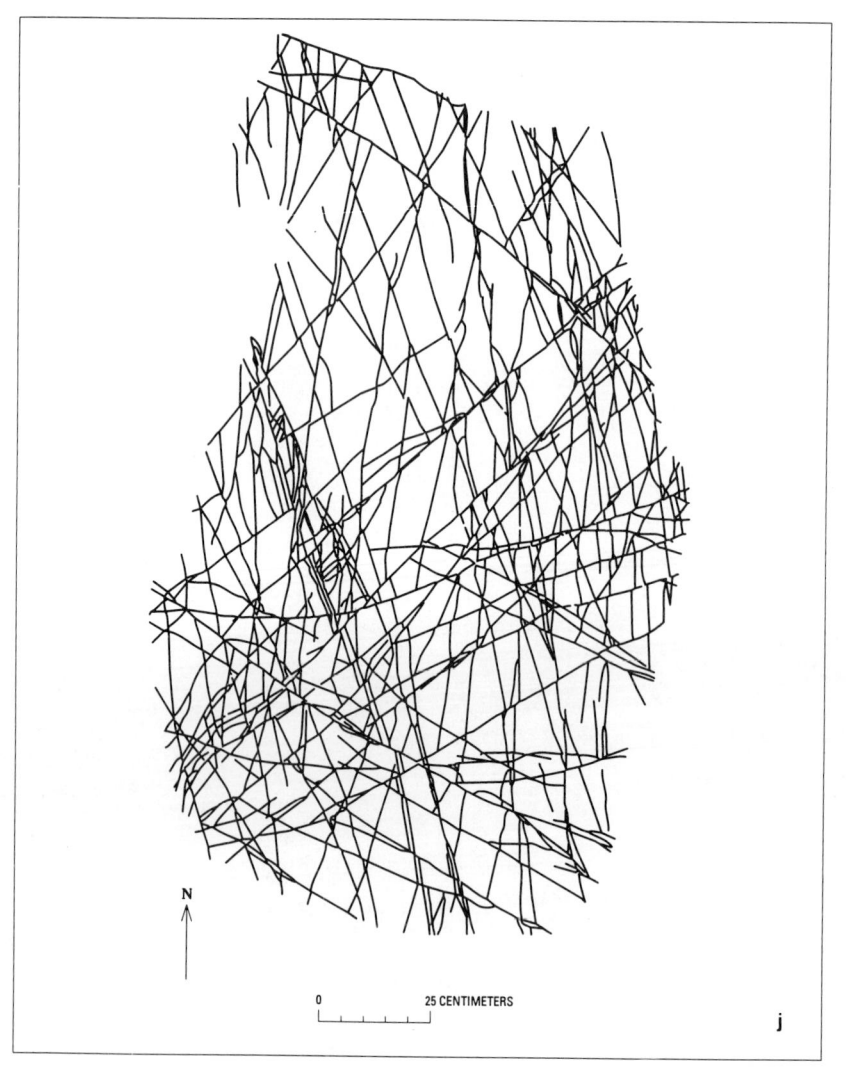

FIGURE 8.7. (*Continued*)

maps in Figs. 8.7a–d, l, and p reveals the same pattern in network evolution. Analyzing maps in Figs. 8.7e–g, i, n, and q reveals that inadequate exposure or mapping of fracture intersections or restarting the evolution by mineral infilling of previous generations renders the approach inoperable. Specifically it is inappropriate to use this approach on intersecting deformation bands (see the maps in Figs. 8.7j and k) and intersecting faults (see the maps in Figs. 8.7m and o) because younger faults truncate older fractures. Note however, that first-generation nonintersecting faults (see the map in Fig. 8.7p) are long and subparallel, as are first-generation joints shown in Fig. 8.9a.

The evolution pattern should begin anew when mineral infillings mechanically heal

8. FRACTAL ANALYSIS OF SCALING AND SPATIAL CLUSTERING OF FRACTURES

FIGURE 8.7. (*Continued*)

previous fracture generations. This predicts that one or more stages of infilling are required to permit development of highly ordered fracture patterns composed of more than one or two generations of fractures. Highly ordered fracture patterns are not observed in the stratigraphic section at Yucca Mountain, for example, where there has been little fracture healing by mineral infilling (Barton and Hsieh, 1989).

The spatial distribution of fractures within the network evolves as fractures are sequentially added to the network. The change in the box-flex fractal dimension during the evolution of the network shown in Fig. 8.9 is plotted in Fig. 8.10. The fractal dimension ranges from 1.29 for the earliest stage of network development to 1.50 for the complete

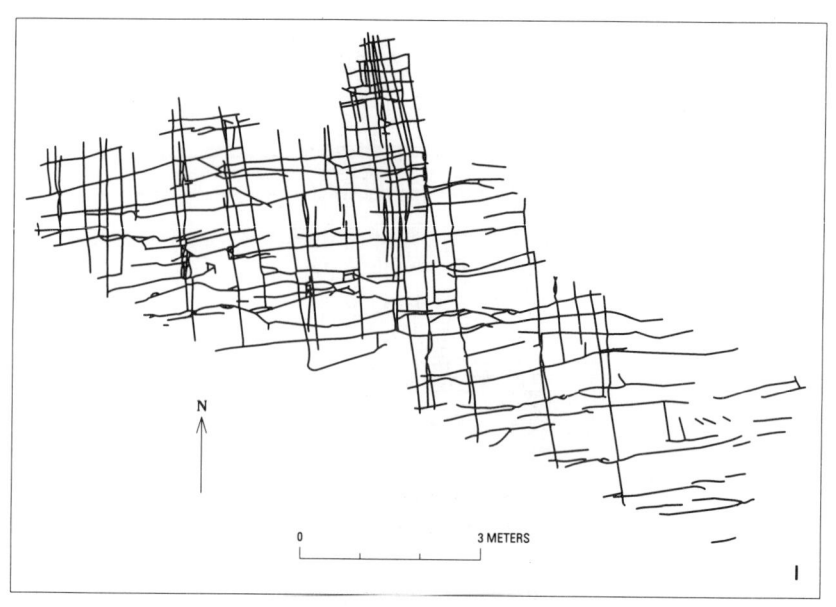

FIGURE 8.7. (*Continued*)

network. Note also that connectivity within the network is low during initial stages of development and increases as more fractures are added.

During network evolution, larger blocks are preferentially broken by subsequent fracturing (see Figs. 8.9b–f); continuing this process reduces the range of block sizes. In fault gouge, comminution by grinding also produces a fractal distribution of particle sizes with a fractal dimension of 1.6 over six orders of magnitude in scale (Sammis and Biegel, 1989), but it does not preferentially break up larger particles. The physics of grinding involves more than fracturing: In grinding the rotation and translation of particles produces large particles mechanically isolated from one another by smaller particles, and this reduces the probability of large particles being further broken up (Sammis and Biegel, 1989).

8.7.1. Percolation Threshold for Fracture Trace Maps

Fluid flow through a fracture trace network requires connectivity across the network. Connectivity for porous media has been studied using site percolation models (for a review, see Feder, 1988). An important property of percolation models is the percolation threshold below which connectivity is confined to a finite region (cluster) and above which connectivity extends across the medium (spanning cluster). For two-dimensional site percolation models, the spanning cluster always has a fractal dimension of 1.89 (Feder, 1988). By analogy to such models, I propose that the fractal dimension at the percolation threshold for fracture trace networks is approximately 1.35, based on the fractal dimension of the network consisting of the first two fracture generations shown in Fig. 8.9b. Note that the fracture trace networks in Fig. 8.7 above the percolation threshold have fractal dimensions greater than 1.35, except Fig. 8.7p, which is below the percolation threshold and has a fractal dimension of 1.32.

8. FRACTAL ANALYSIS OF SCALING AND SPATIAL CLUSTERING OF FRACTURES   167

FIGURE 8.7. (*Continued*)

8.7.2. Generating Synthetic Fracture Networks

The evolution of fracture patterns just described is the basis for our computer generation of synthetic fracture networks (Barton and others, 1987). We have developed computer code to generate synthetic two-dimensional fracture trace networks by randomly selecting values from frequency distributions of fracture trace length, spacing, orientation, crossing, abutting, and dead-end fracture intersection distributions obtained from analyzing our fracture trace maps. The procedure is a two-dimensional analog to generating a disordered Cantor dust, in that spatial correlation is built in. This is fundamentally different from a Poisson process, which is uncorrelated and produces fracture trace patterns with a dimension of 2.

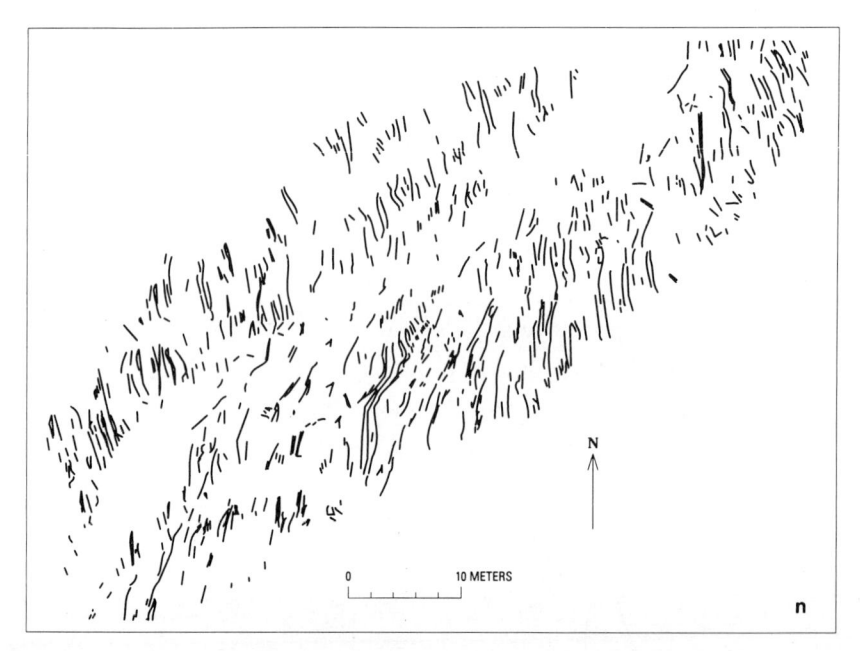

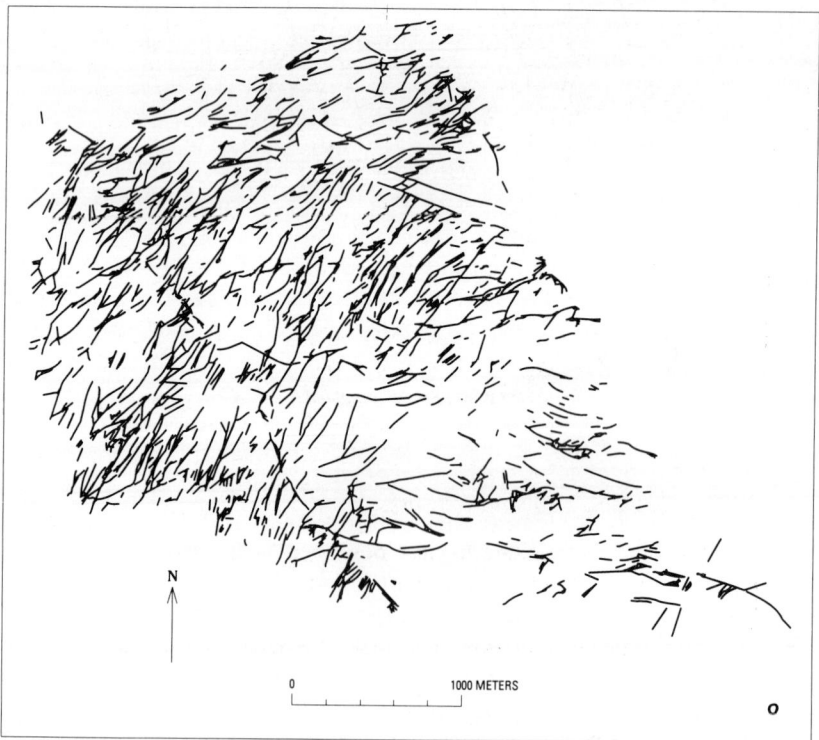

FIGURE 8.7. (*Continued*)

8. FRACTAL ANALYSIS OF SCALING AND SPATIAL CLUSTERING OF FRACTURES

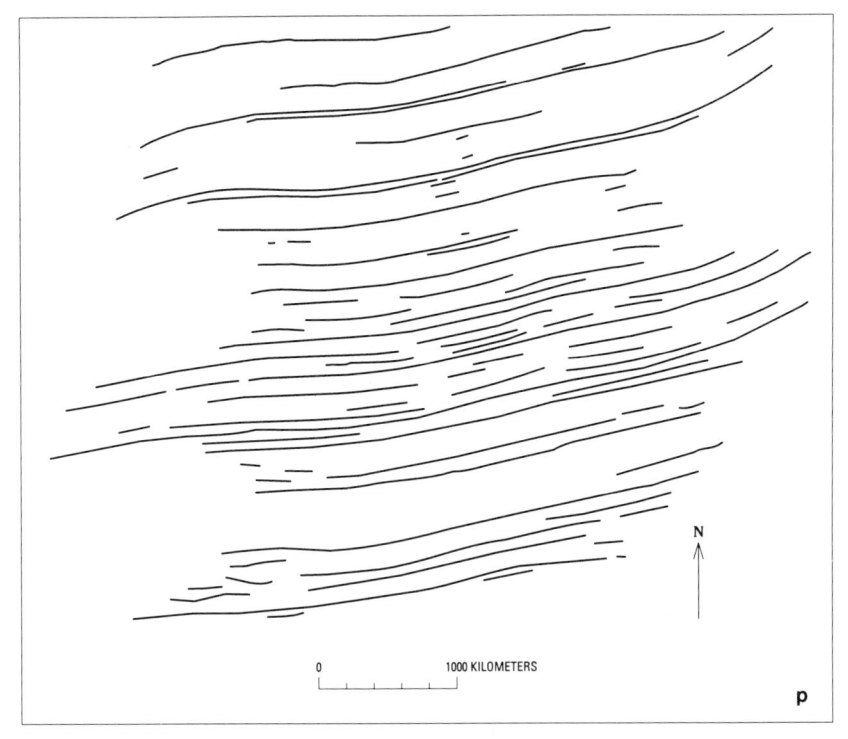

FIGURE 8.7. (*Continued*)

The basis for generating synthetic networks is the observation that fracture networks become more complex with time as younger, more diversely distributed fracture generations are added to previous generations. To define the distribution for each generation, I analyzed the sequence of fracturing in maps in Figs. 8.7a–d, and h. Relative age of the fractures is determined by abutting relations (younger abut older ones), as previously described and illustrated in Fig. 8.9. Our synthetic networks are generated by randomly selecting values of trace length, spacing, orientation, and terminations from those distributions particular to each generation. All generations after the first are initiated along traces of the preceding generation. This procedure produces synthetic networks with appropriate statistical distributions, including fractal scaling and spatial-clustering distributions. This procedure is true to the observed natural evolution of fracture networks and represents an advance beyond the Poisson process used by Dershowitz (1984), Long and others (1985), and Chiles (1988). Box-flex analysis of the synthetic network is used to verify that the fractal dimension of the synthetic network falls within the range of 1.3–1.7, which we find for fracture networks mapped in the field. Refining this approach for generating synthetic fracture networks should include further conditioning the network to observed data.

8.7.3. *Extrapolating from One- and Two-Dimensional Samples*

A most important characteristic of disordered, nonsymmetric, self-similar fractals is that their dimension as measured by volumetric, areal, and linear samples is each,

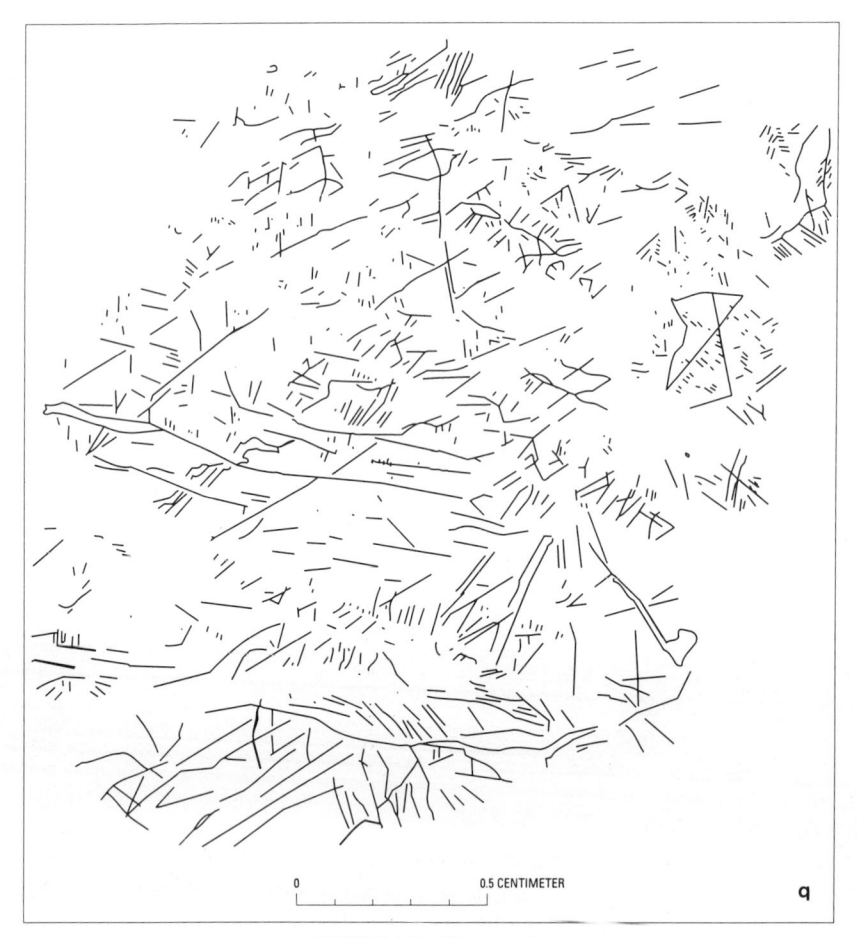

FIGURE 8.7. (*Continued*)

respectively, one less than the former. For example, the fractal dimension of an areal slice through a volumetric fractal of dimension 2.6 is 1.6, while that of a line sample is 0.6. Note that this relation does not hold for ordered fractals; for fractals with symmetry, such as Serpinski carpets, which are areal slices through Menger sponges (Mandelbrot, 1982), and for self-affine fractals.

As previously stated, drill holes, that is, line samples, are the most common mode of sampling scaling and spatial-clustering distributions of fractures in rock. Yet it is two- or optimally three-dimensional characterization of these properties that is needed as input to fractured rock hydrology and mechanical deformation models.

Analyses of one- and two-dimensional samples in Chapter 8 indicate that when fracture networks are shown to be fractal, it is appropriate to extrapolate from a one-dimensional sample to two dimensions. The fractal dimension of 23 linear samples of the spacing of veins at the Perseverence mine in Juneau, Alaska, ranges between 0.41–0.62, with an average of 0.50, while the dimension of an areal sample of the veins (see the map

8. FRACTAL ANALYSIS OF SCALING AND SPATIAL CLUSTERING OF FRACTURES

in Fig. 8.7n) is 1.48—almost exactly an integer dimension more, as expected for disordered, nonsymmetric, self-similar fractal systems. Extrapolating the vein spacing to three dimensions predicts a volumetric fractal dimension of approximately 2.50, but this cannot be verified because there are no three-dimensional maps of the vein networks. Note that this approach averages any directional anisotropy present in the fracture network and therefore should be considered a first-order measure of the scaling properties of fracture networks.

Extrapolating to smaller sizes is also possible. Qualitative observations reveal that the number of fractures in rock continues to increase to the scale of microfractures. Fractal analysis of the smallest areas represented in this study (see the map in Fig. 8.7q) reveals that the fractal dimension ($D = 1.58$) at that size falls well within the range of 1.32–1.70 found over the range of scale of all other maps.

8.8. DISCUSSION

If fractal analyses of future fracture maps at different scales and locations have the same range as results just presented, then it is acceptable to map fractures at one scale and extrapolate spatial and statistical geometric properties to larger and smaller scales.

Once the fractal dimension of a pattern or object in nature is determined, it is possible to model that pattern or object from a single generator. A generator is the fundamental building block from which a fractal pattern or object is produced by iteratively replacing each piece of the generator with a reduced version of the generator. The task of deducing a generator for a particular fractal pattern observed in nature is not easy. One approach is to guess at the generator, as was done by King (1983) for the map pattern of traces of subsidiary faults in the immediate vicinity of large-scale fault bends. To simulate fault patterns, King (1983) proposed a nonoverlapping three-dimensional space-filling generator, but he assumed that it was never fully formed. Crosscutting fractures are common on the maps in Fig. 8.7, which suggests that a proper generator for modeling fracture trace patterns is overlapping. The generator proposed by King (1983) was for faults that did not overlap, and therefore it is not appropriate for patterns of crossing fractures. I have not succeeded in finding a generator for modeling fracture trace maps. A most promising method for deducing a fractal generator is the iterated function systems approach developed by Barnsley and Demko (1985), which systematically deduces a fractal generator for a given fractal object.

Paul Meakin (verbal communication) suggested that the behavior in Fig. 8.8 may not represent fractal behavior but rather a smooth gradual crossover between power law scaling with slopes of 1 and 2. If true, it is not appropriate to extrapolate the fractal dimension beyond length scales measured for each map in Fig. 8.7. I propose that fractures are present everywhere in rock, over many orders of magnitude in length scale, from microfractures to megascopic fractures. Thus the correct dimension for fracture networks in rock is 3 in volume space and 2 in planar (map) section. For the 17 fracture maps analyzed in Fig. 8.7, the fractal dimensions of fracture traces do not exceed 1.7. I suggest that the dimensions are less than 2 because only those fractures with apertures sufficient to render them visible were mapped. I propose that fracture networks are multifractal as a function of aperture, so that as we include fractures of successively smaller aperture, the fractal dimension of a given fracture map approaches 2. Scaling of large-aperture fracture networks will yield fractal dimensions less than 2, and such fractal dimensions correctly characterize scaling for

TABLE 8.2. Fractal Dimensions and Parameters of Fracture Trace Maps

Map Number	Fractal Dimension Box Flex Method	Location	Rock Unit/Type	Age	Map Scale	Length of Shortest Fracture (m)	Length of Longest Fracture (m)	Map Reference
8.7a	1.52	Yucca Mountain, NV	Densely welded upper lithophysal unit of the Tiva Canyon member of the Paintbrush tuff	Miocene	1:54	0.25	12	Barton and Hsieh, 1989
8.7b	1.38	Yucca Mountain, NV	Densely welded upper lithophysal unit of the Tiva Canyon member of the Paintbrush tuff	Miocene	1:104	0.50	17	Barton and Hsieh, 1989
8.7c	1.50	Yucca Mountain, NV	Densely welded upper lithophysal unit of the Tiva Canyon member of the Paintbrush tuff	Miocene	1:65	0.39	15	Barton and Hsieh, 1989
8.7d	1.61	Yucca Mountain, NV	Densely welded upper lithophysal unit of the Tiva Canyon member of the Paintbrush tuff	Miocene	1:205	0.59	42	Chap. 8
8.7e	1.59	Yucca Mountain, NV	Densely welded upper lithophysal unit of the Tiva Canyon member of the Paintbrush tuff	Miocene	1:63	0.23	8.9	Chap. 8
8.7f	1.54	Yucca Mountain, NV	Densely welded upper lithophysal unit of the Tiva Canyon member of the Paintbrush tuff	Miocene	1:74	0.24	13	Chap. 8

8. FRACTAL ANALYSIS OF SCALING AND SPATIAL CLUSTERING OF FRACTURES

8.7g	1.70	Yucca Mountain, NV	Densely welded orange brick unit of the Topopah Spring member of the Paintbrush tuff	Miocene	1:55	0.20	12	Chap. 8
8.7h	1.50	Cedar City, UT	Quartz monzonite	Oligocene	1:500	1.7	110	Kolb and others, 1970
8.7i	1.60	Lannon, WI	Niagaran dolomite	Silurian	1:33	0.091	6.4	La Pointe and Hudson, 1985
8.7j	1.50	Morrison, CO	Lyons sandstone (orientation unknown)	Permian	1:3	0.12	9.7	Chap. 8
8.7k	1.58	Valley of Fire, NV	Aztec sandstone	Triassic?/ Jurassic	1:96	0.2	17	Barton and Hsieh, 1989; Hill, 1990
8.7l	1.52	Alhambra Rock, Mexican Hat, UT	Rico limestone	Penn/Permian	1:5	0.08	4.8	Olson and Pollard, 1989
8.7m	1.49	Yucca Mountain, NV	Paintbrush tuff	Miocene	1:12,000	53	6300	Scott and Bonk, 1984
8.7n	1.48	Juneau, AK	Perseverance slate (subvertical exposure)	Triassic	1:78	0.4	8	Chap. 8
8.7o	1.52	Goldhill, CO	Boulder Creek granodiorite	Precambrian	1:12,000	26	970	Goddard and Lovering, 1938
8.7p	1.32	South Atlantic seafloor	Basalt	Cretaceous–present	$1:4.4 \times 10^7$	9.9×10^4	6.3×10^6	Cande and others, 1988
8.7q	1.58	Timmins, Ontario	Albite, quartz, scheelite	Archean	1:65	5×10^{-4}	1.3×10^{-2}	Chap. 8

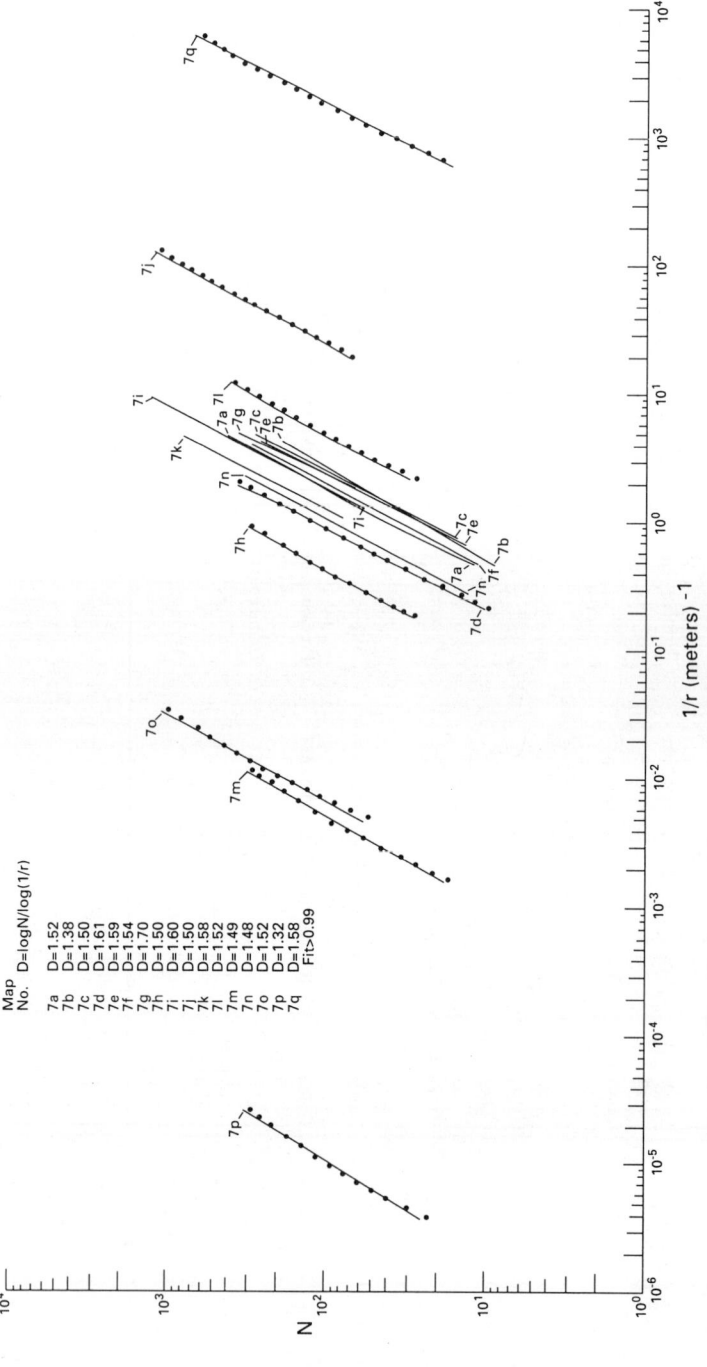

FIGURE 8.8. Fractal plot for fracture trace maps (Figs. 8.7a–q) using the box-flex method. N = number of occupied cells, r = cell size, and D = fractal dimension. The continuous stream of data points generated for each analysis by the box-flex method are not shown; every third data point is shown for analyses that do not overlap on plot.

8. FRACTAL ANALYSIS OF SCALING AND SPATIAL CLUSTERING OF FRACTURES 175

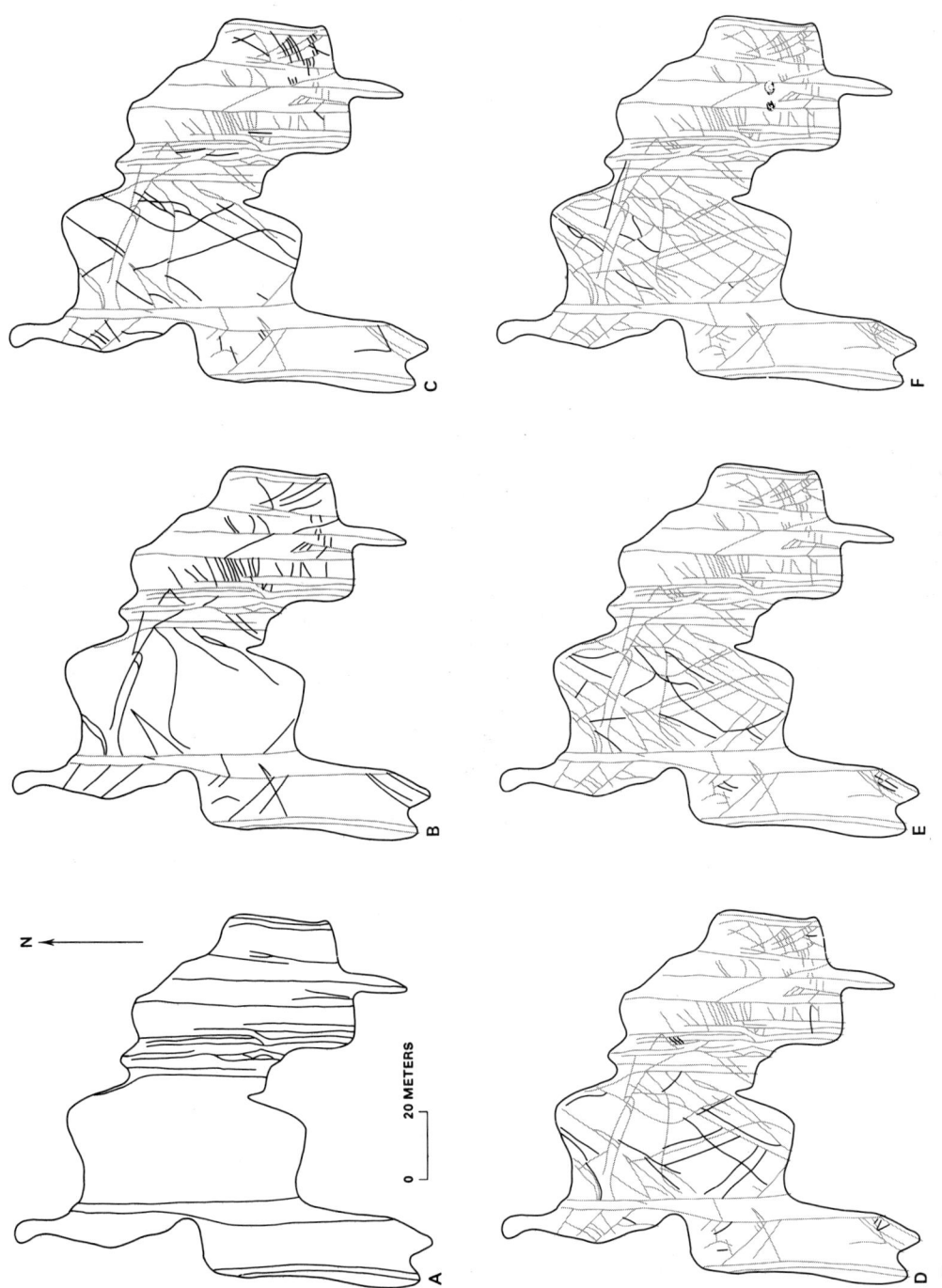

FIGURE 8.9. Fracture trace map in Fig. 8.7h dissected to show fractures grouped into six relative age generations based on abutting relations. (a) First generation fractures; (b–f) fractures added by generation; latest generation shown in black, previous generations shown in gray.

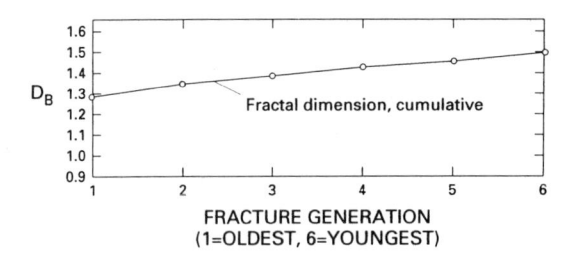

FIGURE 8.10. Plot of the box-flex fractal dimension versus fracture generation for the evolution of the fracture trace map in Fig. 8.7h shown in Fig. 8.9. Curve shows fractal dimension of cumulative pattern as each succeeding generation is added.

studies of fluid flow through fracture networks. However calculations of storativity in fracture networks should assume a dimension of 3 in volume space and 2 in planar section.

8.9. CONCLUSIONS

Fractal dimensions for scaling and spatial-clustering of fractures along drill holes prove to be random Cantor dusts with fractal dimensions ranging from 0.42–0.62 over a range of four orders of magnitude. The fractal dimensions for scaling and spatial clustering of fracture trace maps range from 1.38–1.70 over nearly ten orders of magnitude in length scale. A fracture network sampled by both drill holes and mapping a planar outcrop revealed fractal dimensions of 0.50 and 1.48, respectively, nearly an integer dimension difference, as is expected for disordered, nonsymmetric, statistically self-similar fractal patterns. This result suggests that extrapolating from one-dimensional samples to planar or volumetric dimensional samples is not unreasonable for fracture networks. The change of fractal dimension from the iterative addition of new fractures during the evolution of fracture networks is investigated by disarticulating and then reconstructing the evolution of a fracture trace map. Analysis reveals that the fractal dimension of the network increases with the addition of each successive generation of fractures. Thus the fracture network exhibits multifractal behavior with time. The evolution of fracture networks is proposed as a physical model for constructing synthetic fracture networks. A fractal dimension of 1.35 is found for an evolving fracture trace network at or just beyond the percolation threshold. There is no three-dimensional (volumetric) analysis because of a lack of data.

ACKNOWLEDGMENTS

I am indebted to the following individuals for their help in fracture mapping: Eric Larsen, Thomas A. Howard, and Patricia E. Baechle of Fenix & Scisson, Inc., maps in Figs. 8.7a, c, d, b, and f respectively; William R. Page and Constance K. Throckmorton of the US Geological Survey, maps in Figs. 8.7g and e, respectively; Marc B. Gamble, of Colorado College, the map in Fig. 8.7j; Robin E. Hill, of the University on Nevada, Las Vegas, the map in Fig. 8.7k; and Dominic Channer of the university of Toronto, the map in Fig. 7.8q. I am most indebted to Thomas A. Schutter of the US Geological Survey, who programmed the box algorithms used to produce the fractal plots and calculate the fractal dimensions presented in Chapter 8. William D. Stansbury of the Colorado School of Mines aided in the fractal analyses summarized in Table 8.1. Petro A. Zlatev of the Colorado School of Mines, Paul R. Herring, and William J. Thomas of the University of Colorado,

Boulder aided in the fractal analyses summarized in Table 8.2. Glen Miller of WGM-Echo Bay Mining Company generously provided data for the fractal analysis of vein spacing in the Perseverence mine. The photograph in Fig. 8.1d was taken by Kenneth Bradbury of the Wisconsin Geological and Natural History Survey.

REFERENCES

Baecher, G. B., Lanney, N. A., and Einstein, H. H., *Statistical Description of Rock Properties and Sampling*, Proceedings of the 18th US Symposium on Rock Mechanics, Colorado School of Mines Press, Golden, CO, American Institute of Mining Engineers, 1977, pp. 5c1–5c8.

Barnsley, M. F., and Demko, S., Iterated Function Systems and the Global Construction of Fractals, *Proc. Roy. Soc. London Series A* **399**, 243 (1985).

Barton, C. A., and Zoback, M. D., Self-similar distribution and properties of macroscopic fractures at depth in crystalline rocks in the Cajon Pass scientific drill hole, *J. Geophys. Res.* **97**, **B4**, 5181 (1992).

Barton, C. C., "Systematic Jointing in the Cardium Sandstone along the Bow River, Alberta, Canada," Ph.D. diss., Yale University, 1983.

———, Fractal characteristics of fracture networks in rock, in *Rock Mechanics—Contributions and Challenges*, 31st US Symposium on Rock Mechanics (A. A. Balkema, Rotterdam, 1990).

Barton, C. C., and Larsen, E., Fractal geometry of two-dimensional fracture networks at Yucca Mountain, southwest Nevada, in *Fundamentals of Rock Joints*, Proceedings of the International Symposium on Fundamentals of Rock Joints, Bjorkliden, Lapland, Sweden (O. Stephannson, ed.) (Centek Publishers, Lulea, Sweden, 1985), pp. 74–84.

Barton, C. C., Gott, C. B., and Montgomery, J. R., Fractal scaling of fracture and fault maps at Yucca Mountain, southern Nevada (abs.), *Eos Trans. Am. Geophys. Union* **67**, **44**, 870 (1986).

Barton, C. C., and Hsieh, P. A., *Physical and Hydrologic Flow Properties of Fractures*, Field Trip Guidebook T385, 28th International Geological Congress (American Geophysical Union, Washington, DC, 1989).

Barton, C. C., Page, W. R., and Larsen, E., Pattern of development of fracture networks, *Geol. Soc. Am. Abstr. Programs* **18**, **6**, 536 (1986).

Barton, C. C., Schutter, T. A., Page, W. R., and Samuel, J. K., Computer generation of fracture networks for hydrologic-flow modeling, *Eos Trans. Am. Geophys. Union* **68**, **44**, 1295 (1987).

Barton, C. C., and Scholz, C. H., Consequences of fractal distributions of fracture length and aperture on fluid flow through fractured rock, in press.

Bear, J., *Dynamics of Fluids in a Porous Media* (American Elsevier, New York, 1972).

Brown, S. R., Measuring the dimensions of self-affine fractals: The example of rough surfaces, in this volume, Chap. 4.

Chiles, J. P., Fractal and geostatistical methods for modeling of a fracture network, *Math. Geol.* **20**, 631 (1988).

Conrad, F., and Jacquin, C., Representation d'un reseau bi-dimensionnel de fractures par un modele probabiliste. Application au calcul des grandeurs geometriques des blocks matriciels, *Rev. Institut Fr. Pet.* **28**, 843 (1973).

DeMarsily, G., *Flow and Transport in Fractured Rocks: Connectivity and Scale Effect*, International Association of Hydrologists, Memoirs, Vol. 17, Part 2 (International Association of Hydrogeologists, 1985), pp. 267–77.

Dershowitz, W. S., "Rock Joint Systems," Ph.D. diss., Massachusetts Institute of Technology, 1984.

Feder, J., *Fractals* (Plenum Press, New York, 1988).

Goddard, E. N., and Lovering, T. S., "Geologic Map of the Gold Hill Mining District, Boulder County, Colorado" (US Geological Survey, 1938).

Hausdorff, F., Dimension und ausseres Mass, *Math. Ann.* **79**, 157 (1919).

Hill, R. E., "Analysis of Deformation Bands in the Aztec Sandstone," Valley of Fire State Park, Nevada, M.S. thesis, University of Nevada, Las Vegas, 1990.

Hirata, T., Fractal dimension of fault systems in Japan: Fractal structure in rock fracture geometry at various scales, *Pure Appl. Geophys.* **131**, 290 (1989).

Hohn, M. E., *Geostatistics and Petroleum Geology* (Van Nostrand Reinhold, New York, 1988).

Hudson, J. A., and Priest, S. D., Discontinuities and rock mass geometry, *Int. J. Rock Mech. Min. Sci. Geomech. Abstr.* **16**, 339 (1979).

———, Discontinuity frequency in rock masses, *Int. J. Rock Mech. Min. Sci. Geomech. Abstr.* **17**, 279 (1983).

Journel, A. G., and Huijbregts, C. J., *Mining Geostatistics* (Academic Press, London, 1978).

King, G. C. P., The accommodation of large strains in the upper lithosphere of the earth and other solids by self-similar fault systems: the geometric origin of b-value, *Pure Appl. Geophys.* **121**, 761 (1983).

Kolb, C. R., Farrell, W. J., Hunt, R. W., and Curro, J. R., Jr., "Geological Investigation of the Mine Shaft Sites, Cedar City, Utah," US Army Engineer Waterways Experiment Station, Report MS-2170, 1970.

Korvin, G., Fractured but not fractal: Fragmentation of the Gulf of Suez basement, *Pure Appl. Geophys.* **131**, 290 (1989).

Kulander, B. K., Barton, C. C., and Dean, S. L., "Applications of Fractography to Core and Outcrop Investigations," US Department of Energy METC/SP-79/3, 1979.

La Pointe, P. R., Analysis of the spatial variation in rock mass properties through geostatistics, in *Rock Mechanics—a State of the Art*, Proceedings, 21st US Symposium on Rock Mechanics (D. A. Summers, ed.) (University of Missouri Press, Rolla, 1980), pp. 570–80.

La Pointe, P. R., A method to characterize fracture density and connectivity through fractal geometry, *Int. J. Rock Mech. Min. Sci. Geomech. Abstr.* **25**, 421 (1988).

La Pointe, P. R., and Hudson, J. A., "Characterization and Interpretation of Rock Mass Joint Patterns," Geological Society of America, Special Paper 199, 1985.

Long, J. C. S., Endo, H. K., Karasaki, K., Pyrack, L., MacLean, P., and Witherspoon, P. A., Hydrologic behavior of fracture networks, in *Hydrology of Rocks of Low Permeability*, International Association of Hydrologists, Memoirs, Vol. 17, Part 2 (International Association of Hydrogeologists, 1985), pp. 449–62.

Madden, T. R., Microcrack connectivity in rocks: A renormalization group approach to the critical phenomena of conduction and failure in crystalline rocks, *J. Geophys. Res.* **88**, 585 (1973).

Majer, E. L., McEvilly, T. V., Eastwood, F. S., and Myer, L. R., Fracture detection using P- and S-wave VSP's at the Geysers geothermal field, *Geophysics* **53**, 76 (1988).

Mandelbrot, B. B., *The Fractal Geometry of Nature* (W. H. Freeman, New York, 1982).

Manning, C. E., Fractal clustering of metamorphic veins, *Geology* **22**, 335 (1994).

Olson, J., and Pollard, D. D., Inferring paleostress from natural fracture patterns: A new method, *Geology* **17**, 345 (1989).

Oxaal, U., Murat, M., Boger, F., Aharony, A., Feder, J., and Jossag, T., Viscous fingering on percolation structures, *Nature* **329**, No. 6134, 32 (1987).

Paillet, F. L., Use of geophysical well-logs in evaluating crystalline rocks for siting of radioactive waste repositories, *Log Anal.* **32**, 85 (1991).

Priest, S. D., and Hudson, J. A., Estimation of discontinuity spacing and trace length using scanline surveys, *Int. J. Rock Mech. Min. Sci. Geomech. Abstr.* **13**, 135 (1976).

———, Estimation of discontinuity spacing and trace length using scanline surveys, *Int. J. Rock Mech. Min. Sci. Geomech. Abstr.* **18**, 183 (1981).

Pruess, S. A., Some remarks on the numerical estimation of fractal dimension, in this volume, Chap. 3.

Ramirez, A. L., and Daily, W. C., Evaluation of alternate geophysical tomography in welded tuff, *J. Geophys. Res.* **92**, 8, 7843 (1987).

Robinson, P. C., "Connectivity, Flow, and Transport in Network Models of Fractured Media," Ph.D. diss., Oxford University, 1984.

Sammis, C. G., and Biegel, R. L., Fractals, fault-gouge, and friction, *Pure Appl. Geophys.* **131**, 255 (1989).

Schwartz, F. W., Smith, L., and Crow, A. S., A stochastic analysis of macroscopic dispersion in a fractured media, *Water Resour. Res.* **19**, 1253 (1983).

Scott, R. B., and Bonk, J., "Preliminary Geologic Map of Yucca Mountain, Nye County, Nevada, with Geologic Sections," US Geological Survey Open-File Report 84-491, 1984.

Segall, P., and Pollard, D. D., Joint formation in granitic rock of the Sierra Nevada, *Geol. Soc. Am. Bull.* **94**, 563 (1983a).

———, Nucleation and growth of strike slip faults in granite, *J. Geophys. Res.* **88**, **B1**, 555 (1983b).

Smalley, R. F., Jr., Chatelain, J.-L., Turcotte, D. L., and Prévot, R., A fractal approach to the clustering of earthquakes: Applications to seismicity of the New Hebrides, *Bull. Seism. Soc. Am.* **77**, 1368 (1987).

Thorp, R. K., Watkins, B. J., and Ralph, W. E., Strength and permeability of an ultra-large specimen of granitic rock, in *Rock Mechanics Theory—Experiment—Practice*, Proceedings of the 24th US Rock Mechanics Symposium (C. C. Mathewson, ed.) (Texas A&M University Press, College Station, 1983), pp. 511–18.

Velde, B., Dubois, J., Touchard, G., and Badri, A., Fractal analysis of fractures in rocks, *Tectonophys.* **179**, 345 (1990).

Watanabe, K., Stochastic evaluation of the two-dimensional continuity of fractures in a rock mass, *Int. J. Rock Mech. Min. Sci. Geomech. Abstr.* **23**, 431 (1986).

Witherspoon, P. A., Amick, C. H., Gale, J. E., and Iwai, K., Observations of a potential size effect in experimental determination of the hydraulic properties of fractures, *Water Resour. Res.* **15**, 1142 (1979).

9

Fractal Fragmentation in Crustal Shear Zones

C. G. Sammis and S. J. Steacy

9.1. FRACTAL STRUCTURES IN THE EARTH'S CRUST

Understanding the geometry and evolution of fracture systems in the crust is a fundamental problem in geomechanics. Although the basic elements of brittle fracture, such as the nucleation, growth, and interaction of microcracks from preexisting flaws, have been studied extensively in the laboratory and are reasonably well understood theoretically, patterns of such fractures that develop in zones of crustal deformation is a topic of current research. The reason is that crustal fracture networks are developed over a wide range of scales, and the interaction between structures at different scales is not well understood. Most laboratory experiments involve deformation on only one scale before the effects of sample boundaries become important.

The geometry and evolution of multiscale fracture patterns is central to many practical problems in geomechanics; three examples are (1) the mechanics of earthquake instability, (2) the spatial and temporal distribution of earthquakes, and (3) the flow of geothermal and hydrocarbon fluids in fracture-dominated crustal reservoirs. The earthquake mechanics question centers on how parameters of rate- and state-dependent friction laws measured in the laboratory are to be extrapolated to natural faults. Problems arise because multiscale shear zone geometry is only poorly approximated by the laboratory analog of two surfaces in contact. A deeper understanding of spatial and temporal distributions of regional seismicity also requires a better understanding of the structure and mechanics of the fault systems on which it is developed. Similarly the structure of joint and fissure systems that control fluid flow in many environments can be modeled better if we know how they develop.

C. G. Sammis and S. J. Steacy • Department of Geological Sciences, University of Southern California, Los Angeles, California.

Fractals in the Earth Sciences, edited by Christopher C. Barton and Paul R. La Pointe. Plenum Press, New York, 1995.

There is recent direct evidence that multiscale crustal fracture patterns are self-similar, so they can be described using fractal geometry. Hirata (1989) found that regional fault systems in Japan are self-similar from scale lengths of 2–20 km. with fractal dimensions from 1.05–1.60. Sornette (1991) found that the fault pattern in the Mojave Desert, California, is fractal with dimension D_f = 1.7. Barton and Hsieh (1989) and Barton (1992) measured the fractal dimension of fracture patterns mapped on outcrops over scale ranges from centimeters to tens of meters. They mapped a total of 15 outcrops, obtaining fractal dimensions ranging from 1.58–1.80, with a mean and standard deviation of 1.67 ± 0.07. Sammis and others (1987) and Sammis and Biegel (1989) demonstrated that the cataclastic fragments in the size range of 10 μm–1 cm within the Lopez Canyon fault zone in Southern California form a self-similar pattern with a fractal dimension of D_f = 1.6 ± 0.1 in two-dimensional planar section. Biegel and others (1991) found that cataclastic fragments from the San Gabriel fault zone in Southern California had a fractal size distribution with D_f = 1.6. Turcotte (1986) collected fragmentation data from the literature for a wide variety of processes. For fragmentation produced by underground explosions, the particle size distribution was self-similar, with a fractal dimension near 2.50 in three dimensions. For an isotropic random fractal, this corresponds to D_f = 1.50 in planar section for comparison with other patterns just described.

There is also indirect evidence of a fractal structure in crustal fractures. Wu and Aki (1985) concluded that the backscattering of S-waves is consistent with a fractal distribution of scatterers having dimension D_f = 1.6 at kilometer scale lengths. Observed temporal variations in coda Q^{-1} have led Jin and Aki (1988) to identify scatters with the crustal fracture network. King (1983) pointed out that the observed regional power law relation between the number of earthquakes and their magnitude can be explained by characteristic earthquakes on a fractal network of faults.

Two physical explanations for the observed self-similarity of crustal fracture and fragmentation have been suggested. One, proposed by Sornette and others (1990), is based on long-range screening and growth-enhancing interactions between fractures in a developing network. The other, proposed by Sammis and others (1987), is based on short-range nearest neighbor interactions between adjacent fragments in a well-developed, granulated shear zone.

Sornette and others (1990) proposed that fractal fracture networks are the result of random growth in a critically loaded system controlled by long-range screening and enhancement interactions between the growing fractures. The idea is that the growth of a fracture pattern is a Laplacian growth phenomenon in the same general class as diffusion-limited aggregation (DLA) and dielectric breakdown discussed by Meakin (1988), Hermann (1988), and others (see Stanley and Ostrowsky, 1988). In their model, entire domains of the crust are loaded to a critical state in an extension of Bak and Tang's (1989) fault model to regional faulting. The nature of long-range interactions and how they control the fractal dimension have yet to be worked out. Observations by both Sornette and others (1990) and Barton (1992) of D_f near 1.7, which is predicted by simple DLA models (Meakin, 1988), supports the hypothesis that the sequential growth of fractures may be a member of the general class of Laplacian growth phenomena.

Sammis and others (1987) proposed that D_f = 1.6 observed in the deformation of fragmented materials is the direct consequence of the fracture probability of each fragment being completely determined by the relative size of its nearest neighbors. These researchers present a simple mechanical argument that a fragment is most likely to fracture when it is loaded by other fragments its size. The end result of such a process is a distribution in which no two fragments the same size are nearest neighbors at any scale. Since this is a property of the Sierpinski gasket, or topologically equivalent fractals, which have $D_f = 1.58$, Sammis and others (1987) argued that this explains the observed fractal dimension.

The Sierpinski gasket however is a geometrically perfect figure that is much more regular than the observed gouge patterns. To explore the generation of random fractal fragmentations, Steacy and Sammis (1991) developed a computer automaton that implements the nearest neighbor fragmentation rules. They found that a fracture process controlled by nearest neighbor interactions does indeed produce a random fractal fragmentation but the fractal dimension is determined by the definition of nearest neighbors. Palmer and Sanderson (1991) developed a model for crushing ice that also accounts for the relative size of contacting fragments. In their model, the fractal dimension of 2.5 has the special significance that for such a distribution, fragments of all sizes make an equal contribution to the crushing force.

Both fractal growth and fractal fragmentation mechanisms have been simulated in the laboratory. Davy and others (1990) studied the growth of a fractal pattern of shear fractures in a layer of a cohesionless sand floating on a viscous substratum of silicone putty. They simulated the India–Asia continental collision by the slow advance of a rectangular indenter from the south, with a free boundary to the east. The pattern of conjugate shear fractures that formed to the northeast of the indenter was found to be fractal with $D_f = 1.7 \pm 0.05$, independent of the viscosity of the substrate.

Biegel and others (1989) produced fractal fragmentation by deforming a layer of initially uniform-sized fragments in simple shear. They documented the evolution of a fractal fragment size distribution having $D_f = 1.6$. Marone and Scholz (1989) also observed an artificial gouge evolve to a fractal fragment size distribution with $D_f = 1.6$ in the planar section. Their observations supported the nearest neighbor comminution theory, since the only time they observed two fragments the same size in contact, one of them was shattered. In addition they found that after this distribution formed, deformation localized in shear bands that had a reduced particle size with respect to the bulk gouge.

Differences between these mechanisms and their resultant fractal dimensions may reflect a difference in the total strain at which the two processes operate. The initial development of a fracture network at low strains with D_f near 1.7 may be a fractal growth phenomenon, as proposed by Sornette and others (1990). As strain increases, the fracture network may be expected to grow denser and ultimately to fragment the medium. The local stress state may be increasingly controlled by local fracture geometry and less by remotely applied stress. As fragmentation develops, stress heterogeneity may be controlled by the relative size of adjacent fragments, as proposed by Sammis and others (1987), such that continuing deformation leads to the development of a fractal fragmentation having $D_f = 1.6$. We argue later that further deformation of a fractal fragmentation can be expected to lead to a localization of the shear displacement within the fragmented zone, as observed in laboratory experiments by Marone and Scholz (1989).

Natural examples of fractal fracture networks just discussed can be classed as resulting from either fractal growth or fractal fragmentation. Networks of tensile fractures observed by Barton (1992) may be examples of fractal growth at small strains, since they do not fragment the rock and have a fractal dimension near 1.7. At the opposite extreme, cataclastic fragments in fault zones having D_f near 1.6 can be interpreted as examples of fractal fragmentation at large strains.

In Chapter 9, we focus on the mechanics of fractal fragmentation, beginning with a review of recent studies of the fragment size distribution of gouges in a natural fault zone, where the gouge has a fractal structure (Sammis and others, 1987; Sammis and Biegel, 1989). We then discuss a micromechanical model (Sammis and others, 1987; Biegel and others, 1989; Steacy and Sammis, 1991), which explains why fault gouge has a fractal structure and relates the fractal dimension to the mechanical fragmentation process. We also review a series of laboratory experiments in which initially monosized grain distributions were observed to evolve to fractal distributions with the theoretically predicted dimension (Biegel and others, 1989).

Finally we speculate on the possibility that micromechanics, which produce the fractal granulation observed in fault gouge, also apply to larger scale deformations of the brittle crust. A large-scale fractal fragmentation in crustal shear zones may explain regional power law earthquake statistics (King, 1983) and the generation of local earthquake coda by backscattering from fractal heterogeneities (Wu and Aki, 1985; Jin and Aki, 1988). The idea that the crust may deform by cataclastic flow is not new. Gallagher (1981) pointed out morphological similarities between crustal blocks and deformation belts in China and the microscopic structure of cataclastic rock. In Chapter 9, we extend Gallagher's observation to suggest that shear-induced fragmentation on a crustal scale may also be fractal with a dimension near 2.6 and this fractal geometry is related to the nested hierarchal structure observed in many well-developed crustal shear zones.

9.1.1. The Fragment Size Distribution of a Natural Fault Gouge

Sammis and others (1987) measured the fragment size distribution of fault gouge from the Lopez Fault zone in the San Gabriel Mountains of southern California. Undisturbed samples were impregnated with low-viscosity epoxy and sectioned to produce 2.5-cm diameter thin sections, about 100 μm thick. These sections were photographed in reflected light using optical and electron microscopy at magnifications ranging in factors of 2 from 12.5× to 1600×. Four representative photomosaics are shown in Fig. 9.1; the full set is published in Sammis and others (1987). It is important to point out that the area of each photomosaic was chosen so that the largest fragment in the field was on the order of about one-fourth the frame size. This procedure improves the counting statistics, but it introduces a sampling bias with consequences in tests for self-similarity discussed later.

The fragment size distribution was measured from the photomosaics using a transparent overlay on which five concentric circles were drawn. The diameter of each circle was a factor of 2 larger than the next smaller one. These circles were used to sort fragments on each mosaic into four binary classes as indicated in Table 9.1, which lists the number of fragments per unit area in each size class. A class $n = 1$ fragment is defined to have a diameter between the largest and second largest rings on the 12.5× mosaic. A class $n = 2$ fragment has a diameter between the second and third largest rings, and so on. If we define $N(n)$ as the number of class n fragments in a mosaic of area A, then the fragment density is $N(n)/A$.

9.1.2. Testing the Particle Distribution for Self-Similarity

Two methods were used to test the fragment distribution for self-similarity. The first test quantifies the idea that a self-similar distribution looks the same at all magnifications,

9. FRACTAL FRAGMENTATION IN CRUSTAL SHEAR ZONES

while the second tests the power law relation between fragment density and fragment size expected for a fractal distribution. To understand better how and why these tests work, it is useful to apply them first to the perfect fractal in Fig. 9.2. In this figure, triangular fragments are numbered according to size class to illustrate the sampling procedure. If we arbitrarily assume that the actual area of the entire triangular mosaic shown at 12.5× is 1 cm², then the measurement procedure previously described yields fragment densities given in Table 9.2a. Note that fragment density for a given size class fragment is different at different magnifications; this is an artifact of the sampling bias that was introduced by avoiding larger fragments in each mosaic rather than choosing sample areas at random.

The first test for self-similarity quantifies the idea that if the distribution is fractal, then it is impossible to determine the magnification at which the mosaic was prepared. In this case, the number of fragments with a diameter between a given pair of circles on the template should be the same independent of the magnification. To quantify this scale independence, we introduce a new parameter $N_A(n)$, which is the number of fragments in an area scaled to the mean fragment diameter of the nth class.

$$N_A(n) = \frac{N(n)[\overline{L(n)}]^2}{A} \tag{1}$$

where

$$[\overline{L(n)}]^2 = \frac{[L^2(n) + L^2(n+1)]}{2} \tag{2}$$

is proportional to the mean fragment area in each class. For the ideal fractal in Fig. 9.2, recall that we defined the area of the 12.5× mosaic to be 1 cm². The diameter of Circle 1 on the overlay, which circumscribes the Class 1 fragment, is therefore $L(1) = 0.878$ cm. The diameter of the Circle 2 is then $L(2) = 0.439$ cm, and so on. Equations (1) and (2) have been used to calculate $N_A(n)$ for this ideal fractal, and results are given in Table 9.2b. Because of self-similarity, $N_A(n)$ is the same for each pair of circles, independent of the magnification of the mosaic. Note that $N_A(n)$ decreases for higher order circles; this is another artifact of our biased sampling strategy.

In Table 9.3, $N_A(n)$ was calculated for the Lopez Fault gouge using measured fragment densities in Table 9.1. These data are plotted in Fig. 9.3, where it is apparent that the scaled density is independent of magnification for each pair of circles. By this criterion, the gouge is self-similar.

9.1.3. Measuring the Fractal Dimension

Self-similar distributions can be characterized by a fractal dimension defined as (Mandelbrot, 1982)

$$D_f = \frac{\log N(n)}{\log 1/L(n)} \tag{3}$$

For the ideal fractal in Fig. 9.2, $N(n)$ increases by a factor of 3 for each reduction of $L(n)$ by a factor of 2; hence the fractal dimension is $D_f = \log 3/\log 2 = 1.58$. Rewriting Eq. (3) as

$$\log N(n) = -D_f \log L(n) \tag{4}$$

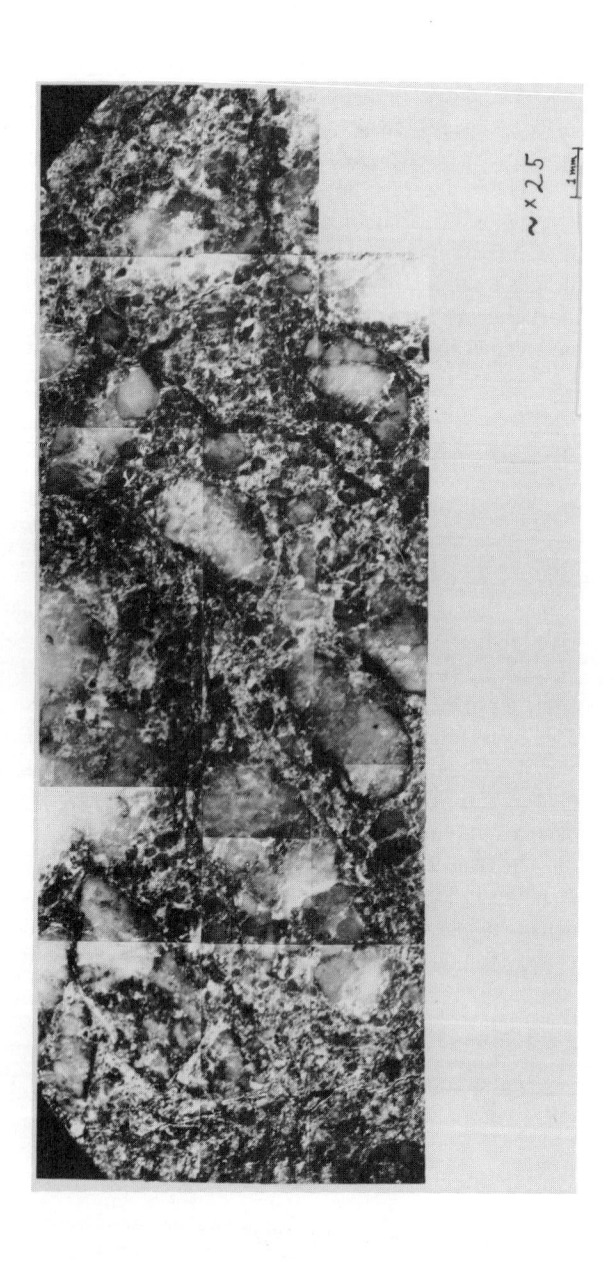

~×2.5

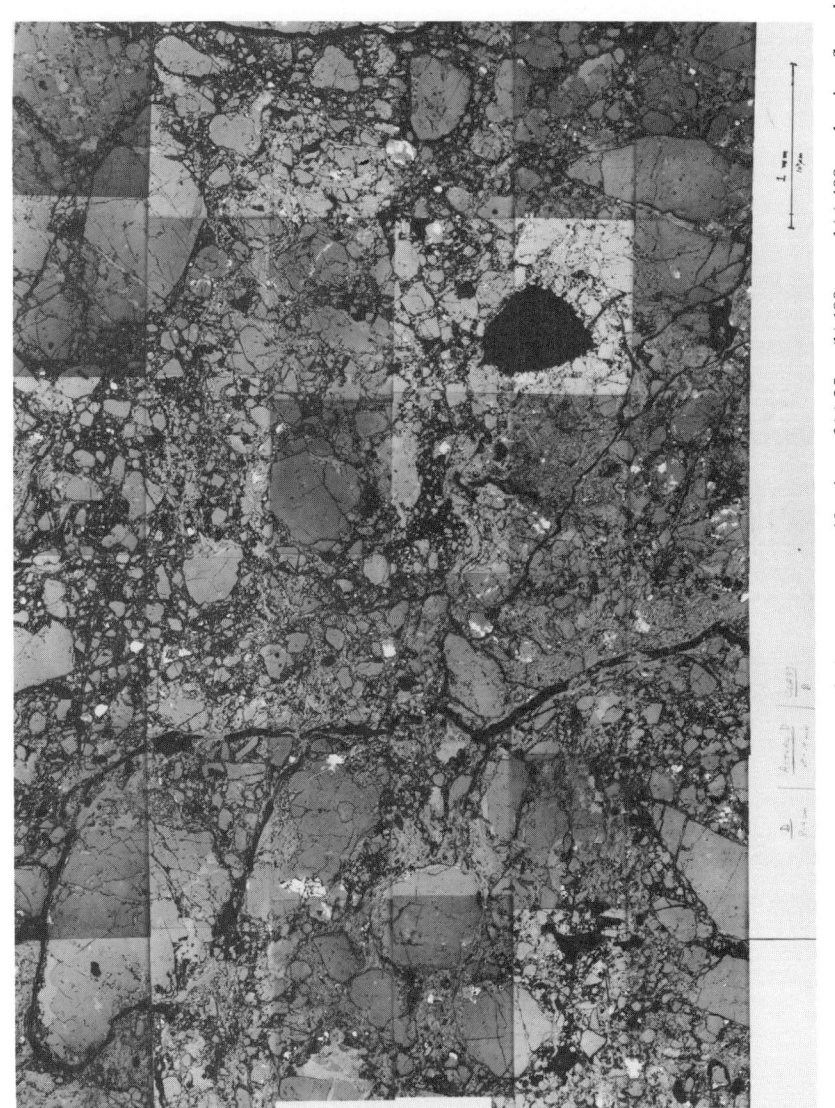

9. FRACTAL FRAGMENTATION IN CRUSTAL SHEAR ZONES

FIGURE 9.1. Photomosaics of the Lopez canyon fault gouge at magnifications of (a) 25×, (b) 100×, and (c) 400× taken in reflected light using an optical microscope and at (d) 1600× using a scanning electron microscope (Sammis and others, 1987). Note that the pictures look similar at all scales except 1600, where long splintery fragments are evident.

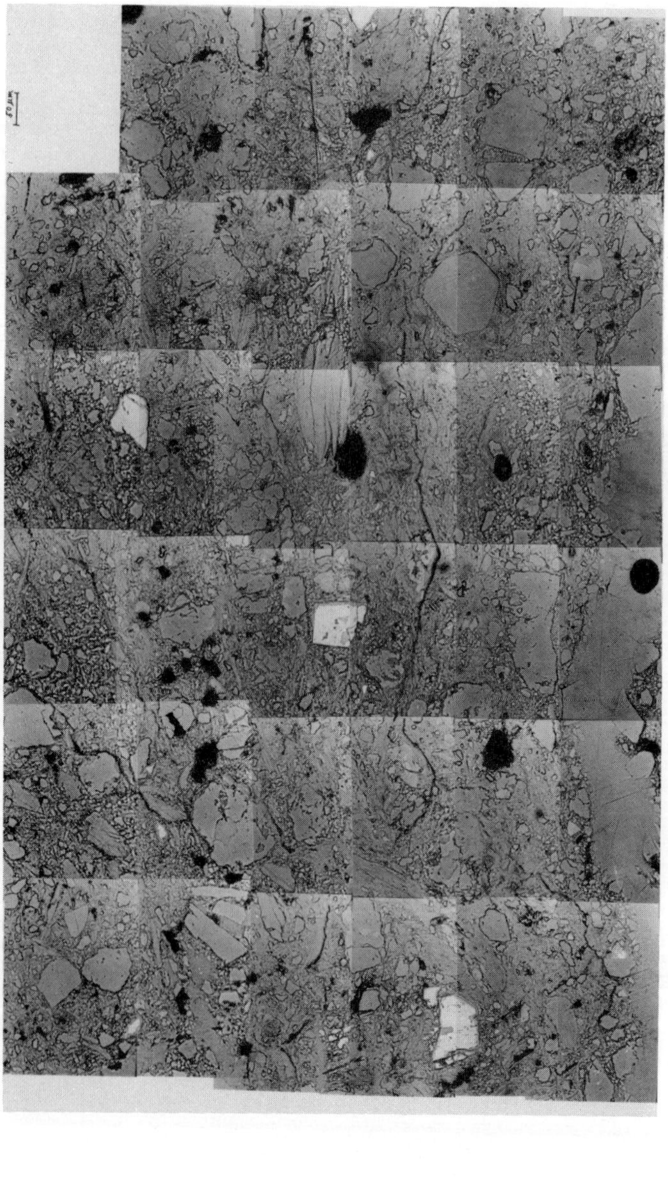

9. FRACTAL FRAGMENTATION IN CRUSTAL SHEAR ZONES

FIGURE 9.1. (*Continued*)

TABLE 9.1. Particle Densities for the Lopez Canyon Fault Gouge

Grain diameter (mm)	\multicolumn{7}{c}{Particle density in mm^{-2} measured at the following magnifications}
	12×	25×	50×	100×	200×	400×	800×	1600×
6.4–3.2	0.0026							
3.2–1.6	0.018	0.029						
1.6–0.8	0.054	0.090	0.040					
0.8–0.4	0.17	0.20	0.35	0.43				
0.4–0.2		0.58	1.14	1.15	1.62			
0.2–0.1			3.59	4.48	5.84	2.80		
0.1–0.05				14.1	16.2	20.0	26.3	
0.05–0.025					46.6	62.4	112.0	139.0
0.025–0.0125						157.0	286.0	300.0
0.0125–0.00625							825.0	993.0
0.00625–0.003125								3.37 × 10^4

it is apparent that the fractal dimension can be found as the negative of the slope on a plot of $\log N(n)$ versus $\log L(n)$. Such plots for the Lopez Fault gouge are given in Fig. 9.4. Note that D_f must be determined independently at each magnification because the sampling bias previously discussed precludes combining data from different photomosaics. Figure 9.5 compares values of D_f measured at each magnification. The mean of all the values is $D_f = 1.60 \pm 0.11$. This dimension is comparable to $D_f = 1.58$ for the ideal distribution in Fig. 9.2; we argue later that this dimension has an important mechanical significance.

In the discussion thus far, we have implicitly assumed that the true three-dimensional

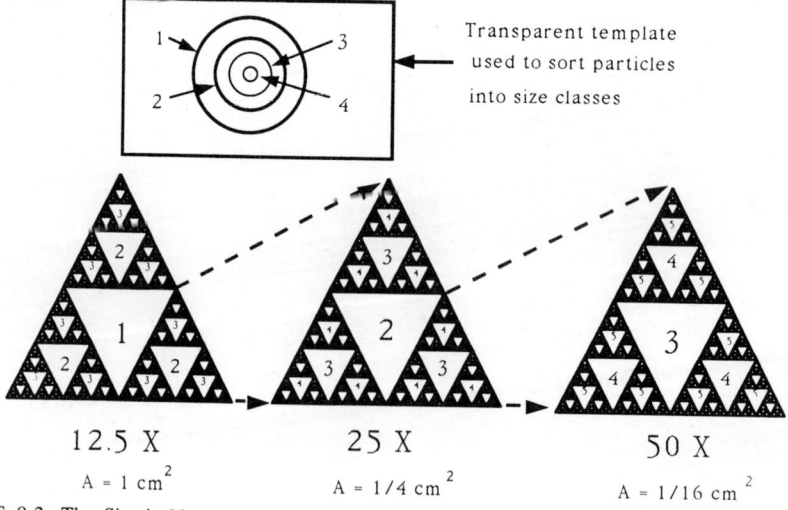

FIGURE 9.2. The Sierpinski gasket (two-dimensional fractal dimension of 1.58) illustrates the sampling technique used to test for self-similarity. The number of triangles that can be covered by a given circle (but not by the next smaller circle) is constant at each magnification for this self-similar figure.

9. FRACTAL FRAGMENTATION IN CRUSTAL SHEAR ZONES

TABLE 9.2.
(a) Particle Densities for the Sierpinski Gasket in Figure 2

Mag.	True area of the mosaic	class n = 1	class n = 2	class n = 3	class n = 4	class n = 5
12.5×	1*	1	3	9	—	—
25×	1/4	—	4	12	36	—
50×	1/16	—	—	16	48	144

(b) Scaled Particle Densities, $N_A(n)$, for the Sierpinski Gasket

			$N_A(n)$		
Mag.	class n = 1	class n = 2	class n = 3	class n = 4	class n = 5
12.5×	0.482	0.363	0.273	—	—
25×	—	0.482	0.363	0.273	—
50×	—	—	0.482	0.363	0.273

*The true area of the 12.5× mosaic is assumed to be 1 square unit.

fragment size distribution can be determined from a two-dimensional analysis of a planar section through the gouge. However a planar section through a random agglomeration of mono-sized spheres produces an apparent distribution of circles with diameters ranging from the true diameter to zero, depending on how the plane cuts each of the spheres (Underwood, 1968). Sammis and others (1987) showed that a planar section through a random, isotropic fractal distribution of spheres of dimension D_f produces a fractal distribution of circles with a fractal dimension one unit less than the dimension of the three-dimensional structure, that is

$$D_f(2D) = D_f(3D) - 1 \qquad (5)$$

TABLE 9.3. Scaled Particle Densities for the Lopez Canyon Fault Gouge

Grain diameter (mm)	Number in scaled area, $N_A(n)$, at the following magnifications							
	12×	25×	50×	100×	200×	400×	800×	1600×
6.4–3.2	0.066							
3.2–1.6	0.12	0.19						
1.6–0.8	0.087	0.14	0.064					
0.8–0.4	0.068	0.080	0.14	0.17				
0.4–0.2		0.058	0.11	0.12	0.16			
0.2–0.1			0.089	0.11	0.15	0.070		
0.1–0.05				0.088	0.10	0.13	0.16	
0.05–0.025					0.073	0.097	0.18	0.22
0.025–0.0125						0.062	0.11	0.12
0.0125–0.00625							0.081	0.097
0.00625–0.003125								0.082

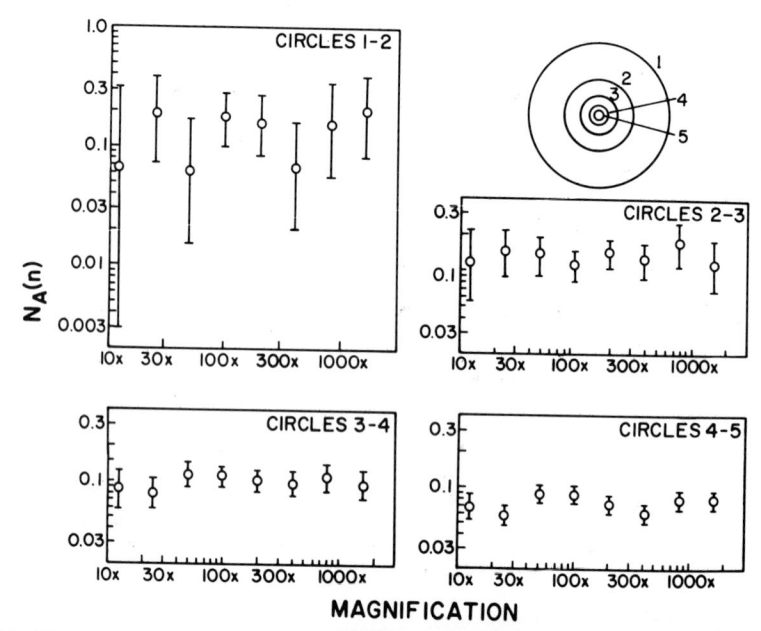

FIGURE 9.3. The number of particles in an area scaled to the particle diameter, defined as $N_A(n)$ in the text. Error bars indicate the 95% confidence level at which the correct particle density was measured, assuming the underlying spatial distribution is Poisonian. Smaller error bars for smaller circles reflect the greater number of smaller particles on each mosaic. Since $N_A(n)$ is constant for each pair of rings, this indicates that the distribution is self-similar over the range of particle sizes samples (Sammis and others, 1987).

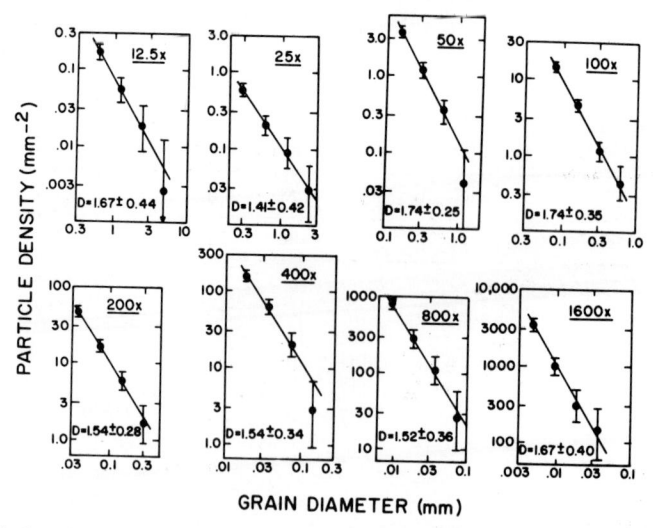

FIGURE 9.4. The logarithm of the particle density as a function of the logarithm of the particle size for the Lopez canyon fault gouge. The slope gives the fractal dimension. Each magnification is analyzed separately because of the sampling bias introduced by avoiding large particles at each scale (Sammis and others, 1987).

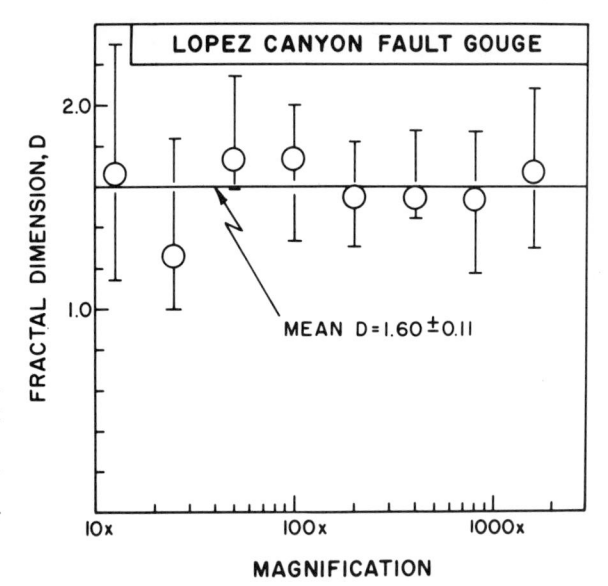

FIGURE 9.5. The fractal dimension determined at each magnification for the Lopez canyon gouge. Error bars represent the maximum and minimum slopes consistent with error bars in Fig. 9.4. The consistency of the fractal dimension measured at each scale is another indication of self-similarity over the range of particles sizes analyzed (Sammis and others, 1987).

As a specific example, consider the fractal distribution of cubes in Fig. 9.6. Note that there are six cubes with dimension L/2 for each cube of dimension L. Hence by Eq. (3), the fractal dimension is

$$D_f(3D) = \frac{\log 6}{\log 2} = 2.58 \qquad (6)$$

In the planar section on one of the faces, there are three squares with dimension L/2 for each square of dimension L. Hence the fractal dimension of a planar section is

$$D_f(2D) = \frac{\log 3}{\log 2} = 1.58 \qquad (7)$$

in agreement with Eq. (5). Note that Eq. (5) is strictly true only for a random isotropic fractal.

Equation (5) can be extended to a one-dimensional sampling of intervals between the fractures that bound the fragments. This distribution is also fractal with a dimension one less than that of the two-dimensional section.

$$D_f(1D) = D_f(2D) - 1 \qquad (8)$$

This can be demonstrated for the fractal distribution in Fig. 9.6 using a statistical approach. Suppose M holes are drilled in random positions on the top of the cube, as illustrated by the schematic drilling rigs in Fig. 9.6. The probability of intersecting a Class 1 cube is one-half, so the number of intervals of length L is $M/2$. Similarly the probability of intersecting just one Class 2 cube is one-fourth, intersecting two Class 2 cubes is one-fourth, and not interesecting Class 2 cube is one-half. The number of intervals of length $L/2$ is therefore $M/4 + 2M/4 = 3M/4$. The fractal dimension of one-dimensional intervals is therefore

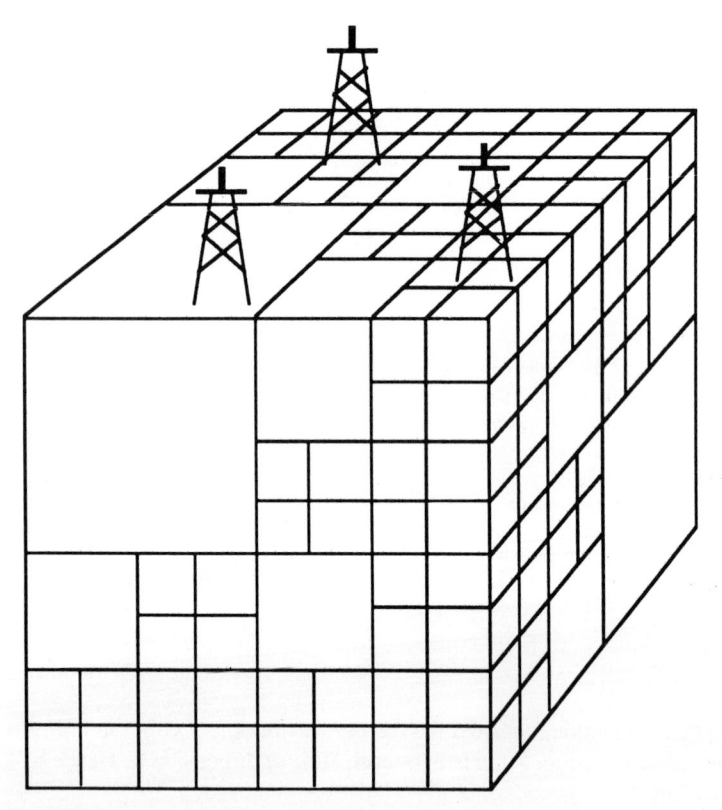

FIGURE 9.6. Fractal cube with dimension $D_f = 2.58$. Since the cube is an isotropic fractal, it has the property that the fractal dimension of any plane through it can be determined by subtracting 1. Hence the fractal dimension of any plane is $D_f = 1.58$. Similarly the fractal dimension of spacings between fractures in any random drill hole is $D_f = 0.58$.

$$D_f(1D) = \frac{\log(3M/4/M/2)}{\log 2} = \frac{\log 3/2}{\log 2} = 0.58 \qquad (9)$$

in agreement with Eq. (8). Again these relations hold only if the three-dimensional fractal is isotropic.

These relations were observed by Barton (1992), who found that the fractal dimension of intervals between fractures in a borehole was in the range of 0.6–0.8, while the dimension of two-dimensional patterns of fractures on related outcrops was in the range of 1.6–1.8.

9.1.4. Determining Fractal Limits

A physical system must have upper and lower size limits beyond which self-similarity breaks down. These are sometimes called the fractal limits. Sammis and others (1987) observed a change in the nature of comminution in the Lopez Fault gouge at fragment diameters on the order of 10 µm, which they identified as the lower fractal limit. Fragments larger than about 10 µm were approximately equant, while those smaller than 10 mm were

9. FRACTAL FRAGMENTATION IN CRUSTAL SHEAR ZONES

long and splintery. Mineral cleavage and intracrystalline porosity appear to play an important role in the comminution at this scale (see Fig. 9.1). It was apparent that further magnification would not produce similar-looking mosaics.

Sammis and Biegel (1989) extended the fragment analysis of the Lopez Fault gouge from a maximum diameter of 0.4 mm to 40 cm to search for the upper fractal limit. Flat circular molds with a diameter of about 15 cm were driven into the gouge formation, then used to make epoxy impregnated castings. These were brought to the laboratory, polished, and used to measure fragment densities over a range of diameters from 1.6–25.6 mm. Densities of fragments larger than 25.6 mm were measured by clearing a flat surface directly on the gouge deposit and counting *in situ*. All fragments were treated as if they were the largest in the field of view by subtracting the area of any larger fragments from the analysis area A. Results of this study are summarized in Fig. 9.7 in which new values of $N_A(n)$ are combined with those between Circles 1 and 2 from the previous study. Note that there is a sharp decrease in the scaled density at a fragment diameter on the order of 1 cm. Sammis and Biegel (1989) identified this as the upper fractal limit.

9.2.5. The Micromechanics of Constrained Comminution

Having established that the gouge from the Lopez Fault is self-similar with fractal dimension $D_f = 1.6$ between fractal limits 10 µm–1 cm, we now turn to the more interesting question of why it is fractal. Fragmentation processes do not necessarily produce a fractal distribution of fragments. Commercial crushing and grinding processes generally produce a Rosin–Rammler distribution of the form

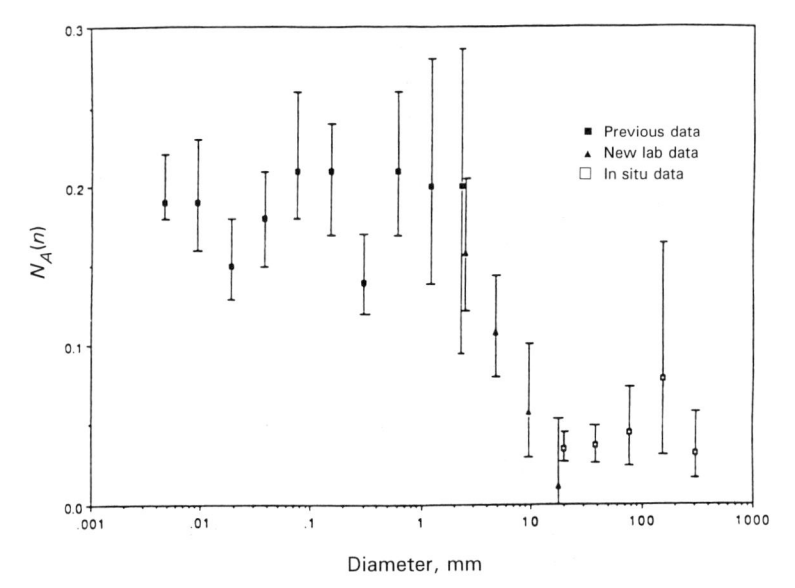

$$P_V(L) = 1 - e^{-L/l} \tag{10}$$

FIGURE 9.7. The scaled density of the Lopez fault gouge for particle diameters between 5 µm and 40 cm. The decrease in scaled density at a diameter of about 1 cm is identified as the upper fractal limit (Sammis and Biegel, 1989).

where $P_V(L)$ is the cumulative fraction of the initial volume corresponding to fragments with linear dimensions up to L (Prasher, 1987). Gilvarry and coworkers (Gilvarry, 1961; Gilvarry and Bergstrom, 1961; Gilvarry and Bergstrom, 1962; Gilvarry, 1964) showed theoretically and experimentally that the Rosin–Rammler distribution results when comminution is controlled by a Poisson distribution of initial flaw sizes in the starting material.

Epstein (1947) investigated the case where comminution is not controlled by the distribution of initial flaws but the fragmentation process itself is scale-independent in the following sense:

Fragmentation can be considered as composed of discrete steps or breakage events.

The probability of breakage of any piece during any step of the process is a constant independent of the size of the piece, the presence of other pieces, and the number of steps prior to a given step.

The distribution of pieces obtained from the application of a single breakage event to a given piece is independent of the dimension of the piece broken.

Epstein showed that under these three assumptions, any initial fragment distribution, no matter how skewed, eventually tends toward a log-normal distribution of the form

$$P_N(L) = \frac{1}{\bar{s}\sqrt{2\pi}} \int_{-\infty}^{f(L)} e^{-u^2/2} \, du \qquad (11)$$

where $f(L) = (\log L - \bar{m})/\bar{s}$, $\bar{m}$ and $\bar{s}$ are the mean and standard deviation, respectively, of logarithms of the fragment size distribution, and $P_N(L)$ is the cumulative fraction of the total number of fragments with linear dimension less than L.

Why does fault gouge follow neither a Rosin–Rammler nor log-normal distribution? The answer appears to lie in the way fragments are loaded. In industrial processes, such as pounding or tumbling, the volume of the fragmented mass is unconstrained. Each fragment is equally likely to suffer an impact, which means that the fracture probability of a given fragment is controlled by its strength. If fragment strength is independent of fragment size, then Epstein's theory is relevant, and a log-normal distribution is produced. If larger fragments are more fragile due to an initial Poissonian flaw distribution, then Gilvarry's theory is relevant, and a Rosin–Rammler distribution is produced. In a fault zone however, the volume of the fragmented mass is constrained. The stress field is compressive, so fragments are not free to change their relative positions without either additional fracturing or significant dilatational work. The stress field in the gouge is inhomogeneously concentrated along stress paths of contacts between fragments (Oda and others, 1982). In this environment, a fragment's fracture probability appears to be controlled by the geometry of its neighboring fragments that supply the load, and not by its size, mineralogy, nor other strength-controlling factor.

To understand how a fragment's fracture probability can be controlled by the geometry of its neighbors, consider the interaction of cylindrical fragments illustrated in Fig. 9.8. In Case a the shaded fragment is being loaded by two fragments the same size. Fragments loaded this way by a compressive force F fail by tensile splitting along the load axis. (This geometry forms the basis of the Brazilian strength test (Jaeger and Cook, 1979). The tensile stress is $\sigma_t = -2F/\pi L$. However at constant loading stress σ_1, the force applied to each

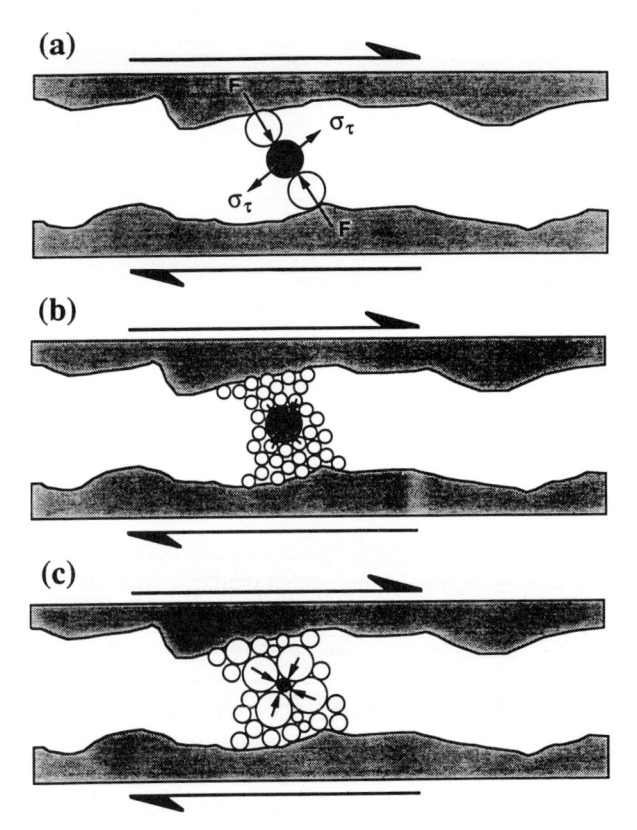

9. FRACTAL FRAGMENTATION IN CRUSTAL SHEAR ZONES

FIGURE 9.8. Schematic diagram illustrating the effect of the size of neighboring particles on the tensile stress developed within a particle. (a) The shaded particle is loaded in compression by two neighbors the same size. Tensile stress σ_τ in the shaded particle is a maximum in this case and independent of particle size. (b) The shaded particle is loaded by smaller particles. Tensile stress is reduced by nonaxial components of the applied force. (c) The shaded particle is loaded by larger particles, and tensile stress is again lowered by nonaxial force components.

fragment is proportional to its diameter L, so that $F = L\sigma_1$ and the tensile stress $\sigma_t = 2\sigma_1/\pi$ is independent of the fragment size. Now consider Case b in Fig. 9.8, the shaded fragment is loaded by smaller neighboring fragments. Although the net force along the stress path is the same, components orthogonal to the stress path thus reduce the tensile stress. The same is true for Case c, in which the shaded fragment is loaded by larger neighbors. Hence tensile stress is a maximum in fragments loaded by neighbors the same size. We propose that this neighbor-size effect overrides a fragment's intrinsic strength to dominate fracture statistics and produce the observed fractal distribution.

There is some experimental evidence to support this hypothesis. Following Schönert (1979), Steacy (1992) broke binary mixtures of glass beads in compression. As shown in Fig. 9.9, the percentage of 3-mm beads broken as the fraction of 1.5-mm beads in the mixture decreases. Since the probability of 3-mm beads being in contact with other 3-mm beads increases when their fraction in the mixture is greater, this supports the hypothesis that fragments are most likely to fracture when loaded by fragments of the same size.

If the fracture probability is maximum for neighboring fragments the same size (regardless of their size), then the distribution evolves toward a geometry in which no two fragments the same size are neighbors *at any scale*. The Sierpinski gasket in Fig. 9.2 and the array of cubes in Fig. 9.6 both have exactly this property: No two triangles (or cubes) the

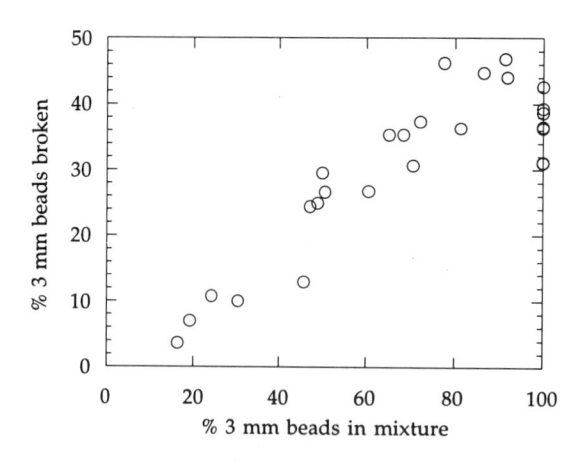

FIGURE 9.9. Percentage of 3-mm beads broken as a function of their fraction in mixtures of 3- and 1.5-mm beads. Note that a larger percentage of the 3-mm beads survive as they become increasingly surrounded by 1.5-mm neighbors (Steacy, 1992).

same size are neighbors (that is, share an edge in two-dimensional space or a face in three-dimensional space) at any scale except the smallest—below the lower fractal limit. These patterns also have the same fractal dimension of $D_f(3D) = 2.58$. Hence a fractal distribution with this dimension has the correct geometry for comminution controlled by the neighbor-size effect.

Although the deterministic fractals just described have the requisite nearest neighbor geometry, it is not obvious that a system which evolves according to a fracture probability based solely on the relative size of nearest neighbors will produce a *random* fractal with the same dimension. Steacy and Sammis (1991) tested whether a fractal dimension of $D_f = 2.6$ is the unique consequence of a fragmentation process that eliminates same-sized nearest neighbors at all scales. They constructed a computer automaton in two and three dimensions, that repeatedly applies simple rules to encode the physics of nearest neighbor comminution. They found that the automaton produced random fractals, but the fractal dimensions varied from 1.1–1.7 in two dimensions and 2.0–2.8 in three dimensions, depending on the packing geometry and the presence of long-range interactions imposed by uniform strain conditions. Figure 9.10 shows a typical random fractal produced by the automaton.

Turcotte (1986) documented many other processes that generate fractal fragment distributions with fractal dimensions in the range of 2.5–2.8. These include underground chemical explosions in mining operations, underground nuclear explosions, and such impacts as asteroid collisions. It can be argued that these are all examples of constrained comminution, since the fragments load each other with limited relative motion.

9.1.6. Generating a Fractal Gouge in the Laboratory

To test the preceding model for constrained comminution, Biegel and others (1989) deformed layers of crushed Westerly granite between sliding blocks of Westerly granite in the double-direct shear friction apparatus described by Dieterich (1978) (see Fig. 9.11). Each of the blocks was given a rough surface by hand lapping with #60 silicon carbide. Artificial gouge was prepared by crushing and sieving scrap material that remained after the blocks were cut. Three sample assemblies were prepared with a 3-mm-thick gouge layer

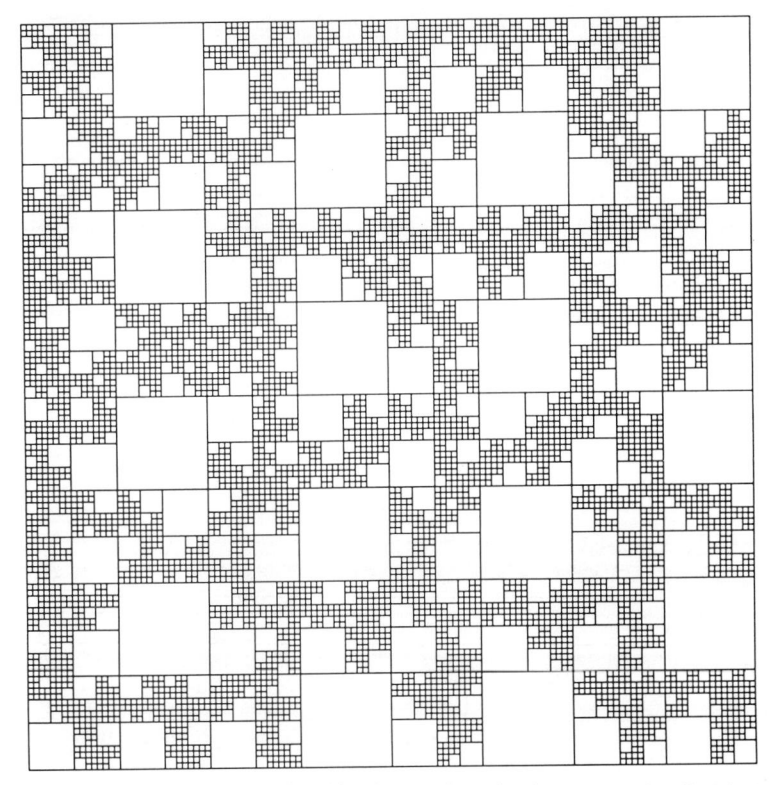

9. FRACTAL FRAGMENTATION IN CRUSTAL SHEAR ZONES

FIGURE 9.10. Typical fractal pattern produced by the two-dimensional automaton described in the text. The automaton began with an 8 × 8 array of identical squares, and fragmentation proceeded though five levels of fragment sizes. The fractal dimension of this random pattern is $D_f = 1.71$ (Steacy, 1992).

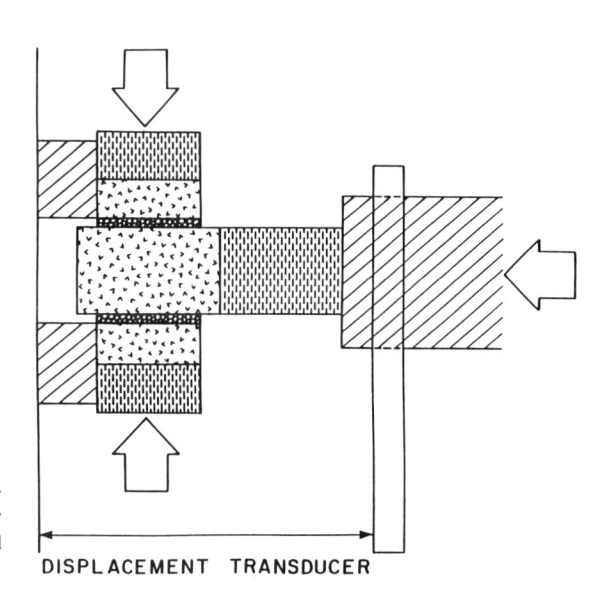

FIGURE 9.11. Double-shear friction apparatus used to produce a fractal gouge by a simple shear deformation (Biegel and others, 1989).

DISPLACEMENT TRANSDUCER

with a uniform initial fragment size distribution diameters and from 360–710 μm. Each sample was deformed at a normal stress of 10 MPa and a sliding speed of 10 μm s^{-1}. The first sample was stopped after a total displacement of 1 mm, the second after 5.5 mm, and the third after 10 mm. At the end of each run, the vertical (shear) load was first removed, then a clamp was tightened around the sample assembly to preserve the gouge structure before the horizontal (normal) load was withdrawn. The clamped three-block sandwich, with gouge intact, was then completely vacuum-impregnated with a low-viscosity epoxy.

Vacuum-impregnated sample assemblies were sectioned in the plane containing the simple shear strain, and thin sections were prepared. Photomicrographs were taken in reflected light and assembled into the photomosaics shown in Fig. 9.12. After 1 mm displacement, Fig. 9.12a shows that significant comminution has already taken place, with many of the largest grains reduced to smaller sizes. Note that many of the fragments appear

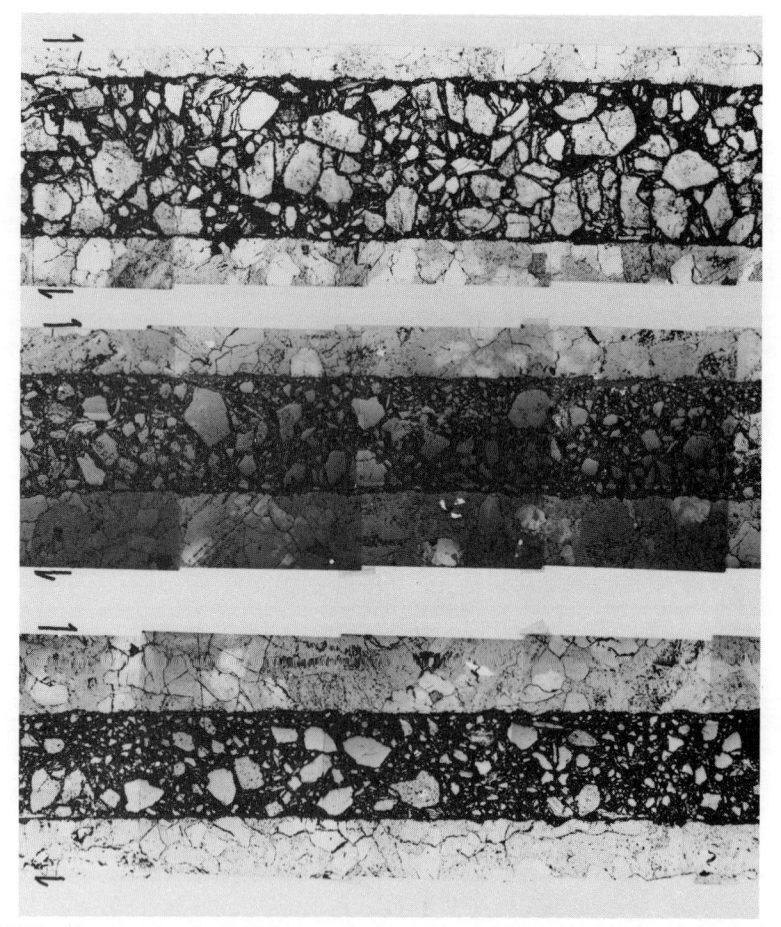

FIGURE 9.12. Photomosaics of an initially monosized gouge layer deformed in right-lateral simple shear. (a) The top layer, which is approximately 1.6 mm thick, was deformed by 1 mm relative displacement of the blocks, (b) the middle by a 5.5-mm displacement, and (c) the bottom, by a 10-mm displacement (Biegel and others, 1989).

9. FRACTAL FRAGMENTATION IN CRUSTAL SHEAR ZONES

to have failed in tension, presumably by loading under compression as previously hypothesized. Many fragments show tensile cracks radiating from compression poles. Smaller fragments formed from these radial cracks contribute to the high percentage of smaller fragments in the gouge layer. Intergranular porosity is still quite large. Figure 9.12b shows the gouge after 5.5 mm of slip. Very few fragments have nearest neighbors the same size, and the void space has been greatly reduced. A layer of smaller fragments is observed near the gouge rock surface; the homogeneity of comminution appears to break down near this interface. Figure 9.12c shows the gouge layer after 10 mm of displacement. Porosity has been further reduced, as well as the density of the largest grains. The fragment size distribution in Fig. 9.12c is plotted in Fig. 9.13. The gouge has a nearly ideal self-similar fragment distribution, with a fractal dimension of 2.6 over the range examined. This is the same dimension observed for the Lopez Fault gouge, and it is also the value predicted by the nearest neighbor comminution model. In fact comparing Fig. 9.12c with the natural fault gouge in Fig. 9.1 supports the hypothesis that the same comminution and deformation mechanisms are operating in these laboratory faults as in natural crustal faults.

9.1.7. Fractal Fragmentation and Shear Localization

Shear localizations are often observed in laboratory fragmentation experiments. For example, Marone and Scholz (1989) observed that shear localizations in triaxial experiments developed after a fractal particle size distribution was established. To see why constrained comminution may lead to shear localization, consider the evolution of a fractal fragmentation such as that illustrated in Fig. 9.12. Fracturing begins with the largest fragments, since all other factors being equal, these contain the largest microfractures and are therefore the weakest. In the beginning, many large fragments have same-sized neighbors, but as one member of each such pair is fractured, the remaining large fragments become increasingly isolated. When no neighboring pairs of large fragments remain, the process shifts to the next largest size fragments for which many neighboring pairs exist,

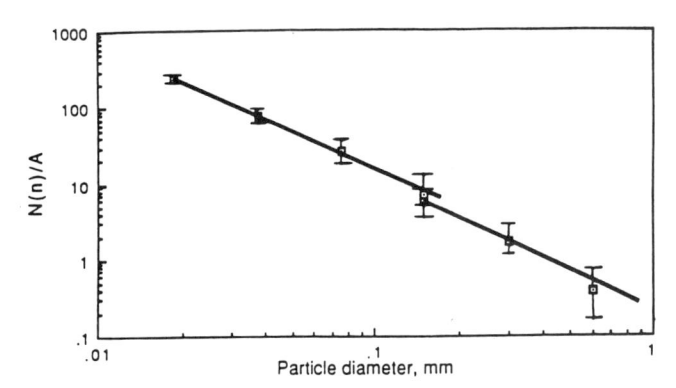

FIGURE 9.13. The logarithm of the particle density as a function of the logarithm of the particle size for the gouge layer in Fig. 9.12c, which was deformed by a 10-mm displacement. The slope gives a fractal dimension of 1.6 with a maximum variation of ±0.3. The offset is caused by using two different photomosaics having slightly different densities, and it is not significant (Biegel and others, 1989).

some of which were produced by fracturing larger fragments in the preceding step. Neighboring pairs at this scale are likewise eliminated until remaining fragments at this size are also isolated. The fragmentation process thus proceeds to ever smaller fragments until some fragmentation limit is reached. In commercial crushing, this is known as the grinding limit (Prasher, 1987). For oxides and silicates, it occurs when fragment size reaches about 3 μm. Smaller fragments are much stronger because they do not contain microfractures required to nucleate fracture.

When this lower limit is reached, fragmentation begins again at the scale of the largest fragments, since they are again the weakest, even though cushioned by smaller fragments. When a largest fragments fails, it produces a cascade of fragmentation at progressively smaller scales as pairs of smaller fragments are formed by larger fragments breaking and then eliminated through nearest neighbor comminution. If this region of enhanced comminution concentrates stress and thereby increases the fracture probability of nearby large fragments with respect to more distant ones, then a shear localization results.

Such a shear localization is shown in Fig. 9.14. A fractal pattern was first produced by the automaton (Steacy and Sammis, 1991) previously described. After same-sized nearest neighbors had been eliminated at all scales, a largest fragment was randomly chosen to

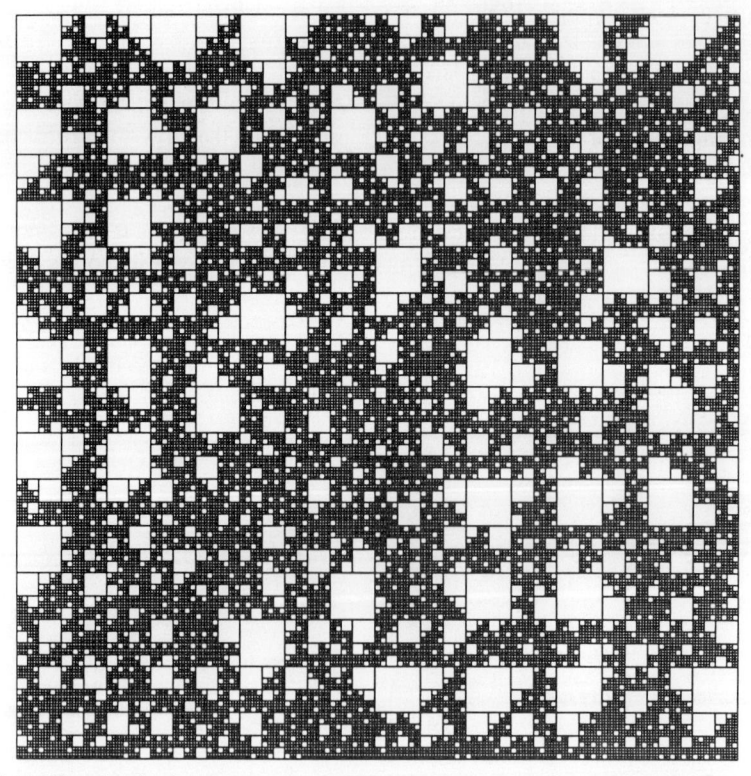

FIGURE 9.14. Shear localization produced in a random fractal using computer automaton. After all same-sized pairs were eliminated, a largest fragment was randomly chosen to fracture. Nearest neighbor comminution then proceeded at all scales. The resultant region of enhanced comminution biased the choice of the next large fragment to break. The end result is a through-going zone of finer fragments in a larger scale fractal fragmentation.

9. FRACTAL FRAGMENTATION IN CRUSTAL SHEAR ZONES 201

FIGURE 9.15. Localization observed in simple shear experiments by Biegel and others (1989). Localizations were observed only when the starting fragment size was very small—possibly because fine fragments were very close to the grinding limit, so that deformation localized at the relatively low strains achieved in the experiment.

fracture. Nearest neighbor comminution then proceeded at all scales, and this region of enhanced comminution biased the choice of the next fragment to break. The end result is a through-going zone of finer fragments in the larger scale fractal fragmentation. Similar localizations were observed by Biegel and others (1989) in experiments with initially very fine-grained gouge. Figure 9.15 shows such a shear band which developed when a 45-μm gouge was deformed in simple shear; note the similarities with Fig. 9.14. It is interesting that these localizations were observed only by Biegel and others (1989) when the starting fragment size was very small. This may be because fine fragments were very close to the grinding limit, so deformation localized at relatively low strains achieved in the experiment.

9.2. DISCUSSION

We now return to the question of whether processes of fractal fragmentation followed by shear localization observed in the laboratory also occur in the crust. Consider the overall structure of a well-developed crustal shear zone like the San Andreas system. Strain is spread over tens to hundreds of kilometers on a nested hierarchy of faults-within-faults. At the largest scale, the crust is broken into microplates or blocks (Anderson, 1971; Suppe and others, 1975; Smith, 1978; Hill, 1982; Sibson, 1986; Weldon, 1986). Regional deformation is dominated by major bounding faults, such as San Andreas, San Jacinto, Elsinore, Hayward, and Calaveras. A closer look at any one of these faults reveals a zone of shattered rock several kilometers wide and within which zones of finer gouge and concentrated slip can be observed (Davies, 1986; Deng and others, 1986; Sibson, 1986; Wallace and Morris, 1986). Gouge zones in turn often contain smaller scale structures of concentrated slip (Chester and Logan, 1986). Each nested localization is characterized by a sharp reduction in the size of its fragments relative to the granulation in which it is developed. Although fault gouge has been found to be fractal, the geometry of larger scale granulations has yet to be determined.

The nested hierarchal structure of shear zones may have important mechanical consequences. For example, although it is common to model faulting in terms of friction between planar surfaces, the complex structure of fault zones suggests that granular mechanics may offer a useful alternative to classical friction for investigating fault stability (Mandl and others, 1977). Although there is an extensive literature on granular mechanics in soil engineering (Scott, 1963; Atkinson and Bransby, 1978), soil mechanics generally takes a continuum approach in which measured stress–strain curves are used to predict critical states at which shear deformation localizes and becomes unstable. Those studies, which do look at the micromechanics on a granular scale, consider only the case when all grains are the same size; they do not consider the possibility of a reduction in grain size by fracture (Oda and others, 1982). An assessment as to whether the granular approach to faulting problems is either appropriate or useful depends on further investigations into the multiscale structure of fault zones and the micromechanics of fragmented materials.

ACKNOWLEDGMENT

This work was supported by NSF grant EAR-9105500.

REFERENCES

Anderson, D. L., *Sci. Am.* **225**, 52 (1971).
Atkinson, J. H., and Bransby, P. L., *The Mechanics of Soils, an Introduction to Critical State Soil Mechanics* (McGraw-Hill, New York, 1978).
Bak, P., and Tang, C., *J. Geophys. Res.* **94**, 15,635 (1989).
Barton, C. C., in *Fractals in the Earth Sciences* (C. C. Barton and P. R. LaPointe, eds.) (Geological Society of America Memoir, this volume, 1992).
Barton, C. C., and Hsieh, P. A., *Physical and Hydrologic-Flow Properties of Fractures* (American Geophysical Union Guidebook T385, 1989).
Biegel, R. L., Chester, F. M., and Evans, J. P., *Eos Trans. Am. Geophys. Union* **72**, 264.
Biegel, R. L., Sammis, C. G., and Dieterich, J. H., *J. Struct. Geol.* **11**, 827 (1989).
Chester, F. M., and Logan, J. M., *Pure Appl. Geophys.* **124**, 79 (1986).
Davies, R. K., *Pure Appl. Geophys.* **124**, 177 (1986).
Davy, Ph., Sornette, A., and Sornette, D., *Nature* **348**, 56 (1990).
Deng, Q., Wu, D., Zhang, P., and Chen, S., *Pure Appl. Geophys.* **124**, 203 (1986).
Dieterich, J. H., *Pure Appl. Geophys.* **116**, 790 (1978).
Epstein, B., *J. Franklin Inst.* **244**, 471 (1947).
Gallagher, J. J., in *Mechanical Behavior of Crustal Rocks*, Geophysical Monograph 24 (N. L. Carter, M. Friedman, J. M. Logan and D. W. Sterns, eds.) (American Geophysical Union, Washington, DC, 1981), pp. 259–73.
Gilvarry, J. J., *J. Appl. Phys.* **32**, 391 (1961).
———, *Solid State Commun.* **2**, 9 (1964).
Gilvarry, J. J., and Bergstrom, B. H., *J. Appl. Phys.* **32**, 400 (1961).
———, *J. Appl. Phys.* **33**, 3211 (1962).
Herrman, H. J., in *Random Fluctuations and Pattern Growth: Experiments and Models*, NATO ASI Series E, v. 157 (H. E. Stanley and N. Ostrowsky, eds.) (Kluwer Academic Publishers, Boston, 1988), pp. 140–60.
Hill, D. P., *J. Geophys. Res.* **87**, 5433 (1982).
Hirata, T., *Pure Appl. Geophys.* **131**, 157 (1989).
Jaeger, J. C., and Cook, N. G. W., *Fundamental of Rock Mechanics* (Halsted Press, New York, 1979).
Jin, A., and Aki, K., *Bull. Seis. Soc. Am.* **78**, 741 (1988).
King, G. C. P., *Pure Appl. Geophys.* **121**, 761 (1983).
Mandl, G., deJong, L. N. J., and Maltha, A., *Rock Mech.* **9**, 95 (1977).
Mandelbrot, B. B., *The Fractal Geometry of Nature* (W. H. Freeman, San Francisco, 1982).
Marone, C., and Scholz, C. H., *J. Struct. Geol.* **11**, 799 (1989).
Meakin, P., in *Random Fluctuations and Pattern Growth: Experiments and Models*, NATO ASI Series E, Vol. 157 (H. E. Stanley and N. Ostrowsky, eds.) (Kluwer Academic Publishers, Boston, 1988), pp. 174–91.
Oda, M., Konishi, J., and Nasser, N., *Mech. Mater.* **1**, 269 (1982).
Palmer, A. C., and Sanderson, T. J. O., *Proc. R. Soc. London A* **433**, 469 (1991).
Prasher, C. L., *Crushing and Grinding Process Handbook* (Wiley, New York, 1987).
Sammis, C. G., and Biegel, R., *Pure Appl. Geophys.* **131**, 255 (1989).
Sammis, C., King, G., and Biegel, R., *Pure Appl. Geophys.* **125**, 777 (1987).
Schönert, K., Proc. 4th Tewksbury Symp. Fracture, 17–18 Feb., U. of Melbourne, Australia 3:1-3:30, University of Melbourne (1979).
Scott, R. F., *Principles of Soil Mechanics* (Addison-Wesley, Reading, MA, 1963).
Sibson, R. H., *Pure Appl. Geophys.* **124**, 159 (1986).
Smith, R. B., in *Cenozoic Tectonics and Regional Geophysics of the Western Cordillera*, Geologic Society of America Memoir 152 (R. B. Smith, ed.) (Geophysical Society of America, 1978), pp. 111–44.
Sornette, D., in *Spontaneous Formation of Space Time Structures and Criticality*, Proceedings of the NATO ASI, Geilo, Norway, April 2–12 (T. Riste and D. Sherrington, eds.) (Kluwer Academic Press, Boston, 1991).
Sornette, D., Davy, P., and Sornette, A., *J. Geophys. Res.* **95**, 17,353 (1990).
Stanley, H. E., and Ostrowsky, N., *Random Fluctuations and Pattern Growth: Experiments and Models*, NATO ASI Series E, Vol. 157 (Kluwer Academic Publishers, Boston, 1988).

Steacy, S. J., "The Mechanics of Failure and Fragmentation in Fault Zones," Ph.D. diss., University of Southern California, 1992.
Steacy, S. J., and Sammis, C. G., *Nature* **353**, 250 (1991).
Suppe, J., Powell, C., and Berry, R., *Am. J. Sci.* **275**, 397 (1975).
Turcotte, D. L., *J. Geophys. Res.* **91**, 1921 (1986).
Underwood, E. E., in *Quantitative Microscopy* (R. T. Delioff and F. N. Rhines, eds.) (McGraw-Hill, New York, 1968).
Wallace, R. E., and Morris, H. T., *Pure Appl. Geophys.* **124**, 107 (1986).
Weldon, R., and Humphreys, E., *Tectonics* **5**, 33 (1986).
Wu, R., and Aki, K., *Pure Appl. Geophys.* **123**, 806 (1985).

10

Fractal Distribution of Fault Length and Offsets: Implications of Brittle Deformation Evaluation— The Lorraine Coal Basin

T. Villemin, J. Angelier, and C. Sunwoo

10.1. INTRODUCTION

The geometrical description of fault patterns in a given rock mass can be made through an estimation of three sets of quantitative parameters:

Local parameters related to a particular point of a given fault: average orientation of the local fault surface, fault slip vector, and so on. These parameters concern not only a single fault, and they are determined locally on its surface. Accordingly in maps or sections, their values change along the trace of the fault.

Comprehensive parameters, which characterize the whole of a given fault: average orientation and average offset, intersection length with a given surface (e.g., the topography), surface affected by fault displacement, shape of equal offset curves, and so on. These parameters, which are estimated for a single fault as the local parameters, may account for the geometry of the fault surface entirely.

Network parameters, which characterize a set of faults: fault density, fractal dimension, and so on. These parameters are estimated on a complete set of faults. This complete set may have been observed along a line (e.g., drill hole or gallery), on a surface (e.g., pavement or seam), or in a volume (e.g., quarry or small-scale sample).

T. Villemin • Laboratoire de Géologie Structurale et Appliquée, Université de Savoie, Chambery, France. *J. Angelier* • Laboratoire de Tectonique Quantitative, Université P. & M. Curie, Paris, France. *C. Sunwoo* • Korean Institute of Energy and Resources, Daejeon, Korea.

Fractals in the Earth Sciences, edited by Christopher C. Barton and Paul R. La Pointe. Plenum Press, New York, 1995.

More accurate definitions of some of these parameters are given later in Chapter 10. Numerous attempts have been made in various geological contexts, to estimate one or more of these descriptive parameters (Cailleux, 1958; Hervé and Cailleux, 1962; Kakimi and Kodama, 1974; Muraoko and Kamata, 1983; Villemin and Sunwoo, 1987; Watterson, 1986; Walsh and Watterson, 1987). Some of these approaches aimed not only at building three-dimensional models of fault networks but also at finding predictive laws (Chilès, 1988, 1989; Thomas, 1986, 1987; Turcotte, 1986a, b). Effectively a major interest of such studies lies in the possibility of extrapolating local descriptions to unknown areas. Results of geometrical descriptions are also used in mechanical approaches to faulting (Ferguson, 1985; Gauthier and Angelier, 1985; Chelidze, 1986; Gibowicz, 1986; Scholz and Cowie, 1990; Marrett and Allmendinger, 1990; Davy and others, 1990).

Chapter 10 presents some distribution laws of faults parameters. These laws are based on the example of the Lorraine coal mines (see Fig. 10.1), where numerous data were collected (Sunwoo, 1988), but comparisons with other sources of data suggest that these laws are of more general value. Some consequences of brittle deformation are discussed. The choice of the study area was based on the following considerations:

An excellent knowledge of the geology of the coal basin, where numerous data from drill holes, seismic lines, and especially underground field observations were available (Pruvost, 1934; Donsimoni, 1981)

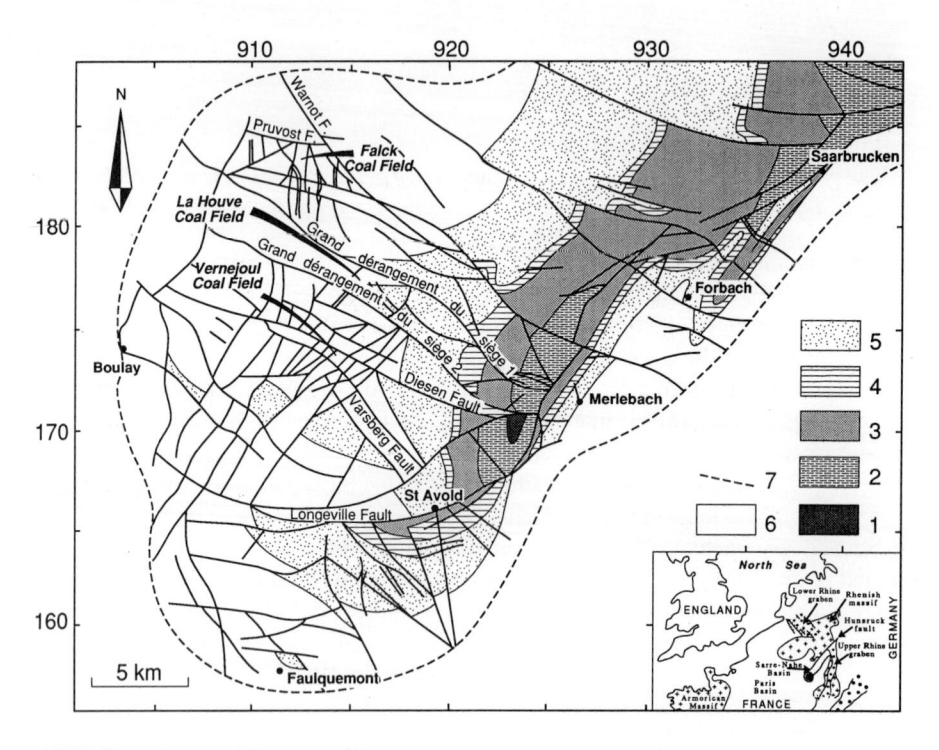

FIGURE 10.1. Structural map of the Lorraine coal basin (top of the carboniferous). (1) B-Westphalian; (2) C-Westphalian (lower part); (3) C-Westphalian (upper part); (4) Merlabach conglomerate; (5) D-Westphalian; (6) Stephanian; (7) limit of worked coal basin.

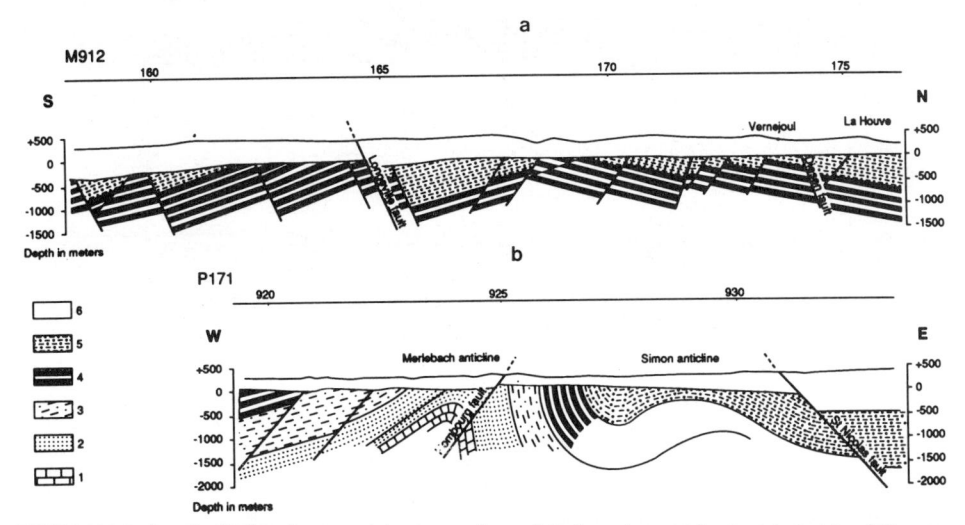

10. FRACTAL DISTRIBUTION OF FAULT LENGTH AND OFFSETS 207

FIGURE 10.2. Longitudinal and transversal cross sections of the Lorraine coal basin, (a) meridian 912 cross section; (b) parallel 171 cross section (see location on Fig. 10.1). (1) B-Westphalian; (2) C-Westphalian (*lower part*: Neunkirchen seams); (3) C-Westphalian (*upper part*: Forbach seams); (4) D-Westphalian; (5) Stephanian; (6) upper Permian and lower Trias.

The presence of relatively simple tectonic features, particularly in the northwestern part of the basin, which corresponds to a gentle monocline cut by large antithetic normal faults (see Fig. 10.2)

Abundant and accurate information on fault patterns observed at different scales during mining exploration and working, because surveyors of the Lorraine Coal Mines Organization carefully recorded structural data in maps (see Fig. 10.3) and sections

We first present the structural context of the Lorraine coal mines and the data collection and second the empirical distributions laws that can be established. In the third part, we discuss the consequences of these laws for understanding the overall brittle deformation.

10.2. GENERALIZED GEOLOGY

The Lorraine coal field (see Fig. 10.1) is the southwestern extension of the late Paleozoic Sarre-Nahe basin, which is widely exposed in adjacent Germany (Purvost, 1934). This basin is filled with Permo-carboniferous clastic sediments more than 5000 m thick in the central part (Falke, 1976; Schäffer, 1986). The regional trend of the deposition area is northeast-southwest, and its northwestern limit is the Hunsruck border fault.

The stratigraphic sequence ranges in age from the Westphalian to the Lower Permian; it includes many coal seams (Teichmüller and others, 1983) interbedded in shales, sandstones, and conglomerates (Schäfer, 1986). These coal deposits characterize rhythmic accumulation during postorogenic Variscan times (Lorenz and Nicholis, 1976). In Lorraine coal-bearing layers are covered by approximately 200–300 m of sandstones, Upper

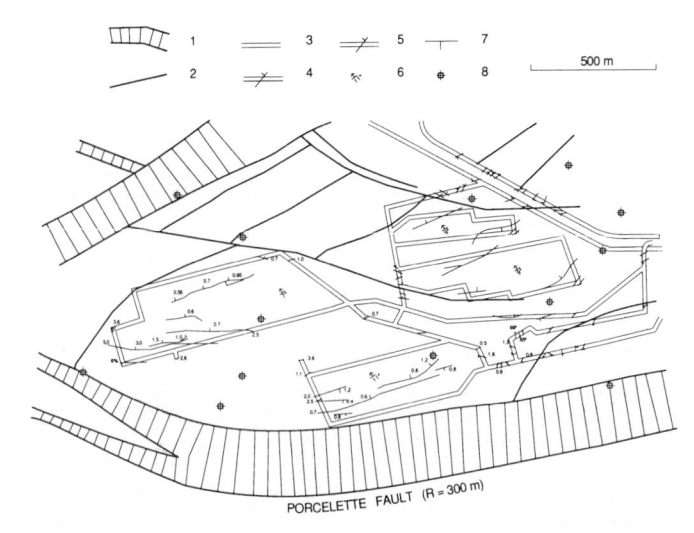

FIGURE 10.3 Faults on mine-detailed maps (*horizontal projection*). (1) Major fault (the hatched area corresponds to the horizontal displacement of the coal seam); (2) local important fault (generally more than a 10-m vertical offset); (2) worked gallery (all worked coal seams are boundered by galleries); (4) fault crossing a gallery with indication of dip direction and vertical offset; (5) fault crossing a gallery with indication of dip direction and dip; (6) dip direction and dip of the coal seam; (7) fault trace on a worked coal seam with indication of dip direction and vertical offset; (8) drill hole.

Permian and Triassic in age (see Fig. 10.2). Contrary to the Sarre-Nahe basin, the Lower Permian is missing; Upper Permian sediments uncomfortably overlie carboniferous formations through a structural and erosional unconformity. Due to tectonism and erosion during the Saxonian, the sedimentary environment of the Lower Permian remains unknown; this period corresponds to the development of the major structures considered in Chapter 10.

The main fold structures in the Lorraine basin trend northeast-southwest: The Simon anticline is the southwestern extension of the Sarre anticline; the Merlebach anticline is located to the southwest as an en echelon extension of the Simon anticline (see Figs. 10.1 and 10.2). In contrast in the western portion of the Lorraine basin, an extensional structure prevails (see Fig. 10.1): Gently dipping faulted-tilted blocks are bounded by large normal faults (see Fig. 10.2a; Villemin, 1987). Data used in Chapter 10 were collected in this western domain (see Fig. 10.1).

The tectonic history of the Lorraine basin is characterized by two contrasting states of stress that prevailed during the Early Permian (Donsimoni, 1981):

A northwestern-southeastern compression, which is probably related to the Saalian event of the Variscan Orogeny, the fold development in the southeastern part of the basin occurred during this event.

A north-south extension related to a brittle deformation with moderate but widespread block faulting and tilting; the stretching factor in the basin, as estimated from balanced cross-sections and other tectonic observations, averages 10% (Villemin, 1987). It is worthwhile to note first that the tectonic features quantitatively analyzed in Chapter 10 developed during this north-south extension and second these extensional phenomena played a major role in Western Europe during the Permo-carboniferous (Lorenz and Nicholis, 1976).

10.3. DATA COLLECTION

In the Lorraine coal basin, information on fault networks is available on maps in two different types of scale, typically 1/1000–1/5000 and 1/10,000–1/50,000 (Villemin, 1987).

Mining plans are first drawn by surveyors at the 1/1000 scale; an example of such a plan is shown in Fig. 10.3. These plans contain complete information on underground works as projected in the horizontal plane (galleries, coal seams with work in progress or worked out). This geological information includes two complementary data sources: the geometry of faults intersected by gallery axes (maps indicate the positions of faults as well as their dip directions, dips and offsets where they cross galleries) and the complete fault traces for each zone of coal seam worked out:

Both types of data provide access to the orientation, offset, and density of faults. Note that because numerous galleries are rather long (500 m or more), spacing distributions are conveniently studied along galleries, whereas fault lengths are estimated only in mapped seams.

Mapped coal seams correspond to areas where the amount of deformation remains low (so that mechanized mine working is technically possible). As a result, all offsets observed along fault traces are small (0.1 to a few meters), so this information is biased. In contrast galleries are not selective with respect to faults; effectively they cross large faults as well as small ones (0.1–100 m of vertical offset).

Second structural maps have been made at scales of the entire Lorraine basin (1/50,000; see Fig. 10.1) or the coal field (1/10,000). These maps are partly interpretative, but they combine accurate observations from the field, drillholes, and seismic lines. Contrary to local maps discussed before, these maps show major faults.

For this study, data acquisition in mining plans and sections was made using both digitizer and computer. The complete data set was recorded on several files linked by database relationships. Special programs were created to analyze these data files (Sunwoo, 1988).

The following three mining areas were examined in detail, and these provided excellent opportunities for data collection (see Fig. 10.1):

The Falk coal field is located in the northwestern part of the basin. Only a few coal seams, Stephanian in age, have been exploited. Numerous normal faults affect this field, which is bounded by the Warndt and Siège 1 major faults.

The La Houve coal field, probably the most extensively worked area, is located between and on both sides of the Siège 1 and Siège 2 normal faults. More than ten coal seams have been worked out until 1968, at depths from 200–1,000 m, mostly in the Westphalian D.

The Vernejoul coal field is presently exploited near the Diesen and Varsberg normal faults.

10.4. DATA DISTRIBUTION AND RELATIONSHIPS AND THEIR IMPLICATIONS

10.4.1. Orientation of Faults

The complete set of faults observed in each detailed study area is easily split into two sets of conjugate normal faults, with westnorthwest-eastnortheast and southwest-northeast strikes (see Fig. 10.4). In addition, underground field observation showed that most slickenside lineations have a strong dip slip component.

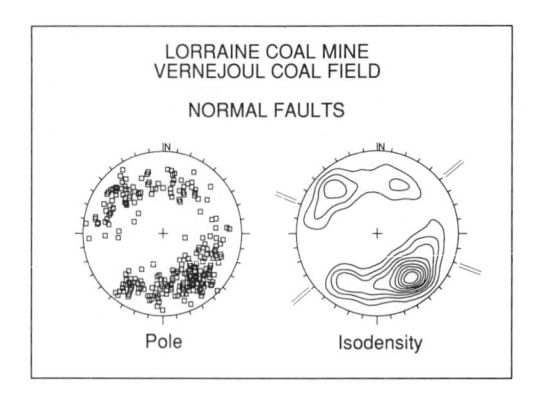

FIGURE 10.4. Poles to normal faults systems (Schmidt projection, lower hemisphere). All plotted data were collected in the Vernejoul coal field.

These two systems of conjugate normal faults can be interpreted either in terms of two distinct extensional tectonic stresses, with northwest-southeast and northnortheast-southsouthwest trends, respectively, or as a single pattern of orthorhombic normal faults, as described by Reches (1978). In the latter case, the trend of extension is approximately north-south.

Chronological relationships between the two conjugate systems have been examined (Villemin, 1987): The northeast-southwest striking normal faults were sometimes found to be older, sometimes more recent than the westnorthwest-eastsoutheast ones. For this reason, we consider that a single north-south distension probably dominated during the Permian.

In any case, the analysis of fault mechanisms related to fault slip orientations demonstrates that the complete data set corresponds to no more than two paleostress regimes, which belong to a single major extensional event. Due to this relative simplicity, the total data set can be considered rather homogeneous.

10.4.1. Fault Model

Here we discuss the two-dimensional shape of a single planar fault and the distribution of net separations within its surface to define a simple planar fault shape model. Such models are generally based on ideal schemes (Watterson, 1986; Walsh and Watterson, 1987); actual reconstructions of fault-offset geometry are scarce due to practical problems in data collection (e.g., Cruden, 1977; Muraoka and Kamata, 1983; Archuleta, 1984; Rippon, 1985; Barnett and others, 1987). Because in some cases a single fault was mapped at various levels (superposed seams), we can reconstruct the model in Fig. 10.5, which we consider representative at least in the Lorraine coal basin.

The most characteristic features of this model are (1) the rectangular shape of the approximately planar fault surface and (2) the rapid increase in offset near the fault limits, in contrast with the relative invariability of offset within most of the fault surface. Surprisingly this offset was found to be more or less constant and close to the maximum offset value inside the 90% displacement contour (see Fig. 10.5).

Note first that this model has been defined at the scale of faults mapped in mining plans (typically 10–100 m) and second outside of the 0% displacement contour line, the offset is not null but simply smaller than a minimum value related to the accuracy of the mining surveys (i.e., approximately 10 cm; see also Walsh and Watterson, 1987), for correction of

10. FRACTAL DISTRIBUTION OF FAULT LENGTH AND OFFSETS

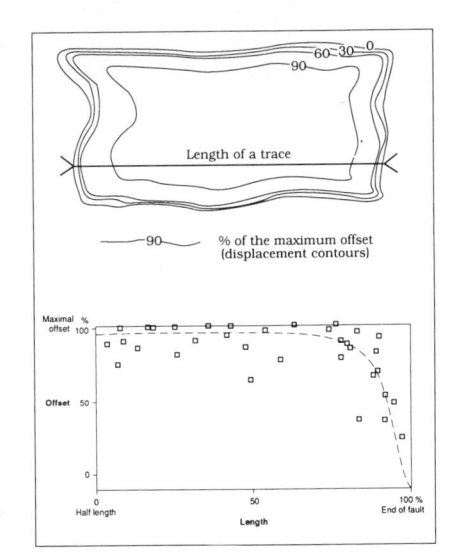

FIGURE 10.5. Schematic contoured displacement curves on a fault surface (see explanations in text).

fault length). Note also that the model accounts for dip slip normal faults in a conspicuously layered rock mass and may not be valid in different geological settings (e.g., magmatic rocks).

10.4.2. Lengths Distribution

The simple model in Fig. 10.5 implies that fault length does not change significantly according to the observation depth, so that the resulting bias in data collection for fault lengths can be considered negligible.

The cumulative distribution of fault lengths is shown on Fig. 10.6. In this diagram, the Y-axis indicates the number $N_L(L)$ of fault lengths larger than the value L on the X-axis. Data plotted in Fig. 10.6 correspond to observations at different scale. The fault length distribution obviously depends on the scale of the map where the faults are drawn. For each scale, the following parameters are defined:

Two limit lengths (lower and upper Ψ' and Ψ); in the studied domain, no fault with its length smaller than the lower limit length was recorded. This choice was made by mine surveyors according to the minimum size of faults they were interested in, also size of the studied domain ascribed a maximum size to faults (see Fig. 10.6).

Two critical lengths (lower and upper Φ and Φ'); on their maps, mine surveyors represented some faults whose lengths are smaller than the lower critical length but not all. In contrast they represented all longer faults. Also because some faults have extensions outside of the mine-worked area, their exact lengths are unknown, and lengths actually observed should be considered with care in statistics. For this reason, we must define an upper critical length.

The lower critical length is called Φ, and the upper limit length is called Ψ hereafter. These two limits, which respectively depend on the survey accuracy and the size of the mapped area, play a major practical role in our quantitative analysis (see Fig. 10.6).

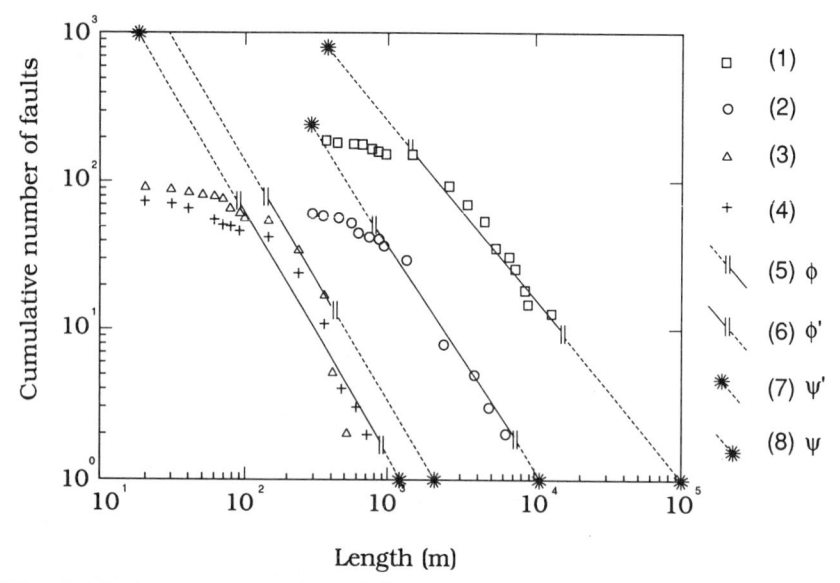

FIGURE 10.6. Distribution of fault lengths. Data from the general map of the (1) Lorraine coal basin; (2) the Vernejoul coal field; (3) the La Houve coal field, seam Marie; (4) and seam François; (5) lower critical length Φ; (6) upper critical length Φ'; (7) lower limit length Ψ'; (8) upper limit length Ψ.

Figure 10.6 shows an example of length–number relationships for about 500 faults (four sets at three different scales). In this case as well as in others not described here (Villemin and Sunwoo, 1987), we can establish the following relationship with a good correlation rate within the range of critical lengths (the straight portion of the curve ranges between Φ and Φ'):

$$N_L(L) = A_L L^{-C} \tag{1}$$

where A_L and C are constants and $N_L(L)$ is the number of faults with length greater than L. The deviation from linearity can be explained by inadequacies in sampling. Figure 10.6 also shows that within the complete range of fault lengths under investigation (i.e., 10–10,000 m) the parameter C is constant (i.e., the slopes of rectilinear segments are similar for the four sets; C averages 1.4). We conclude that the distribution of fault lengths is self-similar for the scales and length range considered.

The parameter A_L has a local significance as an increasing function of the area and the fault density. This parameter is given as a function of a reference length L (Curl, 1986) provided that the value of $N_L(L)$ is accurately known (this is the case for Ψ or Φ)

$$A = \frac{N_L(L)}{L^{-C}} \quad \text{or} \quad A = \frac{N_L(\Phi)}{\Phi^{-C}} \quad \text{or} \quad A = \frac{1}{\Psi^{-C}} \tag{2}$$

Combining Eqs. (1) and (2), we obtain

$$N_L(L) = N_L(\Phi)\left(\frac{L}{\Phi}\right)^{-C} \tag{3}$$

Equation (3) can be extrapolated for lengths smaller than Φ provided we consider the total area where $N_L(\Phi)$ and Φ have been estimated. This relationship however should neither be transferred to other areas nor applied to smaller subareas, due to related variations in parameter A_L.

Let us apply Eq. (3) for Ψ. [Note that $N_L(\Psi) = 1$]. Thus we obtain Eq. (4) between Φ and Ψ

$$\Psi = [N_L(\Phi)]^{-C}\Phi \tag{4}$$

Using Eq. (4), we can determine the length of the longest probable fault totally or partly included in the area studied.

Differentiating Eq. (3) results in Eq. (5)

$$n_L(L) = \frac{-dN_L(L)}{dl} = B\frac{N_L(\Phi)}{\Phi}\left[\frac{L}{\Phi}\right]^{-(C+1)} \tag{5}$$

where $n(l)$ is the fault length probability density function for length l in the area studied. Thus the total length of faults whose lengths lie between L_1 and L_2 is given by Eq. (6)

$$L_T(L_1,L_2) = \int_{L_1}^{L_2} Ln(L)\,dL = \frac{C}{C-1}N_L(\Phi)\Phi^C[L_1^{1-C} - L_2^{1-C}] \tag{6}$$

Substituting $L_1 = \Phi$ and $L_2 = \Psi$ (with $\Psi \gg \Phi$) into Eq. (6) yields Eq. (7) for $1 < C < 2$

$$L_T(\Phi,\Psi) = \frac{C}{C-1}N_L(\Phi)\Phi \tag{7}$$

However, by differentiating and subsequently integrating Eqs. (6–7), which has been used by Scholz and Cowie (1990), we may underestimate the total fault length. As discussed by Marret and Allmendinger (1991), this underestimation results from the assumption that N_L is a continuous function, which is not the case. According to the technique proposed by Marret and Allmendinger (1991) for earthquake seismic moments, a stable estimate of the total fault length is given by Eq. (8)

$$L_T(\Phi,\Psi) = N_L(\Phi)^{-C}\Phi \sum_{1}^{N_L(\Phi)} \frac{1}{(i)^{1/C}} \tag{8}$$

We observe that both techniques resulting in Eqs. (7) and (8) are based on strong assumptions: continuity of N in the first case (Eqs. 5 and 6 are approximations and assume that dL is infinitesimally small) and more constraining distribution of lengths in the second case. A third technique does not involve such assumptions, but it cannot provide an analytical formula, it consists of calculating the total length by a purely numerical method directly from each set of length data.

Equation (7) or (8) enables us to determine the average length $L_M(\Phi)$ of faults whose lengths are greater than Φ [see Eq. (9)], as well as the corresponding fracture density $D(\Phi)$ in the area S under investigation [see Eq. (10)]. Note that we define fracture density as total fracture length $L_T(\Phi,\Psi)$ divided by area S.

$$L_M(\Phi) = \frac{C}{C-1}\Phi \tag{9}$$

$$D(\Phi) = \frac{C}{C-1} N_L(\Phi) \frac{\Phi}{S} \tag{10}$$

From Eqs. (2) and (10), we obtain the parameter A_L defined in Eq. (1) as a function of $D(\Phi)$ and S.

$$A_L = \frac{C-1}{C} \Phi^{C-1} D(\Phi) S \tag{11}$$

Eq. (11) shows that A_L increases linearly with the area studied and the fracture density. As a result, Eq. (1) determined in a given area can be extrapolated to other areas if the value of A is modified according to Eq. (11).

10.4.3. Offsets Distribution

Before analyzing the distribution of offsets, it is worthwhile to examine how far the collected offsets, measured where faults are visible, reflect the average offsets of these faults. As discussed before and suggested in Fig. 10.5, local offset is generally close to the maximum offset (and also to the average offset), so that the bias is minor (see also Marrett and Allmendinger, 1990). In particular the observed and average displacements where a gallery cuts a fault do not significantly differ in a statistical way.

Offset values recorded in mining plans describe the local vertical separation, based on identifying layers on both sides of each fault. Because most faults are normal dip slip in the area studied, there is little difference between separation values thus recorded and actual vertical offsets.

The cumulative distribution of fault offsets is shown in Fig. 10.7. As for fault length data previously discussed, offset data have been collected at different scales. At each scale, we similarly define the two following parameters:

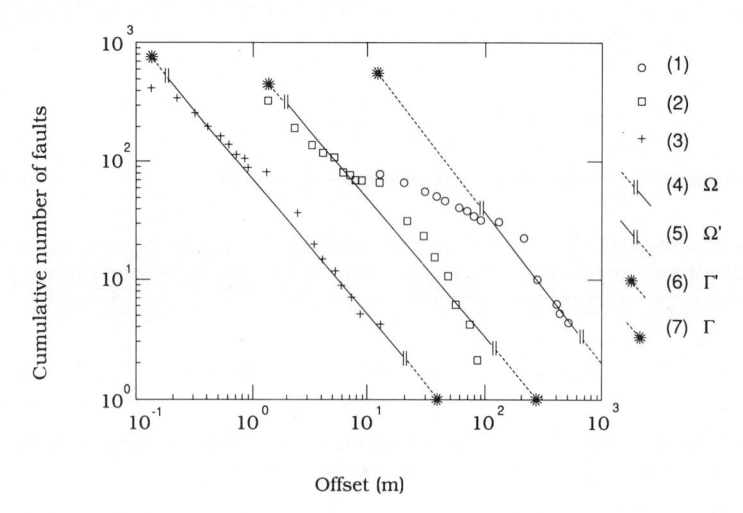

FIGURE 10.7. Distribution of fault offsets (see explanations in text). Data from the general map of the (1) Lorraine coal basin; (2) the Vernejoul coal field map; (3) the detailed mine planes of the Falck coal field; (4) lower critical offset Ω; (5) upper critical offset Ω'; (6) lower limit offset Γ'; (7) upper limit offset Γ.

10. FRACTAL DISTRIBUTION OF FAULT LENGTH AND OFFSETS

Lower and upper limit offsets Γ' and Γ; no offset smaller than Γ' or larger than Γ was recorded.

Lower and upper critical offsets Ω and Ω', between which sampling is complete.

Let $N_D(D)$ be the number of faults whose offset is larger than D. For each set of data, the distribution of $N_D(D)$ fits Eq. (12) satisfactorily (see Fig. 10.7).

$$N_D(D) = A_D D^{-C_1} \qquad (12)$$

where A_D and C_1 are constants. As for fault length distribution, there is no significant difference between the values of C_1 independently determined in several areas and at several scales (in Table 10.1 C_1 averages 1.35). As discussed with regard to length, the parameter A_D has a local significance and can be considered a function of reference offset D if the related value of $N_D(D)$ is accurately known. Especially with the lower critical value Ω, we obtain

$$A_D = \frac{N_D(\Omega)}{\Omega^{-C_1}} \qquad (13)$$

Equation (11) can be used for fault offsets smaller than Ω if the same study area is considered. As for Eq. (5), we define a fault-offset probability density function $n_D(D)$

$$n_D(D) = C_1 \frac{n_D(\Omega)}{\Omega} \left[\frac{D}{\Omega}\right]^{-(C_1+1)} \qquad (14)$$

From Fig. 10.7 and Eq. (11), we conclude that the distribution of fault offset is partly self-similar for the scales and offset range considered.

10.4.4. Fault Size Versus Fault Offset

For a limited number of faults (27; see Fig. 10.8), it was possible to analyze the relationships between size and offset. Fortunately such observations were made at different scales within the ranges of length and offset values already studied in detail; as a consequence, Fig. 10.8 supports a simple proportionality relationship between length and offset as follows

TABLE 10.1. A_D, C_1, and Ψ Parameters from Eq. (12) in Different Locations

Location	Number of Offsets	A_D	C_1	Ψ
La Houve coal field	1615	5.395	1.385	782.4
Vernejoul coal field	488	4.477	1.335	421.7
Falck coal field	502	4.418	1.351	340.5
La Houve seam 1	361	4.112	1.128	441.7
La Houve seam 2	259	4.354	1.228	352.2
La Houve seam Albert	275	4.518	1.334	244.0
La Houve seam Theodore	157	4.263	1.233	286.5
La Houve seam François	271	4.578	1.359	343.9
La Houve seam E	226	4.411	1.342	193.5
All data	2605	14.290	1.369	606.3

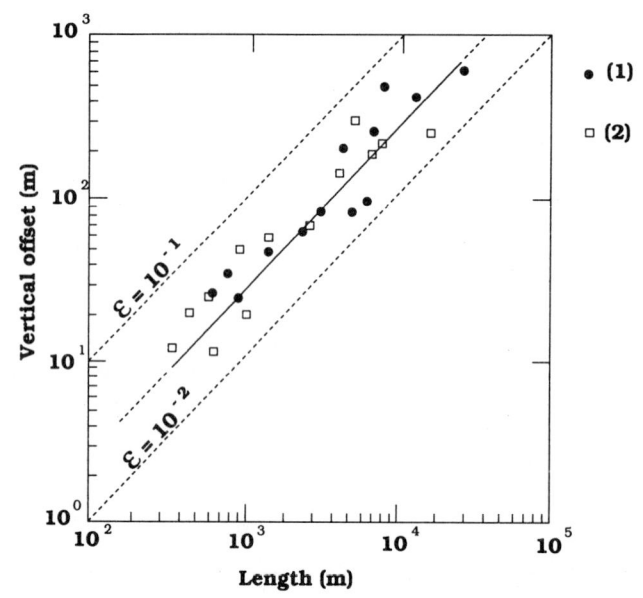

FIGURE 10.8. Fault size versus fault offset. Data from the general map of the (1) Lorraine coal basin and (2) the Vernejoul coal field map.

$$D = \varepsilon L^{-C_2} \qquad (15)$$

Watterson (1986), Walsh and Watterson (1987) and Childs and others (1990) obtained another value for C_2, that is, 1.5; Scholz and Cowie (1990) argued for a value of 1 for C_2. This estimation is more consistent with our data presented on Fig. 10.8.

Equation (14) with $C_2 = 1$ implies that parameters C and C_1 [see Eqs. (1) and (12), respectively] should be equal. In other words, the distributions of fault lengths and fault offsets reveal a common power law and correspond to common fractal dimensions (as defined by Mandelbrot 1977, 1982) within the range of scales considered here. Determining areal extension ratios related to normal faulting is simplified by taking into account the previously described distribution laws. The increase in horizontal surface related to the $n_L(L)$ normal faults of length L is given by

$$s(L) = \frac{Ln_L(L)D}{\tan(\Theta)} \qquad (16)$$

where Θ is the fault dip (assumed constant). Replacing the offset r and the number $n_L(L)$ by corresponding functions of L [given by Eqs. (14) and (5), respectively], we directly obtain the expression of $s(L)$ as a simple function of L, with C, A_L, ε, and Θ as parameters in Eq. (17)

$$s(L) = \frac{\varepsilon A_L C}{\tan \Theta} L^{1-C} \qquad (17)$$

Integrating Eq. (17) between L_1 and L_2, we obtain the total surface $S(L_1,L_2)$ added by normal faulting for all faults whose lengths range from L_1–L_2

10. FRACTAL DISTRIBUTION OF FAULT LENGTH AND OFFSETS

$$S(L_1, L_2) = \int_{L_1}^{L_2} s(L) \cdot dL = \frac{\varepsilon A_L C}{(2 - C)\tan \Theta}[L_2^{2-C} - L_1^{2-C}] \quad (18)$$

Adopting $L_2 = \Psi$ and $L_1 = 0$, we easily estimate the total surface S_M added as a result of the whole normal faulting (see also Scholz and Cowie, 1990)

$$S_M = \frac{\varepsilon \cdot A_L \cdot C}{(2 - C) \cdot \tan \Theta} \Psi^{2-C} \quad (19)$$

With Eq. (2) and $L = \Psi$, the parameter A_L becomes Ψ^C. We finally obtain a simple expression of S_M as a function of C and ε (from distribution laws), Θ (observed), and Ψ (related to the observation scale).

$$S_M = \frac{\varepsilon C}{(2 - C)\tan \Theta} \Psi^2 \quad (20)$$

Equation (20) can be used to obtain rapid estimates of areal extension ratios integrating all fault sizes. In the case of the Lorraine coal basin, we thus obtained values that average 1.2 for the fields studied.

10.4.5. Earthquakes and Fractal Organization of Faults

Earthquake magnitude distribution is reliably described by the following equation:

$$N_M(M_0) = aM_0^{-B} \quad (21)$$

where $N_M(M_0)$ is the number of earthquakes of seismic moment larger than M_0, a variable in space and time. The exponent B has a universal value of $2/3$.

Several attempts were made to relate seismic moment M_0, displacement on the fault, and corresponding active fault surface (Scholz, 1968; Thatcher and Hanks, 1973; Okubo and Aki, 1987; Scholz, 1990). The following estimates are proposed by Scholz and Cowie (1990):

$$M_0 \sim DL \quad (22a)$$
$$M_0 \sim DL^2 \quad (22b)$$

Equation (22a) is applied in case of faults that span the brittle crust and Eq. (22b) in case of faults that do not. Combining Eqs. (21) and (22a and b), we obtain

$$B = \frac{C}{C_2 + 1} \quad (23a)$$

$$B = \frac{C}{C_2 + 2} \quad (23b)$$

Using values of C and C_2 estimated in the Lorraine coal basin (1.4 and 1.0, respectively), Eq. (23a) gives a value of B identical to the theoretical value of $2/3$. As a major consequence of this comparison, we conclude that (1) the distribution law of earthquake moments probably controls in a direct way the distribution of fault lengths (and offsets) in the area studied and (2) at each scale of fault, there is a corresponding brittle layer entirely spanned by faults limited by décollement and/or rapidly decreasing offset (see Fig. 10.5).

10.5. FAULT DEFORMATION AND SELF-SIMILARITY

10.5.1. Self-Similarity of Fault Networks

The concept of fractal was introduced by Mandelbrot (1977, 1982); one of its most remarkable features is the description and understanding of self-similarities. In two dimensions, a set of lines may correspond to a fractal distribution (this set of lines may simply describe the network of fault traces in a map). In this case, a fractal dimension can be determined (within the range of 1–2 for planar networks).

Barton and Larsen (1985) proposed and applied techniques that enable us to characterize the fractal organization of a planar network of fractures and to compute the corresponding fractal dimension. Schematically their method is based on repeated determinations in grids of the number $N_S(c)$ of meshes that contain at least one segment of fracture of the map, where S is the total area under investigation (as shown in the fracture map) and c is the side of a square mesh. If $N_S(c)$ fits a power function of c according to Eq. (24) within a certain range of scales, the network of fracture traces has fractal or self-similar properties within this range, and its fractal dimension f can be determined.

$$N_S(c) \sim c^{-f} \qquad (24)$$

We applied this technique to accurate underground fault trace maps made at different scales in the Lorraine coal basin. Examples of results are shown in Fig. 10.9. Distributions thus characterized in log-log diagrams display a clearly linear portion between two values of c, which are called E and E' (lower and upper bound, respectively)

For $c < E$, meshes are too small relative to the map accuracy (small fractures were not mapped); as a result, the slope of the curve decreases.

For $c > E'$, meshes are too large with respect to map size (all meshes fractures); as a result, the slope of the curve increases.

Obviously the shape of the curve can be considered significant within the range E–E' exclusively, so that taking into account data accuracy and map size, we can simply determine a fractal dimension within a limited range.

In the case of fault networks in the Lorraine coal basin (e.g., Fig. 10.9), our analysis led to the following conclusions:

Each fault trace network was found fractal within certain limits (E and E' in each diagram). This fractal character was determined at several contrasting scale ranges (typically from 1–10,000 m). The fractal dimension at each scale ranges from 1.25–1.5. Smallest values of f are obtained for smallest box sizes.

Fractal dimensions thus determined for different fault networks at different scales do not significantly differ. No significant change in network organization through changes in scale could be identified. As a result, we consider that network of fractures whose length and offset are larger than 1 m and 0.1 m, respectively, probably represents a single fractal network with fractal dimension of about 1.5 in the area under investigation.

10.5.2. Distribution and Principal Direction of Deformation

In the preceding section, we demonstrated that the distribution of fault lengths and offsets is self-similar for the scales of observation considered. As a consequence, it is

10. FRACTAL DISTRIBUTION OF FAULT LENGTH AND OFFSETS

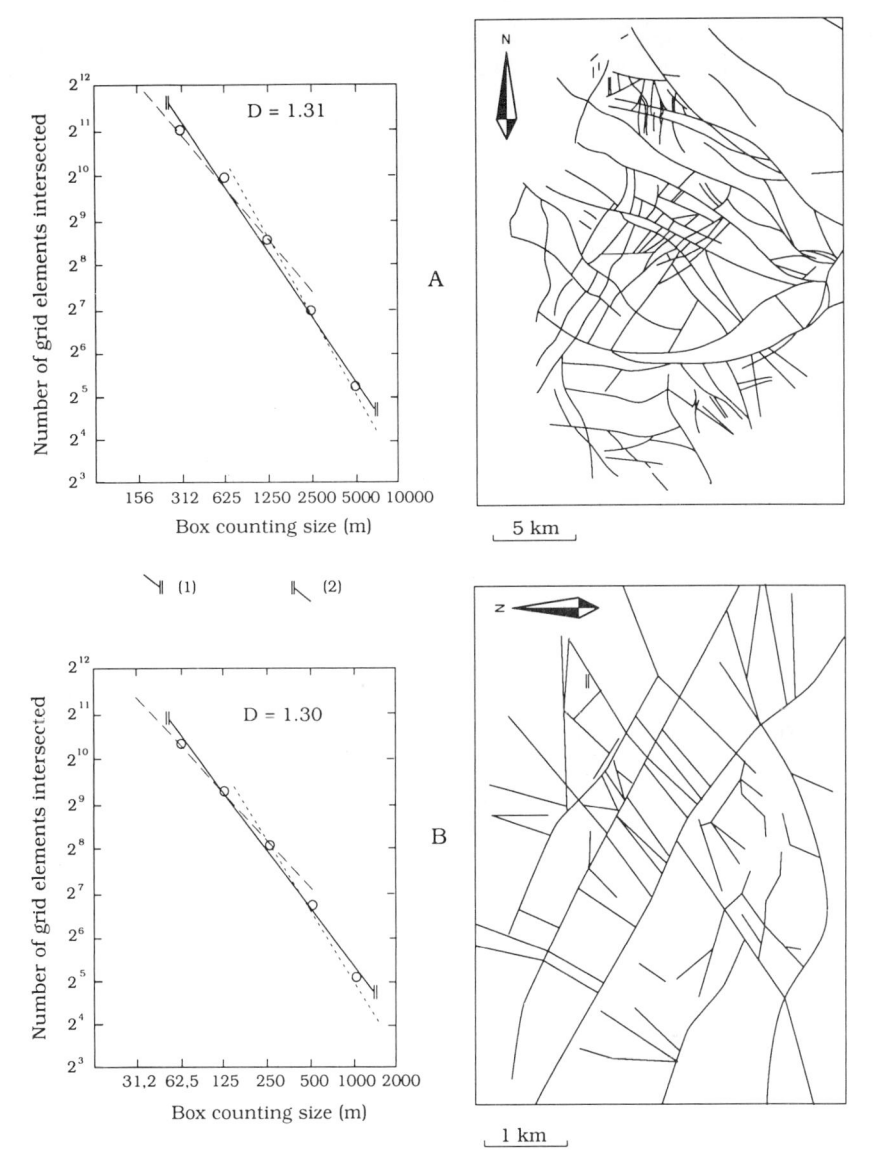

FIGURE 10.9. Fractal analysis of fault networks. (a) Western part of the Lorraine coal basin; (b) Vernejoul coal field at a 500-m depth; (1) upper bound of relative size E'; (2) lower bound of relative size E.

possible to obtain estimates of the total deformation in a given rock volume by analyzing subsets instead of complete fault populations. One can reliably determine deformation rates by analyzing fault data within a certain range of length and/or offset. This kind of determination, based on taking self-similarity properties of fault networks into account, has been discussed. Conclusions from such partial analyses can be extrapolated to other scales of observations through proportionality relationships, provided that self-similarity properties are identified.

We studied the deformation related to faults by using a simple geometrical model of

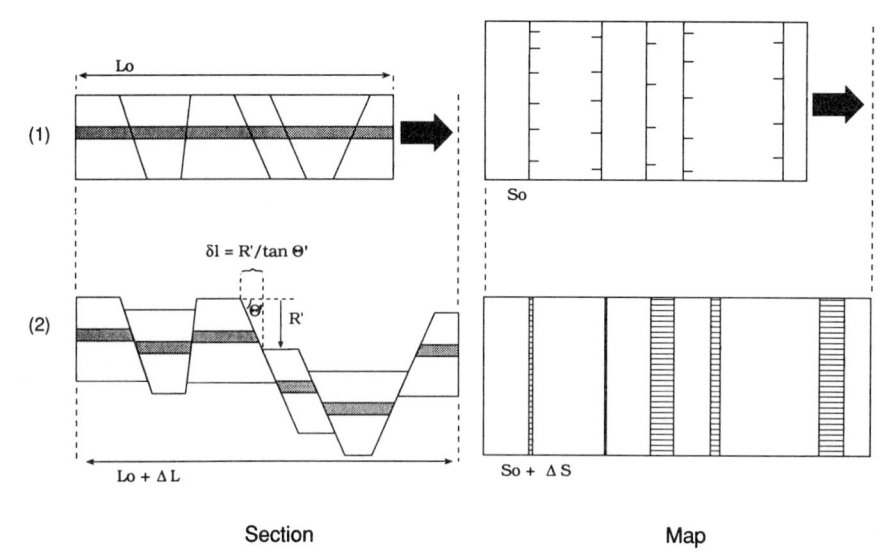

FIGURE 10.10. Model of brittle extensional deformation. From this model, linear lengthening factor ε_L and surfacic deformation factor ε_S can be estimate (see explanation in text).

extension (Sunwoo, 1988) shown in Fig. 10.10. Figure 10.11 shows an example of results. Note that this model does not account for the effect of block tilting. Because amounts of tilt due to extension remain small in the Lorraine coal basin (10° average), results are valid as a first approximation. A different method for computing the total deformation related to faulting in a rock mass has been described by Gauthier and Angelier (1985) and applied to extensional faulting in the same basin.

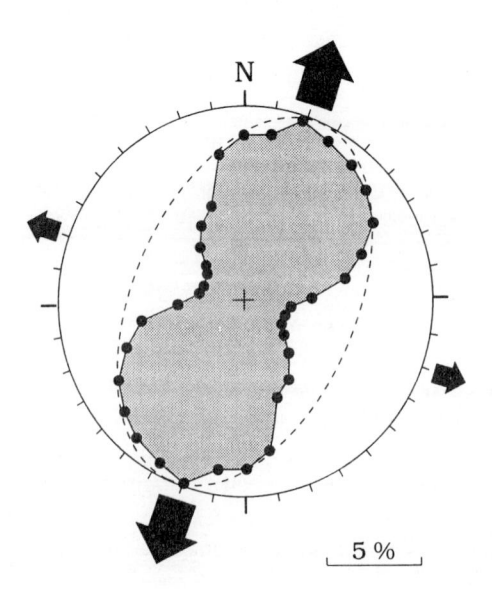

FIGURE 10.11. Rose diagram of linear-lengthening factor. This rose diagram was obtained from master faults in the western part of the Lorraine coal basin in all directions from a central point. Radii are proportional to the linear-lengthening factor ε_L. Large and small arrows indicate, respectively, the direction of maximum and minimum horizontal deformation (ε_M and ε_m).

10. FRACTAL DISTRIBUTION OF FAULT LENGTH AND OFFSETS

The model we used to determine deformation (see Fig. 10.10) enables us to compute a local linear lengthening factor ε_L in each direction. This factor ε_L, defined along a vertical section of length L that cuts N faults, is given by Eq. (24), where Θ is the dip of the fault and τ the angle between vertical cross section and direction of the fault

$$\varepsilon_L = \frac{\Delta L}{(L - \Delta L)} \qquad \Delta L = \sum_1^N \delta L_i, \quad \delta L_i = R_i/(\tan \Theta_i \sin \tau_i) \qquad (25)$$

Considering sections in all directions from a given point and determining all lengthening rates, we reconstruct the total horizontal deformation. The result in this case of extensional deformation in the Lorraine coal basin is illustrated in Fig. 10.11. There are two principal directions of horizontal deformation which correspond to the maximum and minimum ε_M and ε_m, respectively. In this case, $\varepsilon_M = 10\%$ and $\varepsilon_m = 3\%$. For extensional deformation, we can simply represent the trends of maximum lengthening thus determined, which results in the map in Fig. 10.12. From the regional point of view, we thus observe that extensional trends are north-south in the southern area, northnortheast-southsouthwest in the northern area, and variable in a narrow intermediate zone.

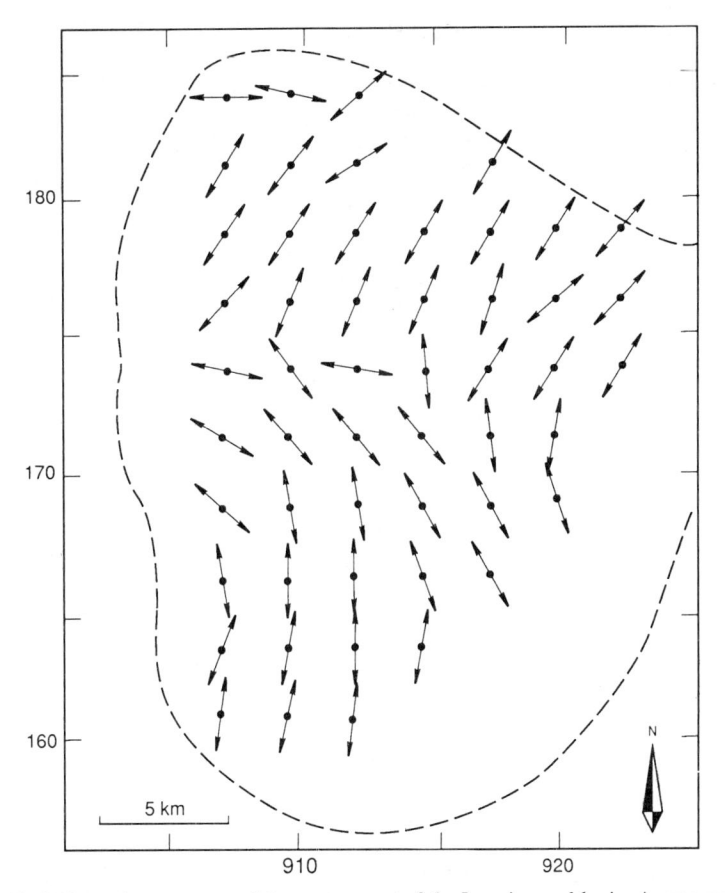

FIGURE 10.12. Deformation axes map of the western part of the Lorraine coal basin. Arrows represent local maximum lengthening trends ε_M.

Similar determinations have been made in terms of horizontal surface at the scale of the basin. We consider a regular grid in the map and compute the local deformation for each mesh by taking into account all horizontal surfaces added by each fault slip.

Surface deformation factors can be studied in the same way as linear ones previously mentioned in Eq. (24). They are defined as follows

$$\varepsilon_S = \frac{\Delta S}{(S - \Delta S)}, \quad \Delta S = \sum_1^N \delta S_i, \quad \delta S_i = L_i R_i / \tan \Theta_i \quad (26)$$

Based on such determinations, local surface deformation factors are mapped as shown in Fig. 10.13. These factors range from 5–25% in the basin studied. Periodical distributions are observable in this map: The first one corresponds to northnortheast-southsouthwest trends (N 20° E) with a wavelength of about 5 km; the other one corresponds to eastsoutheast-westnorthwest trends (N 120° E) with the same wavelength.

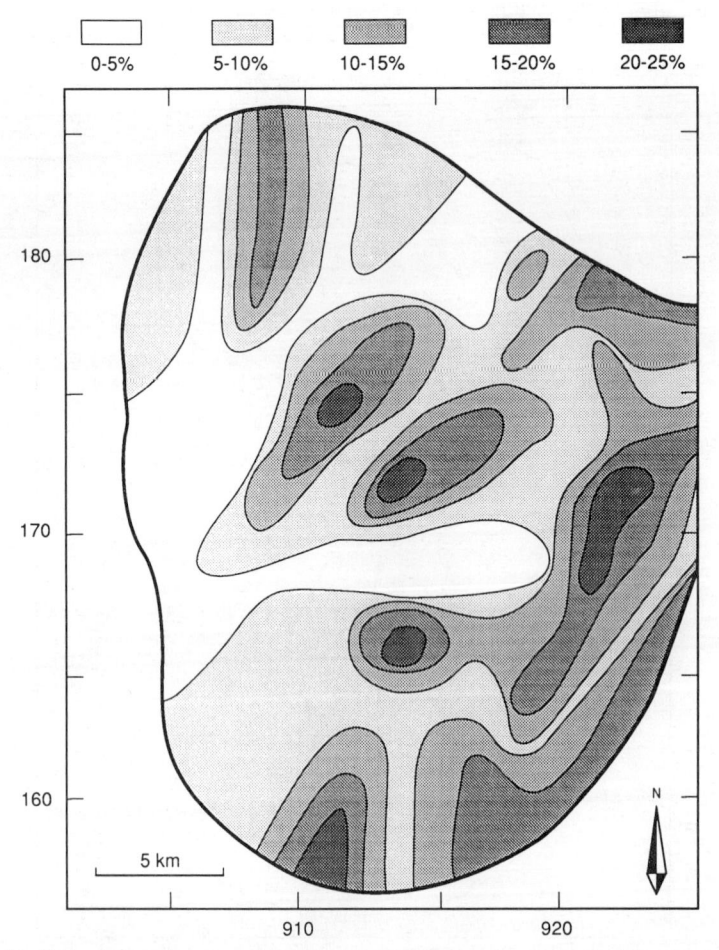

FIGURE 10.13. Deformation values map of the western part of the Lorraine coal basin. This map represents local values of the surfacic deformation factor ε_S.

10. FRACTAL DISTRIBUTION OF FAULT LENGTH AND OFFSETS

Such periodical distributions correspond to a succession of narrow elongated zones with alternating large and small amounts of extensional deformation. These zones strike parallel to the two major normal fault trends of the basin. The wavelength (usually 5 km, with a maximum of 8 km) is remarkably consistent in first approximation with the total thickness of the stratified sedimentary cover (Brand and others, 1976). Because this sedimentary cover is entirely affected by the faults studied and because there is probably a basal décollement (at the top of the Devonian rock units), we suspect that this periodicity of normal faulting at the regional scale reflects a boudinage phenomenon (Cloos, 1947; Ramberg, 1955; Masuda and Kuriyama, 1988) at large scale.

10.6. CONCLUSIONS

Based on detailed analyses of fault networks in the western portion of the Lorraine coal basin, we conclude that

1. The cumulative distribution of fault lengths and the cumulative distribution of fault offsets clearly fit simple power laws within the range of observation scales (typically respectively 10–10,000 m and 0.1–10 m). These relationships are conveniently described by equation $N(x) = Ax^{-C}$, where $N(x)$ describes the number of faults and x describes the variable considered (L or D), while A and C are parameters.
2. The values of C (1.35–1.4) remains constant through scale changes (within the same range), whereas the value of A varies linearly with the size of the area considered and local fracture density.
3. The value of C is the same (within the range of uncertainties) for both distributions considered (fault lengths and fault offsets).
4. Fault offset is directly proportional to the length of the fault.
5. The fault network, as observed in maps at different depths and with different accuracies (within the range of scales considered), displays a fractal organization with a fractal dimension of about 1.5.
6. Based on self-similarity relationships thus established, determining total deformation due to faulting in a rock mass does not require a complete knowledge of faults at all scales, and extrapolating results obtained within a certain range of scales (especially faults larger than a given size) can be used.
7. In the Lorraine coal basin, the extensional deformation thus analyzed is characterized by a maximum lengthening horizontal axis trend perpendicular to the largest normal fault lines.
8. In the same basin, large and small amount of horizontal extension (in linear and surfacic terms) are distributed in narrow parallel elongated zones, which probably reflect large-scale boudinage of the stretched sedimentary cover above a décollement surface despite the small amount of extension.

APPENDIX

In this Appendix, the main original data sets used in this study are presented and illustrated in a raw way. Figure 10.3 in the text illustates basic data collection in detailed mining plans. Table 10.2 (hereafter) provides data sets used in Figure 10.6 (in the text) to describe fault length distribution. Likewise Table 10.3 contains the data sets relative to fault-offset distribution, as described in Fig. 10.7 in the text. All values mentioned in Tables 10.2 and 10.3 are given in metres and ordered by increasing values. In each table, columns correspond to different data subset, collected in different areas and at different scales; a columns numbers in Tables 10.2 and 10.3 refer to the same number in figure captions for Figs. 10.6 and 10.7, respectively.

TABLE 10.2.

Length	(1)	(2)	(3)	(4)	Length	(1)	(2)	(3)	(4)	Length	(1)	(2)	(3)	(4)
5	—	—	—	—	400	185	58	5	4	7500	21	1	—	—
10	—	—	—	—	450	182	54	2	4	8000	18	1	—	—
15	—	—	91	73	500	181	53	2	3	8500	17	0	—	—
20	—	—	90	71	550	181	49	0	3	9000	15	—	—	—
25	—	—	89	69	600	180	45	—	3	9500	14	—	—	—
30	—	—	88	65	650	172	43	—	2	10,000	13	—	—	—
35	—	—	86	63	700	168	43	—	2	15,000	5	—	—	—
40	—	—	83	62	750	166	42	—	1	20,000	1	—	—	—
45	—	—	81	59	800	162	42	—	1	25,000	1	—	—	—
50	—	—	81	55	850	158	41	—	1	30,000	0	—	—	—
55	—	—	81	53	900	155	38	—	1	35,000	—	—	—	—
60	—	—	77	52	950	154	34	—	0	40,000	—	—	—	—
65	—	—	73	51	1000	152	30	—	—	45,000	—	—	—	—
70	—	—	66	49	1500	117	13	—	—	50,000	—	—	—	—
75	—	—	63	46	2000	94	9	—	—	55,000	—	—	—	—
80	—	—	62	46	2500	83	7	—	—	60,000	—	—	—	—
85	—	—	60	45	3000	70	6	—	—	65,000	—	—	—	—
90	—	—	58	45	3500	61	6	—	—	70,000	—	—	—	—
95	—	—	56	44	4000	55	4	—	—	75,000	—	—	—	—
100	—	—	55	43	4500	48	4	—	—	80,000	—	—	—	—
150	—	—	42	32	5000	35	3	—	—	85,000	—	—	—	—
200	—	—	33	24	5500	33	3	—	—	90,000	—	—	—	—
250	—	61	25	15	6000	31	3	—	—	95,000	—	—	—	—
300	190	60	17	11	6500	30	3	—	—	100,000	—	—	—	—
350	188	59	12	9	7000	26	2	—	—					

TABLE 10.3.

Offset	(1)	(2)	(3)	Offset	(1)	(2)	(3)	Offset	(1)	(2)	(3)
0.05	—	542	540	0.6	—	239	156	2.5	—	96	29
0.1	—	537	495	0.65	—	210	134	3	—	84	21
0.15	—	504	457	0.7	—	207	125	3.5	—	81	19
0.2	—	483	414	0.75	—	192	112	4	—	80	15
0.25	—	410	340	0.8	—	192	107	4.5	—	71	13
0.3	—	388	303	0.85	—	183	100	5	—	70	12
0.35	—	329	252	0.9	—	180	98	5.5	—	67	10
0.4	—	316	232	0.95	—	174	93	6	84	67	9
0.45	—	276	195	1	—	173	90	6.5	81	64	7
0.5	—	266	188	1.5	—	131	51	7	81	64	7
0.55	—	244	161	2	—	114	39	7.5	81	59	7

(Continued)

TABLE 10.3. (*Continued*)

Offset	(1)	(2)	(3)	Offset	(1)	(2)	(3)	Offset	(1)	(2)	(3)
8	81	58	5	60	42	16	—	400	7	—	—
8.5	81	57	5	65	42	15	—	450	7	—	—
9	81	55	5	70	40	14	—	500	6	—	—
9.5	80	55	4	75	38	14	—	550	5	—	—
10	80	55	4	80	37	14	—	600	5	—	—
15	73	45	3	85	35	13	—	650	1	—	—
20	67	39	1	90	34	12	—	700	0	—	—
25	59	35	0	95	33	11	—	750	—	—	—
30	56	32	—	100	32	11	—	800	—	—	—
35	53	29	—	150	27	10	—	850	—	—	—
40	50	26	—	200	22	7	—	900	—	—	—
45	47	22	—	250	19	3	—	950	—	—	—
50	46	21	—	300	11	1	—	1000	—	—	—
55	42	17	—	350	7	0	—				

ACKNOWLEDGMENTS

We thank the Houillères du Bassin de Lorraine for permission to work with their data and specialy M. P. Schröeter and his geological staff for their assistance in extracting data from coal mine plans. Reviews by Rick Allmendinger and Patience Cowie were very useful.

REFERENCES

Archuleta, R. J., *J. Geophys. Res.* **89**, 4559 (1984).
Barnet, J. A., Mortimer, J., Rippon, J. H., Walsh, J. J., and Waterson, J., *Am. Assoc. Pet. Geol. Bull.* **71**, 925 (1987).
Barton, C. C., and Larsen, E., *Fractal Geometry of Two-Dimensional Fracture Networks at Yucca Mountain, Southwestern Nevada*, Proceedings of International Symposium on Fundamentals of Rock Joints Bjorkliden, Lapland, Sweden (1985).
Brand, E., *Geol. J.* **27**, A (1976).
Cailleux, A., Etude quantitative de failles: *Revue de Géomorphol. Dyn.* **9**, 129 (1958).
Childs, C., Walsh, J. J., and Watterson, J., *A Method for Estimation of the Density of Fault Displacements below the Limit of Seismic Resolution in Reservoir Formations*, North Sea Oil and Gas Reservoirs. Norwegian Institute of Technology (Graham and Trotman, London, 1990).
Chiles, J. P., *Math. Geol.* **20**, 631 (1988).
———, in *Geostatistical, Sensitivity and Uncertainty Methods for Groundwater Flow and Radionuclide Transport Modeling* (Buxton Battelle Press, Columbus, 1989), pp. 361–85.
Cloos, E., *Am. Geophys. Union Trans.* **28**, 626 (1947).
Cruden, D. M., *Int. J. Rock Mech. Min. Sci. Geomech. Abstr.* **14**, 133 (1977).
Curl, R. L., *Math. Geol.* **18**, 765 (1986).
Davy, P., Sornette, D., and Sornette, A., *Nature* **348**, 56 (1990).
Donsimoni, M., *Mém. BRGM* **117**, (1981).
Falke, H., ed. *The Continental Permian in Central, West, and South Europe* (Dordrecht, Boston, 1976).
Ferguson, C. C., *Math. Geol.* **17**, 403 (1985).
Gauthier, B., and Angelier, J., *Earth Plan.* **74**, 137 (1985).
Gibowicz, S. J., *Pure Appl. Geophys.* **124**, 614 (1986).
Herve, J. C., and Cailleux, A., *Cahiers Geol.* **68–69**, 733 (1962).
Huang, J., and Turcotte, D. L., *Earth Plan.* **91**, 223 (1988).
Kakimi, T., and Kodama, K., *Bull. Geol. Surv. Jap.* **25**, 31 (1974).

Lorenz, V., and Nicholis, I. A., in *The Continental Permian in Central, West, and South Europe* (H. Falke, ed.) (D. Reidel Publ. Co., Boston, 1976), 313.
Mandelbrot, B. B., *Fractal: Form, Chance and Dimension: Geometry of Nature* (W. H. Freeman, San Francisco, 1977).
———, *The Fractal Geometry of Nature* (W. H. Freeman, San Francisco, 1982).
Marrett, R. A., and Allmendinger, R. W., *J. Struct. Geol.* **13**, 6, 735 (1991).
———, *J. Struct. Geol.* **12**, 976 (1990).
Masuda, T., and Kuriyama, M., *Tectonophys.* **147**, 171 (1988).
Muraoka, H., and Kamata, H., *J. Struct. Geol.* **5**, 483 (1983).
Okubo, P. G., and Aki, K., *J. Geoph. Res.* **92**, 343 (1987).
Pruvost, P., *Bassin houiller de la Sarre et de la Lorraine*, t. 3: *Description géologique*, Etude Gites minér (Danel, Lille, France, 1934).
Ramberg, H., *J. Geol.* **63**, 512 (1955).
Reches, Z., *Tectonophys.* **47**, 109 (1978).
Rippon, J. H., *Proc. Yorkshire Geol. Soc.* **45**, 147 (1985).
Schaffer, A., *Mainzer Geowiss. Mitt.* **15**, 239 (1986).
Scholz, C. H., *Bull. Seism. Soc. Am.* **58**, 399 (1968).
———, *The Mechanics of Earthquakes and Faulting* (Cambridge University Press, Cambridge, England, 1990).
Scholz, C. H., and Cowie, P. A., *Nature* **346**, 837 (1990).
Shi, Y., and Bolt, B. A., *Bull. Seism. Soc. Am.* **72**, 1677 (1982).
Sunwoo, C., *Mém. Sc. Terre Univ. Curie*, Paris, 88–28, 195 p (1988).
Teichmüller, M., Teichmüller, R., and Lorenz, V., *Z. D. Geol. Ges.* **134**, 153 (1983).
Thatcher, W., and Hanks, T. C., *J. Geophys. Res.* **78**, 8547 (1973).
Thomas, A., *C. R. Acad. Sc. Paris* **303**, 225 (1986).
———, *C. R. Acad. Sc. Paris* **304**, 181 (1987).
Turcotte, D. L., *Tectonophys.* **132**, 261 (1986a).
———, *J. Geoph. Res.* **91**, 1921 (1986b).
Villemin, T., *Geodinamica Acta Paris* **1**, 147 (1987).
Villemin, T., and Sunwoo, C., *C. R. Acad. Sc. Paris* **305**, 1309 (1987).
Walsh, J. J., and Watterson, J., *J. Struct. Geol.* **10**, 239 (1987).
Walsh, J. J., Watterson, J., and Yielding, G., *Nature* **351**, 391 (1991).
Watterson, J., *Pure Appl. Geophys.* **124**, 365 (1986).

11

Fractal Dynamics of Earthquakes

P. Bak and K. Chen

11.1 INTRODUCTION

Many objects in nature, from mountain landscapes to electrical breakdown and turbulence, have a self-similar fractal spatial structure (Mandelbrot, 1982). This is by no means a trivial observation, since it implies that systems are correlated over large distances. Much effort has been put into computer simulation and characterization of these objects. However the empirical geometrical observation and characterization do not by themselves serve as a physical explanation. It seems obvious that to understand the origin of self-similar structures, we must understand the nature of the dynamical processes that created them: Temporal and spatial properties must necessarily be completely interwoven.

This is particularly true for earthquakes, which have a variety of fractal aspects, as discussed in this volume. The distribution of energy released during earthquakes is given by the Gutenberg–Richter (1956) power law. The distribution of epicenters appears to be fractal with dimension $D \approx 1-1.3$ (Kagan and Knopoff, 1980). The number of after shocks decay as a function of time according to the Omori (1894) power law. There have been several attempts to explain the Gutenberg–Richter law by starting from a fractal distribution of faults or stresses (Kagan and Knopoff, 1987; Huang and Turcotte, 1990; Turcotte, 1989). But this is a hen-and-egg approach: To explain the Gutenberg–Richter law, we assume the existence of another power-law—the fractal distribution.

The Gutenberg–Richter law extends over several orders of magnitude. For instance Johnson and Nava (1985) present data on the New Madrid seismic zone indicating a power law over almost 5 decades. The upper limit is probably due to the fact that measurements were necessarily limited to a period of 167 years, from 1816–1983. Since a human lifetime

---

The submitted manuscript has been authored under contract DE-AC02-7600016 with the Division of Materials Sciences, U. S. Department of Energy. Accordingly, the U. S. Government retains a nonexclusive, royalty-free license to publish or reproduce the published form of this contribution, or allow others to do so, for U.S. Government purposes.

P. Bak and K. Chen • Department of Physics, Brookhaven National Laboratory, Upton, New York.

Fractals in the Earth Sciences, edited by Christopher C. Barton and Paul R. La Pointe. Plenum Press, New York, 1995.

cannot play an essential role for earthquakes, there is no reason to believe that the distribution cannot be extended beyond earthquakes of size $m = 7$ to earthquakes of size 8, 9, and 10, etc., if a geological time period were available for the measurements.

The observation of power laws is of tremendous importance in physics, since it indicates the existence of an underlying scale-invariant mechanism. The Gutenberg–Richter law indicates that the mechanism of small earthquakes is essentially the same as the mechanism for large earthquakes, since otherwise their relative frequency cannot be expected to obey a simple law. Actually the quality of data for earthquakes is excellent compared with other areas of physics, where usually not more than 3 decades are available: Scaling over 8 decades is unheard of. We argue that this is due to the fact that the upper length and time scales for Earth dynamics are much larger than for any system set up by humans.

Recently it has been recognized that many interacting dynamical systems naturally evolve into a self-organized critical state, with avalanches of all sizes—large and small (Bak and others, 1987, 1988; Tang and Bak, 1988; Bak and Chen, 1989). The discovery suggests a rather general dynamical mechanism for the emergence of scaling behavior (including fractal structure) in nature. Shortly after the discovery, it became clear that the simplest and most direct application of this idea might be to earthquakes: The Gutenberg–Richter law, the fractal spatial distribution of epicenters, and other power laws in earthquakes are all manifestations that the crust of the earth operates at a self-organized critical state. Indeed several authors (Bak and others, 1988; Bak and Tang, 1989; Ito and Matsuzaki, 1990; Sornette and Sornette, 1989; Carlson and Langer, 1989) have taken up the idea and presented supporting theoretical evidence, although the Carlson–Langer model fails to reproduce the scaling observed for large earthquakes.

The concept of self-organized criticality is most easily visualized in terms of the prototypical example: a pile of sand. Consider a situation where the pile is built by slowly and uniformly by adding sand, one grain at the time, to a large flat surface with edges where the sand slides off. In the beginning, the sand remains close to the position where it lands. After a while, the pile achieves a slope, and now and then there are small avalanches when the slope somewhere becomes too steep. Avalanches can be thought of as generated by a chain reaction or branching process. Following the initial instability, the falling particle may either stop falling, continue falling, or induce two or more falling particles. Later each falling particle may again stop, continue falling, or induce more falling particles, and so on. The total number of falling particles during this process is a measure of the size of the avalanche. As the process of adding sand continues, the pile becomes steeper and steeper, and larger and larger avalanches appear. Eventually the pile reaches a statistically stationary state where the amount of sand added in average is balanced by the amount of sand falling off the edges, and the growth of the slope stops. The chain reaction for avalanches becomes critical, and avalanches of all sizes occur; this is the self-organized critical state. The frequency of avalanches of different sizes follows a power law distribution similar to the Gutenberg–Richter law (Bak and others, 1987, 1988; Bak and Chen, 1990).

The basic principle of self-organized criticality is that large interactive dynamic systems naturally organize themselves into a state that is perpetually critical. Dynamic forces inevitably carry the system to the critical state without fine tuning external forces. In contrast to critical, for chaotic few-degrees-of-freedom systems (which can be described by a few variables), the self-organized critical state is robust with respect to any change in local microscopic mechanisms for the system. For example, in terms of the sandpile picture, if we try to prevent avalanches by building snow screens, then for a while there will be fewer and smaller avalanches. But eventually the slope adjusts to the new situation, and the critical state is resumed: The critical state is a global *attractor* of the dynamics. This resiliency is important for representing real dynamics in nature. The scaling laws of this critical state are properties of whole systems with many degrees of freedom, and they cannot be deduced by studying local properties. It makes no sense to try to explain large events with a detailed microscopic-engineering approach.

The characterization of the Earth's crust as a system operating at the self-organized critical state is in complete contrast to the view that the crust is a low-dimensional chaotic system. In fact as we demonstrate later, the self-organized critical state is not chaotic at all but operates perpetually at the border of chaos.

We present results of a simple stick slip model of earthquakes, which evolves to a self-organized critical state. Our emphasis is on demonstrating that empirical power laws for earthquakes indicate that the Earth's crust is at the critical state, with no typical time, space, or energy scale. Of course the model is tremendously oversimplified; however in analogy with equilibrium phenomena we do not expect criticality to depend on details of the model (universality).

11.2. MODELS AND SIMULATIONS

In 1956 Gutenberg and Richter observed that the number Q of earthquakes of magnitude greater than m is given by the relation

$$\log_{10} Q = c - bm \tag{1}$$

where b is a universal constant with a value approximately unity, $0.8 < b < 1.2$. The researchers also estimated that the energy E released during an earthquake increases exponentially with m

$$\log_{10} E = c' + dm \tag{2}$$

where d is not known very accurately but generally assumed to be in the range of $1.5 < d < 2.5$. Combining those two relations, we realize that the Gutenberg–Richter law is essentially a power law for the distribution of energy release

$$N(E) = \frac{dQ}{dE} \propto E^{-1-b/d} \equiv E^{-1-\beta} \qquad 0.4 < \beta < 0.6 \tag{3}$$

However despite the universality of the relation, there has been no explanation of this power law behavior. Note that most of the uncertainty lies in relating m to E; there is little doubt that we are indeed dealing with a power law. Kagan (1990) finds the exponent β to lie between 0.5–0.6 from analyzing the Harvard earthquake catalog for large earthquakes.

It is generally assumed that earthquake dynamics are due to a stick-slip mechanism involving the Earth's crust sliding along faults (Burridge and Knopoff, 1967; Otsuka, 1972; Stuart and Mavko, 1979; Sieh, 1978; Mikumo and Miyatake, 1978, 1979; Choi and Huberman, 1984). When a slip occurs at some location, the strain energy at that position is released, and the stress propagates to the near environment. While this picture is rather well-established, no connection between stick-slip models and the actual spatial and temporal correlations has been demonstrated.

The situation that we want to describe is shown in Fig. 11.1a. Two segments of material, representing tectonic plates, are slowly pressed against each other, causing them to slip along their interface. A scaled-down laboratory experiment has actually been performed by Bobrov and Lebedkin (1989), who used aluminum and niobium rods. A fault region was generated as pressure increased, causing a transition from elastic flow (where the rod returns to its original shape once pressure is released) to ductile flow (where compression is irreversible). The researchers indeed observed earthquakes along the fault with a power law distribution independent of the slip material and mechanism (believed to be different for the two materials). In the present context, blocks are tectonic plates grinding against each other along a fault or a fault system. Now and then, parts of the plates slip relative to each other; these slips are ruptures of the crust in earthquakes.

Figure 11.1b shows a one-dimensional model of a single fault. For simplicity one plate is assumed to be rigid and the other to be an elastic medium represented by an array of blocks at positions $x_1, x_2, x_3 \ldots$ connected by springs. The blocks interact with the rigid plate by means of static and dynamic friction forces. We assume that the array is open at one end and extends infinitely in the other direction. Whenever the spring force on a particular block exceeds the critical static friction force, it slides until interaction forces have been reduced below the critical dynamical friction. In the aluminum rod experiment, the process may be dislocation motion caused by an atomic bond shifting. During this process, potential energy is first converted into kinetic energy, then dissipated (radiated) when the blocks are decelerated by the frictional forces.

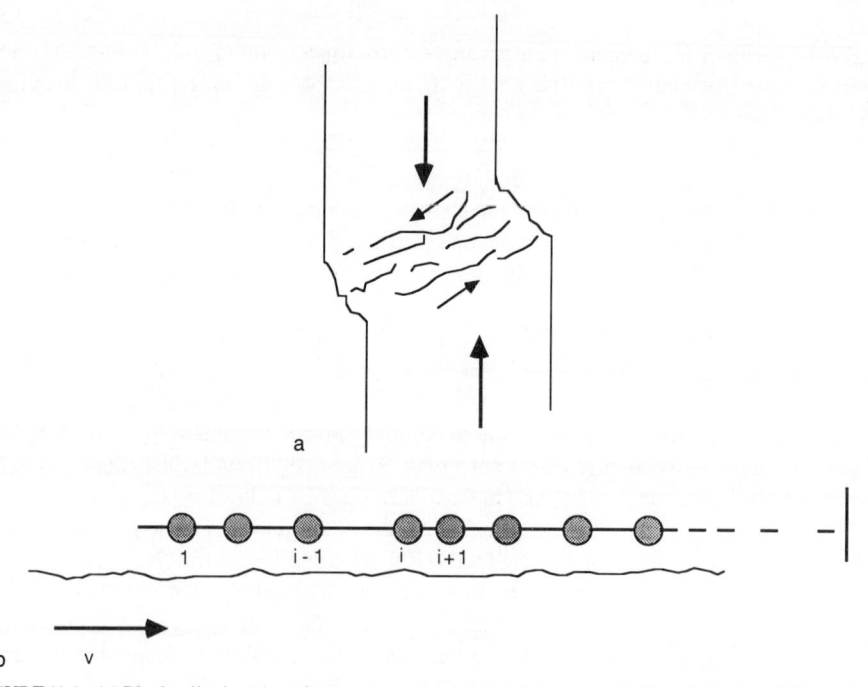

FIGURE 11.1. (a) Blocks slipping along fault system when subjected to stress. (b) One-dimensional illustration of our model. The block-spring chain is pushed along a rough surface with a low velocity v.

11. FRACTAL DYNAMICS OF EARTHQUAKES

Of course since the blocks are at rest between slips, the total force on each block is zero; thus spring forces exactly balance friction forces. When a block slides, the friction force on the block is reduced and so is the spring force on that block. There must be exact conservation of friction forces (or equivalently spring forces) at the individual sliding event since the blocks are at rest both before and after the event. For simplicity we assume that the force is redistributed evenly among nearest neighbors. Note that while forces are conserved, the density of blocks is not: There can be wide fluctuations of the local density of blocks. This distinguishes the model from leaf-spring models (Burridge and Knopoff, 1967) where the average distance between blocks is fixed by leaf springs hooked to a rigid rod. We believe that this model is not physical and introduces a characteristic length into the model. This length leads to deviations from the Gutenberg–Richter law (dominating characteristic large events) not found in nature (Carlson and Langer, 1989). We must explicitly include the perpendicular to the fault system.

The model is driven by slowly pushing the rigid surface relative to the other surface. The time scale set by the pushing is a geological one, and it can be viewed as infinitely large compared with a realistic observation time, so that there is no typical time scale. This is essential for generating power laws and fractal scaling for spatial and temporal correlation functions.

Let us monitor the friction force z_i (which equals minus the spring force) on the ith block. Initially a random distribution of subcritical forces is chosen. The z_is grow at a small rate p until somewhere z reaches the critical value, and a slip event takes place. Without loss of generality, the critical friction force is chosen to be an integer Z_{cr}, and the reduction of friction force is taken to be two units, so

$$z_i \to z_i - 2 \qquad z_{i\pm1} \to z_{i\pm1} + 1 \qquad \text{when } z_i > Z_{cr} \qquad (4)$$

Bak and Tang (1989) present a model driven by letting $z_i \to z_i + 1$ at random positions. This has the advantage that all operations are integers; i.e., the model is a random cellular automaton. In contrast the present model is completely deterministic, with all randomness entering through the initial condition. The model appears to be more physical, since no external random forces are needed. Nevertheless the results, including critical exponents, fractal dimensions, etc., remain the same.

The process initiated by the event in Eq. (4) transfers force to neighbors, allowing for a chain reaction. This chain reaction is the earthquake. As the process continues, the forces z_i generally increase, causing larger and larger earthquakes. Eventually the system is pumped up to a minimally stable state where all forces are near the critical value; that is Int$(z_i) = Z_{cr} - 1$ for all i at this state. The next instability is propagated throughout the system until the excess force is released at the boundary and the system is back to a minimally stable state. Thus statistically stationary state has been reached. We assume that the Earth's crust has had sufficient time to reach a stationary state, so we are generally concerned with this state only.

The dynamics of this one-dimensional model is rather trivial, and the preceding discussion should be viewed as a pedagogical exercise only. To achieve nontrivial critical behavior, it is sufficient to generalize the model to include next-nearest neighbor interactions (Kadanoff and others, 1989). Here we generalize the model to two and three dimensions, keeping in mind that the Earth's crust, and particularly the fault region, is a higher dimensional system. The generalization is rather trivial: Blocks are situated on a d-dimensional lattice, and each block is connected with its $2d$ nearest neighbors. Now z

denotes the component of the friction force along the sliding direction. Again forces increase at a small rate p until the critical value is reached. For simplicity the force is redistributed evenly among neighbors, so no distinction is made between sliding and perpendicular directions in this model. (Later we discuss a more realistic model with an anisotropic long-range redistribution of forces.) The important thing is that forces are conserved, hence,

$$z_i \rightarrow Z_{cr} - 2d \qquad z_{nn} \rightarrow z_{nn} + 1 \qquad \text{when } z_i > Z_{cr} \qquad (5)$$

We might expect the system to be pumped up to the minimally stable state as before, but this turns out not to be the case. It is not difficult to see that the minimally stable state is unstable, since a local instability is amplified by the higher connectivity of the blocks. Thus the state collapses from a domino effect to some other state with smaller forces.

As the system builds up, earthquakes become larger and larger. Eventually a stationary state is reached with earthquakes of all sizes limited only by the size of the system. The size S (= energy radiation) of an earthquake equals the total number of slidings caused by a single initial instability. This is the self-organized critical state whose properties we study later. First let us convince ourselves that indeed the system is critical. Figure 11.2 shows the distribution of earthquakes found by a continuous simulation of the model in two dimensions. The straight line in the log-log plot indicates that the energy distribution is indeed a power law

$$N(s) = s^{-a} \qquad a \approx 1.1 \qquad (6)$$

in accordance with the Gutenberg–Richter law. Geophysicists tell us that the Earth's crust (at least for moderate earthquakes) should be viewed as a three-dimensional system. In three dimensions we find $a \approx 1.3$, that is $\beta = 0.3$ which is in somewhat better agreement with observations. For large earthquakes extending over more than the thickness of the crust, we may expect a cross over from three-dimensional to two-dimensional exponents.

Our exponents should be compared with those of the seismic moment (energy) distribution. It appears that the exponents are smaller than those of earthquakes, indicating

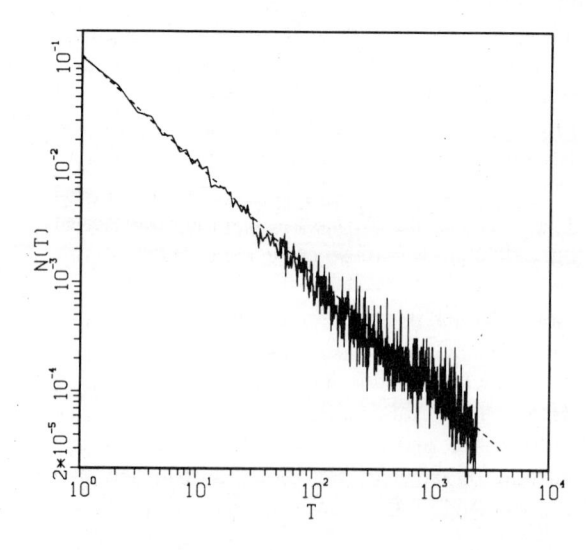

FIGURE 11.2. Distribution of earthquake sizes for 50 × 50 system driven at a rate of $p = 0.00002$. The straight line over several decades indicates a Gutenberg–Richter power law with an exponent slightly greater than unity. $N(T)$ is the number of earthquakes involving T sliding events.

that the model is not in the same universality class as earthquakes. What is important is that there are power laws indicating that the Earth's crust is at a critical state, with no typical time, space, and energy scale.

11.3. DISCUSSION AND CONCLUSIONS

As discussed in the introduction, the critical state is robust with respect to randomness, etc. It is trivial to see that a random distribution of critical forces has no effect, since the model can be transformed into the uniform one by simply shifting the variable z (a gauge transformation). We have also studied models where a fraction of the springs were randomly removed (Bak and others, 1988) and a critical state with the same critical exponents was reached. If the properties of the system are changed during the simulation (due to some external event), the system returns to the self-organized critical state after a transient period. The critical state is a global *attractor* of the dynamics. This resiliency is important for self-organized criticality to apply to a wide range of natural phenomena.

Systems with few degrees of freedom (like the Feigenbaum map, coupled oscillators, circle maps, etc.; for reviews, see Hao, 1984) may also exhibit critical points with power law correlations. Since these have no spatial degrees of freedom, then can not possibly have fractal power law spatial correlations. However critically requires fine tuning some parameter, and the critical point, separating regular from chaotic states, has no robustness at all. Thus any small perturbation throws the system off the critical point by destroying long-time memory effects. Attempts to explain the complicated behavior of earthquakes as low-dimensional (few degrees of freedom) chaos must be considered fundamentally misguided, since chaos implies exponentially decaying correlations, not power laws. The belief that there may be a connection between low-dimensional chaos and fractals is without mathematical foundation. Our model cannot be reduced to a few degrees of freedom at the critical state. Sooner or later, information from far away affects the dynamics of any given point.

Figure 11.3 compares the number of blocks slipping versus time in a simulation where z increases by $p = 0.00002$ per unit time for a system of the size 50×50. Note the irregular evolution of the individual events. The outcome of a single earthquake is quite unpredictable, since it depends on minor details far removed from the initial point of instability.

Forecasting individual earthquakes in such a system is quite impossible, since accurate global information is needed. How do we characterize this unpredictability? Usually the unpredictability of dynamic systems is characterized by the Lyapunov exponent, which defines the amplification of small differences in the initial condition as the system evolves. A positive Lyapunov exponent indicates chaos. We have simulated systems at the critical stationary state that initially differ by a small force f_i, where f_i is a random number from $-q$–q' and q is a small number of the order 10^{-5}. Figure 11.4 shows the average difference of z per site as a function of time (the Hamming distance). The straight line indicates power law behavior. Hence the Lyapunov exponent is zero, and the system is at the border of chaos. Nevertheless the fact that the power is positive indicates the uncertainty of the state of the system grows, albeit much less dramatically than for chaotic systems. The situation for predicting earthquakes is less desperate than for fully chaotic systems, although there is the added complexity of having to deal with many degrees of freedom. We denote such systems as weakly chaotic. Since many dynamic systems are expected to be self-organized critical, we expect weak chaos to be quite ubiquitous in nature.

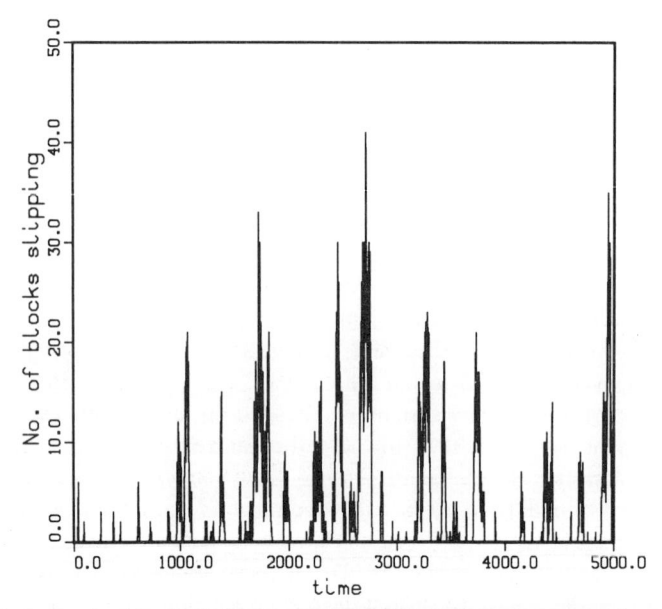

FIGURE 11.3. Evolution of activity, including several earthquakes, for 30 × 30 system driven at a rate of $p = 0.0001$. The plot shows the number of sliding blocks versus time.

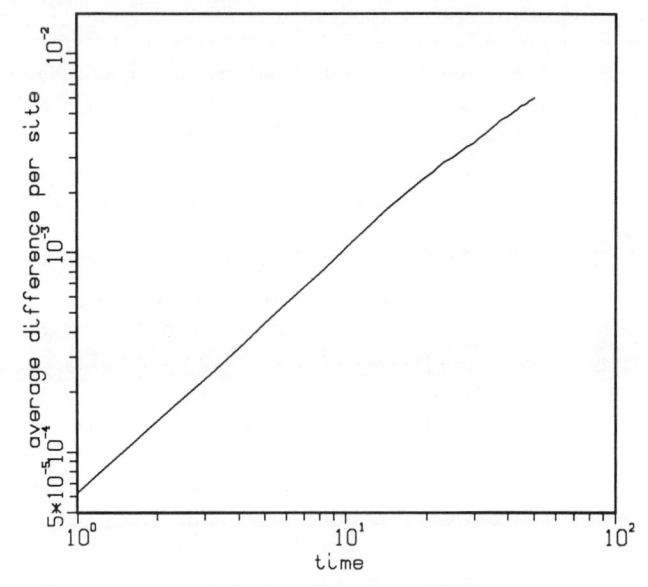

FIGURE 11.4. Power law growth of a small random difference in the initial condition (weak chaos). The plot shows the evolution of the Hamming distance, which is the sum of absolute values of the difference between a system in the critical state and the same system with a small initial, random perturbation versus time.

Once the existence of the self-organized critical state has been established, it is not so difficult to derive other exponents, such as the fractal dimension, characterizing different correlation functions (Tang and Bak, 1988). In particular Ito and Matsuzaki (1990) have generalized our model by adding a random disturbance to sites just subjected to an earthquake. They obtained a spatial clustering of epicenters with a fractal dimension of 1.1. They were also able to obtain a power law distribution of aftershocks (Omori's law). Sornette and Sornette (1989) have shown the existence of $1/f$ noise in the time gap between large earthquakes. A number of other works applying the principle of self-organized critically to earthquakes have been performed. In collaboration with S. Obukhov, Chen and others (1990) have proposed a crack propagation model of earthquakes, which includes realistic features of a long-range redistribution of elastic forces following local ruptures. The model evolves to a self-organized critical state with exponent β close to the observed one. We also notice that Brown and others (1990) have studied a spring block model of earthquakes similar to the one discussed in that paper. Their study confirms the general picture just presented.

ACKNOWLEDGMENT

Thus work was supported by the Division of Materials Science, Office of Basic Energy Sciences, US Department of Energy, under contract DE-AC02-76CH00016. We are grateful to Chris Barton for numerous constructive suggestions for the presentation.

REFERENCES

Bak, P., and Chen, K., *Phys. D.* **38**, 4 (1989).
——, *Sci. Am.* **264**, 46 (1990).
Bak, P., and Tang, C., *J. Geophys. Res. B* **94**, 15,635 (1989).
Bak, P., Tang, C., and Wiesenfeld, K., *Phys. Rev. Lett.* **59**, 381 (1987).
——, *Phys. Rev. A* **38**, 364 (1988).
Bobrov, W. S., and Lebedkin, M., private communications, 1989. Their results were kindly communicated to us by S. Obukhov.
Brown, R. S., Rundle, J. B., and Scholz, C. H., *A Simplified Spring-Block Model of Earthquakes*, Preprint.
Burridge, R., and Knopoff, L., *Bull. Seism. Soc. Am.* **57**, 341 (1967).
Carlson, J. M., and Langer, J. S., *Phys. Rev. Lett.* **62**, 2632 (1989).
Chen, K., Bak, P., and Obukhov, S., *Phys. Rev. A* **43**, 15 (1990).
Choi, M. Y., and Huberman, B. A., *J. Phys. C* **17**, L673 (1984).
Gutenberg, B., and Richter, C. F., *Ann. Geofis.* **9**, 1 (1956).
Hao, B. L., Chaos: World Scientific, Singapore, 1984.
Huang, J., and Turcotte, D. L., *Geophys. Res. Lett.* **232**, 223.
Ito, K., and Matsuzaki, M., *J. Geophys. Res. B* **95**, 6853 (1990).
Johnston, A. C., and Nava, S. J., *J. Geophys. Res. B* **90**, 6737 (1985).
Kadanoff, L. P., Nagel, S. R., Wu, L., and Zhou, S., *Phys. Rev. A* **39**, 6524 (1989).
Kagan, Y. Y., and Knopoff, L., *Geophys. J. R.* **62**, 303 (1980).
——, *Science* **236**, 1563 (1987).
Mandelbrot, B., *The Fractal Geometry of Nature* (W. H. Freeman, San Francisco, 1982).
Mikumo, T., and Miyatake, T., *Geophys. J. R.* **54**, 417 (1978).
——, *Geophys. J. R.* **59**, 497 (1979).
Omori, F., *J. Coll. Sci. Imp. Univ. Tokyo* **7**, 111 (1984).

Otsuka, M., *Phys. Earth Planet. Inter.* **6**, 311 (1972).
Sieh, K. E., *J. Geophys. Res.* **83**, 3907 (1978).
Sornette, A., and Sornette, D., *Europhys. Lett.* **9**, 192 (1989).
Stuart, W. D., and Mavko, G. M., *J. Geophys. Res.* **84**, 2153 (1979).
Tang, C., and Bak, P., *Phys. Rev. Lett.* **60**, 2347 (1988).
———, *J. Statist. Phys.* **51**, 797 (1988).
Turcotte, D. L., *Fractals in Geology and Physics* (Cambridge University Press, Cambridge England, 1992).

12

Mineral Crystallinity in Igneous Rocks
Fractal Method

A. D. Fowler

12.1. INTRODUCTION

The interpretation of the textures of igneous rock, that is the description and analysis of crystal morphology, crystal faces, crystal intergrowths, zoning, and crystal distributions, is essential to understanding rock fabric. Rock and mineral textures reflect the interaction of complex growth and dissolution processes controlled by physical and chemical processes that are at best poorly understood. Mineral morphology or habit is described strictly in qualitative terms (e.g., acicular, equant, ramified), as is mineral crystallinity (euhedral, etc.). The record of chemical variation within crystals (zoning) is also characterized in very qualitative terms.

Although much has been accomplished in quantifying rock and mineral chemistry, comparatively little effort has been made to quantify texture. There are numerous publications quantifying the distribution of crystals in igneous and metamorphic rocks (e.g., Serra, 1966; Agterberg, 1967; Kretz, 1969). However the descriptive approach to documenting the mineral texture of igneous rocks remains essentially unchanged from the days of early nineteenth-century workers. This has in part been due to problems of imaging mineral texture in ~30-μm thick sections and to the fact that there has been no pressing economic incentive to quantify the texture of igneous rocks. Also until recently, appropriate mathematical techniques were unavailable. For instance, although Fourier series transformations can be successfully used to analyze smooth, single-valued continuous curved lines (i.e., differential curves), such transformations are inappropriate for continuous jagged curved lines (i.e., nondifferentiable, continuous multivalued curves). Fractal geometry is uniquely suited to this task. In Chapter 12, as a first step to better documenting texture, I demonstrate

---

A. D. Fowler • Ottawa–Carleton Geoscience Centre, University of Ottawa, Ottawa, Canada.

Fractals in the Earth Sciences, edited by Christopher C. Barton and Paul R. La Pointe. Plenum Press, New York, 1995.

fractals can be used to quantify crystal outlines (crystallinity) and how simulations of fractal growth processes can lead to further understanding of some textures.

12.2. FRACTALS

Although the precise definition of the term fractal is elusive, fractals can easily be grasped by considering some simple geometric forms (deterministic fractals) before proceeding to their natural analogues, random fractals. Figure 12.1 shows a triadic Koch curve constructed on each side of the unit triangle, constructed by dividing the sides into three equal parts, called initiators (Mandelbrot, 1982). The sides are replaced by curves shaped thusly: ⌐⌐. These sides are composed of four equal parts whose length is one-third the original side length; these sides are generators. Each of these four segments is then divided in a third and replaced by a similarly shaped generator composed of elements one-ninth the length of the original side, and so on. The curve has some interesting properties: It is everywhere continuous but nondifferentiable (i.e., a tangent line cannot be drawn everywhere). With successive iterations, the length of the curve goes to infinity, whereas the area bounded by the curve converges to a limit. Note that the curve is self-similar or scale-invariant; that is, each generation is copied onto the previous one, so that with a change of scale, the object looks identical. Mandelbrot (1982) calls such curves fractal; i.e., they have a fractional dimension, not a whole number value. A straight line has dimension 1, a plane has dimension 2, whereas highly ramified curves on the plane have a fractal dimension d, where $1 < d < 2$, depending on how much of the plane they cover. The fractal dimension of the triadic Koch curve can be computed by considering the change in length of the curve with recursion, formulating a statement that 3 raised to the power d equals four ($3^d = 4$), and solving for d

$$d = \frac{\log 4}{\log 3} \qquad (1)$$

See Barton (Chapter 8) for a rigorous proof. As an aid to quick computation of the fractal dimension of deterministic fractals, note that in Eq. 1 the numerator represents log growth factor and the denominator log, scaling factor. There are numerous simple geomet-

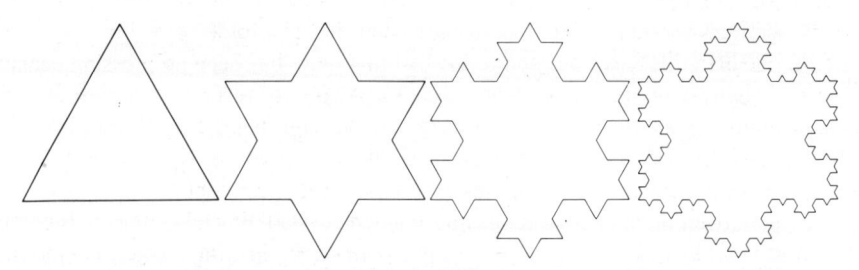

FIGURE 12.1. The first four steps in the production of a triadic Koch curve (see Mandelbrot, 1982). The curve is constructed by dividing the sides of the unit triangle in three and replacing them with a ⌐⌐ shaped line composed of four segments, each one-third the original side length. The process is repeated, resulting in a fractal curve of $\log 4/\log 3 = 1.262$.

ric constructs for two- and three-dimensional space and disconnected sets of points with fractal dimension $d < 1$.

The construction of natural objects is far more complicated than the simple rules for the curves just described. In nature random events or noise often perturb the system. In fact the Koch curves can be made into appealing models of the boundaries of natural objects (e.g., coastlines) by including randomness. One method of limited utility involves introducing into the algorithm a method of randomly choosing between different non-self-intersecting generators (e.g., Mandelbrot, 1982).

Richardson (see Mandelbrot, 1982) showed that if we measure a coastline perimeter (a natural fractal) from a map at a given scale, the measured perimeter becomes larger as the caliper diameter is reduced, because more and more of the fine detail of the curve is included. He also noted a power law relationship between perimeter length and scale of measurement. Many objects in nature are convex (i.e., all cords connecting perimeter points lie entirely within the object), bounded by perimeters that are fractal (e.g., coastlines), and they have compact mass distributions. Measuring their perimeters (p) with calipers of different apertures yields

$$\log p = b + m \log a \tag{2}$$

where b = constant, a = caliper aperture, and m = slope. Mandelbrot (1982) showed that the slope is of great significance and $d = (1 - m)$. Although the method is tedious, reliable results can be obtained by using this technique on crystal outlines.

In addition to compact structures that are more or less convex and bounded by fractal curves, there are objects that are neither compact not convex but have a fractal mass distribution. If we consider the amount of material in a two-dimensional section of a constant density object, it scales to the second power of radius; in other words, the area occupied by the material is proportional to r^2. In contrast the amount of material m contained in fractal objects scales to a smaller power, the fractal dimension, i.e., $m \propto r^d$, where $1 < d < 2$. This means that the amount of material within a given area of the object decreases proportionately as its size increases. Thus fractal objects are dilation symmetric, and very often, they tend to be composed of similar-looking parts over large ranges in scale of observation; examples are snowflakes (Nittman and Stanley, 1986) and some disequilibrium silicate crystals (Fowler and others, 1989). In many cases, such examples are instantly recognized at an empirical level, since they are composed of a hierarchy of similar branches.

12.3. EXPERIMENTAL TECHNIQUES

Methods of measuring d for both sections of fractal objects and curves include the correlation function and box methods. The correlation function (see Fig. 12.2) technique starts by having the computer pick a pixel as an origin that is part of the digitized object under consideration. From this pixel, a series of concentric subshells is constructed at radii r from the origin, and the ratio of pixels that is part of the texture to the total number in the shell [the correlation function $C(r)$], is computed for all r. The process is repeated until all texture pixels have been chosen as origins. The data are then averaged and plotted on a graph $\log C(r)$ versus $\log r$. If the object under consideration is fractal, a straight line relationship with slope $m = d - 2$ is observed. Note that for an item of constant mass

FIGURE 12.2. The correlation function plot. Developed by sequentially choosing each pixel that is part of the branching texture, then constructing concentric subshells at radii r from it. The ratio of pixels that are part of the texture to the total number within the shell $C(r)$ are computed at each site. Averages are taken, and a double logarithmic plot of $C(r)$ versus r yields the fractal dimension. Fig. 12.2 schematically shows two nests of subshells at r.

distribution, such as a compact item, $m = 0$; hence $d = 2$, the dimension of the embedding space or Euclidean dimension.

The box method algorithm (see Fig. 12.3) inscribes a box around the digitized image of the curve or object under consideration. At each step, the box-side length b is reduced by one-half, producing $(1/b)^2$ new boxes nested within the original one. Each time b is reduced, the number of boxes $N(b)$ that contain an element of the texture is counted. A double logarithmic plot of $N(b)$ versus $1/b$ has slope $= d$ (Barnsley, 1988). For this research, the box method was implemented on an ATARI ST MEGA-4 computer by means of a program written in BASIC. The initial box size was 400 × 400 pixels, so that at the last step, there were 16×10^4 boxes of length 1/400. Kaye (1978), Orford and Whalley (1983) showed that analyzing these plots may result in several straight line segments. Kaye (1978) demonstrated that these line segments were due to large-scale (shape) and small-scale (edge) effects. This is entirely reasonable (see Mandelbrot, 1982), since the dimension of an object at a particular scale is the reflection of processes operating at that scale. For example, a plot of coastline detail over the range of scales of interest to a geomorphologist and a surface scientist would likely yield more than one straight line segment. The coastline at a scale of 10^2–10^4 m may be smoothly rounded and contain small bays nested in larger ones, as

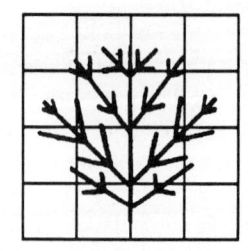

FIGURE 12.3. The box method: The idea is to inscribe a box around the texture and progressively decrease the box side length b. A double logarithmic plot of the number of boxes $N(b)$ that contain some element of the texture versus $1/b$ each scale yields the fractal dimension.

opposed to a clay particle (also part of the coastline) at a scale of 10^{-6} m composed of jagged, randomly oriented plates. In others words, dimension is bounded within the scale of observation, which must be reported with measurements of dimension. Because of the difficulties associated with resolving small objects in thin sections 30 μm thick, the lower bound reported in this study is approximately 5 μm. Since the purpose of this research is to examine mineral crystallinity (i.e., the perimeter), the approach taken is to collect data from roughly two orders of magnitude in scale and to distribute the data collection points equally across this range, so that information gathered at any particular scale is not biased.

Clearly the box method provides far more information as the scale of observation is reduced. All coefficients of linear correlation are > 0.99, and slopes are reported to two significant digits, sufficient for describing the perimeter geometry over the range of observation. The goodness of fit indicated by the correlation coefficient is likely to be overestimated because data are log-transformed (e.g., Krohn and Thompson, 1986). However the homogeneous distribution of data collection points in part compensates for this. Images of crystals for the study were photographed as transparencies, then enlarged by projection to a length of approximately 16 cm. These were entered into the computer using a digitizing tablet. The images were processed so that pixel boundaries were continuous and everywhere only 1-pixel in width. Because the crystals are embedded in solid rock they have only been analyzed in two dimensions in thin sections. Thus no attempt has been made to study potential self-affine fractal properties.

12.4. CRYSTALLINITY

To appreciate mineral growth textures and their significance, prior to examining actual crystals, I describe the large-scale environments of growth in basaltic systems and relate them to crystal growth processes at the microscopic scale. Basaltic magma, the most voluminous lava on the Planet, is produced in many different tectonic regimes. In general it consists of approximately 50% SiO_2, 15% Al_2O_3, 10% CaO, 10% MgO, 10% FeO and other oxides. It is multiply saturated, meaning that numerous mineral phases crystallize from it to produce solid rock. In general these are plagioclase $(Ca,Na)(Al,Si)AlSi_2O_8$, clinopyroxene $Ca(Mg,Fe)Si_2O_6$, orthopyroxene $(Fe,Mg)SiO_3$, olivine $(Fe_2Mg_2)SiO_2$, quartz, or glass. It has long been recognized that basaltic magma is generated within the Earth's mantle and percolates to magma chambers that may be thousands of cubic kilometers in volume and a few kilometers deep. Under near equilibrium conditions (i.e., slow cooling within these chambers), the temperature at the appearance of the first crystals (liquidus temperature) depends on the composition, and it is generally in the range of 1100–1200°C (2000–2200°F). Because the melts are composed of linked (i.e., polymerized) silica and alumina tetrahedra, they are viscous (10^{1-2} Pa s), and diffusion rates are slow. Nonetheless, slowly cooling these chambers over periods of 10^{4-5} years can produce plutonic rocks characterized by relatively well-faceted crystals varying in size from a few millimeters to a few centimeters.

Magma chambers are frequently tapped, thereby leading to the migration of lava to high levels or eruption on the surface. These lavas may entrain crystals formed within the magma chamber, and they likely cool in a manner different from the chamber, thereby producing very different mineral and rock textures. Pillow basalts afford an excellent opportunity for studying a variety of crystal morphologies. Pillows are formed when basaltic lava is extruded into the ocean floor. Although the liquid is immediately quenched, the glass remains hot and plastic, and it can be inflated with magma by the pressure of the flow, so that a lava sac is produced.

These sacs, called pillows, are frequently on the order of a meter in size, and they cool from the margin inward to form a variety of different crystal morphologies. In general the outer margin is characterized by a ~1-cm-thick glass layer formed by the quench. The glass margin may preserve within it crystals grown within the magma chamber prior to eruption, and near its inner boundary, disequilibrium crystals grown during the rapid cooling. In addition because the glassy zone acts as an insulator, heat loss from the pillow interior is relatively slower than at the margin, consequently observed crystal morphology across pillows is often variable. This results because the cooling environment controls the rate of crystal growth in a manner that can affect the geometry of individual minerals.

Above the liquidus temperature, the energy imparted to the crystal by colliding atoms in the liquid is sufficient to break the bonds holding surface atoms to the crystal and as a consequence, cause dissolution. Just below the liquidus, liquid atoms are not sufficiently energetic to cause disattachment and growth may occur. Here crystal growth is governed by atom selection at the crystal–liquid interface. Of the myriad of collisions at the interface, only a few atoms have the correct velocity, orientation, charge, or composition for attachment. Growth is most favored at those sites where the maximum amount of energy is expended in attachment, i.e., where the maximum possible number of bonds are formed (see Fig. 12.4). Growth occurs only infrequently and preferentially where there are steps (i.e., dislocations), because here the new growth atom can coordinate with more atoms than on a plane surface (see Fig. 12.4). Thus growth proceeds by the regular attachment of atoms at dislocations, so that a regular in-filling of the planes occurs, thereby producing a well-faceted crystal.

In contrast to this, at temperatures well below the liquidus (i.e., far from equilibrium), growth kinetics are not governed by choice at the interface but by diffusion within the liquid. As the temperature of the system decreases, many more of the Si-Al tetrahedra share their oxygen atoms, thereby producing a more polymerized viscous melt and reducing rates of diffusion. Thus in comparison to near-liquidus growth, few atoms are able to diffuse to the interface, and those that do are of sufficiently low energy that growth can occur at virtually any site. Once a few protuberances have built up on such crystals, the chance of atomic diffusion into their interstices without collision and attachment is small. There, smooth in-filling growth is not favored; hence crystals tend to form branching arrays. This type of growth is nicely illustrated by the diffusion-limited aggregation algorithm discussed

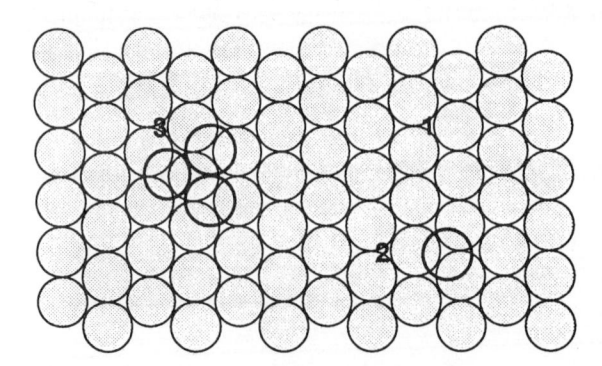

FIGURE 12.4. Conceptual illustration of crystal growth at equilibrium. Because a growth atom landing at Site 3 would coordinate with five atoms, i.e., more than one arriving at sites labeled 2 or 1, Site 3 is favored for growth. Under equilibrium conditions near the liquidus, the crystal selects only those atoms of correct energy and geometry; thus growth proceeds in a stepwise fashion at dislocation sites.

12. MINERAL CRYSTALLINE IN IGNEOUS ROCKS 243

in the following paragraphs (see Fig. 12.5). Crystals that have been transported from an environment of near-liquidus growth to quick-cooling volcanic environments often show well-faceted planar faces with wispy or branching growth from their corners. This occurs because the corners of the earlier grown well-faceted crystal subtend more volume in the liquid than do the faces; hence under diffusion-limited conditions, they become preferred sites of growth.

Figure 12.6 shows two images of plagioclase from an Eocene pillow basalt collected near Omapere, South Island, New Zealand. Here the basalts, which are very well preserved, form a part of a tholeiitic sequence of pillow basalt, basaltic agglomerate, and tuff interlayered with chalks and marl. The pillow margins consist of glass and crystals of plagioclase (labradorite) and olivine. Within the ~1-cm-thick glassy margin, the basalts

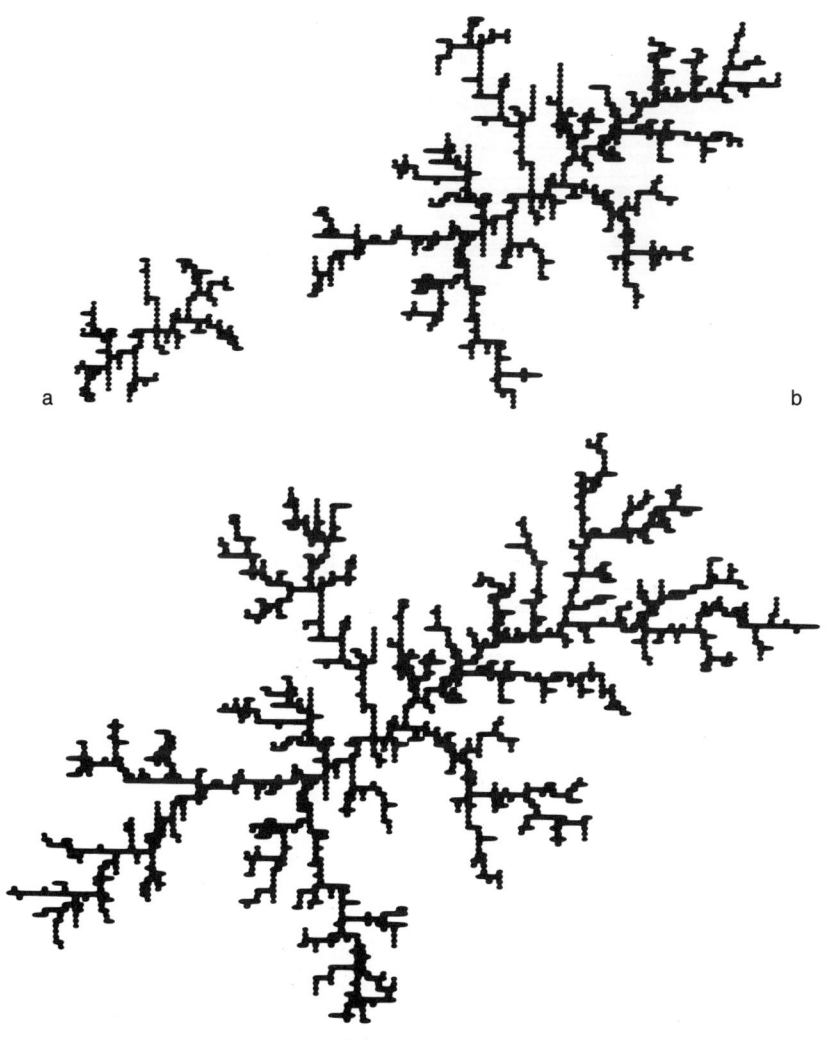

FIGURE 12.5. DLA simulation. Figures 5a, 5b, and 5c show the attachment over time intervals. Note that few random walkers penetrate the pre-existing structure to cause growth.

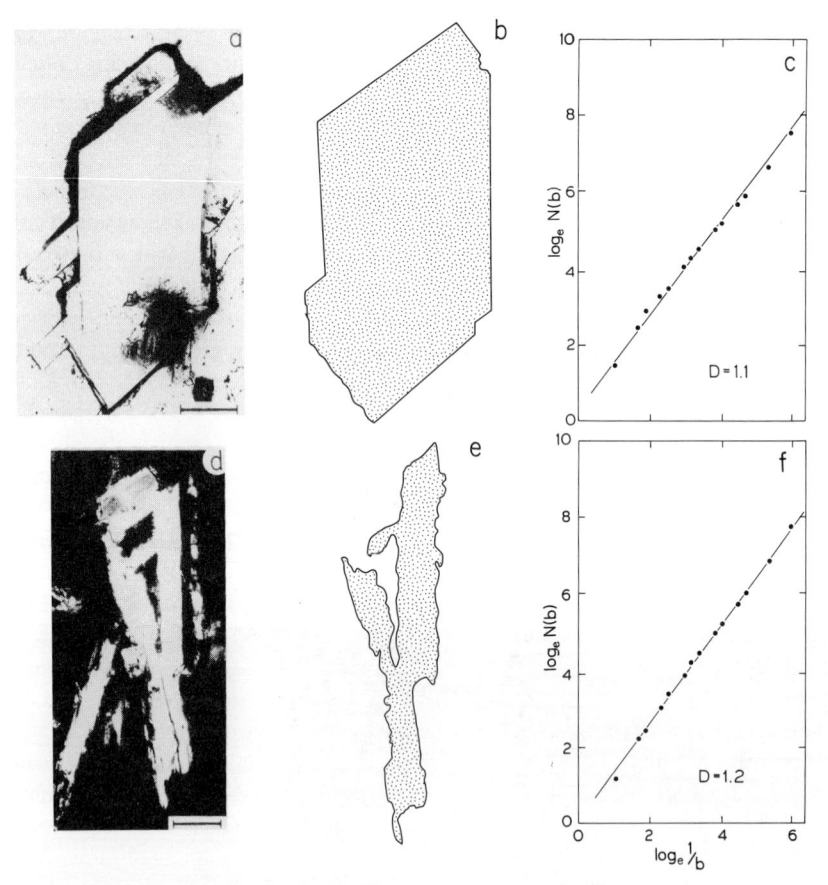

FIGURE 12.6. Plane polarized light (ppl) photomicrograph (a) of a plagioclase crystal from the quench glassy margin of an Eocene pillow basalt from Oamaru South Island, New Zealand. The drawing (b) is the actual crystal perimeter digitized into the computer, and the box-method graph (c) indicates that the fractal dimension of the perimeter of the crystal is 1.1 over the range of 0.005–0.65 mm. (d) A ppl image of plagioclase crystals from the same section as in (c). Here the crystals are within an aphanitic groundmass closer to the pillow core that cooled slower than the pillow margin, resulting in posteruption crystal modification. (e) Analyzed single crystal. (f) It has a fractal dimension of 1.2 over 0.005–0.49 mm.

contain well-faceted euhedral equant and lath-shaped plagioclase crystals, which presumably record a period of preeruption growth in a magma chamber. Typical of these are the crystal outlines in Figs. 12.6a and b. It is dominated at the scale of observation (optical microscope) by straight boundaries, and it has a fractal dimension of 1.1.

Closer to the pillow interiors, plagioclase crystals are more elongate, and these would likely be termed euhedral or subhedral, depending on the observer. Unlike those of the quench margin, the edges of these crystals reflect a period of modification during posteruption cooling. Some of these crystals are well-faceted, and they have so-called swallow tail terminations due to later growth at the crystal corners because of rapid cooling. Other crystals have edges that are rounded and embayed (see Figs. 12.6d and e). I interpret the embayed character of some of these crystals as resulting from partial dissolution.

12. MINERAL CRYSTALLINE IN IGNEOUS ROCKS 245

Dissolution is obviously also in part a diffusion process, and it produces again for geometric reasons, rounded edges and curved embayments. Clearly the two crystals imaged have very different crystallinities, as reflected by the fact that there is a significant difference in the fractal dimensions of their perimeters, 1.1 versus 1.2 (see Figs. 12.6c and f).

Figure 12.7 shows plagioclase crystals from an extremely well-preserved Archean pillow metabasalt in Holloway Township, Abitibi Greenstone belt near Kirkland Lake, Ontario. These rocks are greater than 2.7 gigayears in age, and they have a truly remarkable degree of textural preservation. Here the only evidence of a period of pre-eruption magma chamber growth are small (75-μm) plagioclase microlites preserved in some of the altered glasses. These also served as nuclei for some of the plagioclase spherulites (Fowler and others, 1987). The dominant crystal morphology of these rocks, so-called variolites, is disequilibrium plagioclase spherulites (see Figs. 12.7a, d, and g). The spherulites are characterized by radiating sheafs of needlelike plagioclase crystals. They are not observed in altered glass fragments between pillow margins. They first appear in the altered glass margin a few millimeters from the crystalline basalt. From this cooling contact in the center of the pillow, the plagioclase morphology changes in a manner very similar to that observed in the plagioclase-undercooling experiments of Lofgren (1974). Fowler and others (1987) showed that the morphology changes from small (1–2 mm) close-spaced time fibroradial spherulites to larger ones ($\sim$ 1 cm), then to spherulites with coarser fibers and broader spacing, to branching forms, and finally to dendritic forms within a few centimeters of the quenched glassy margin.

Figure 12.7 shows examples of these disequilibrium textures in section, and they are fractal objects. As we expect, those grown far from equilibrium (see Figs. 12.7d and g), the closed-spaced spherulites, have a higher fractal dimension, 1.6–1.7 (see Figs. 7f and i) than the fan-shaped one in Fig. 12.7a, $d = 1.3$, where it in turn has a larger fractal dimension than the more equilibrium-type forms documented above from New Zealand. The branching forms of such minerals as these are not included in classifications of mineral crystallinity, and they are the subject of a rich and with the exception of a few specialists, an inconsistently applied nomenclature all their own. Clearly the fractal dimension is a more quantitative, but not unique, index of mineral crystallinity that may prove useful along with the qualitative terms euhedral, subhedral, anhedral, ameboid, variolitic, etc.

12.5. FRACTAL GROWTH SIMULATION

Reviews of a microscopic model of growth that has been successfully applied to simulate many fractal objects, including crystals grown far from equilibrium are provided by Meakin (1989) and Vicsek (1989). The algorithm designates a pixel in the center of the computer screen as a growth site or seed. Other pixels are released to move randomly until they intersect the seed and are frozen to it. Once a few pixels stick protuberances preferentially grow because they shield their intervening embayments. These continue to grow, and after many tens of thousands of growth events, a branching morphology is formed because the probability of a random walker descending into the void between branches without collision to cause in-filling growth is nil (see Fig. 12.5). The DLA simulates far from equilibrium growth, since the random walk of the pixels mimics that of species in a liquid under diffusion-limited conditions and each collision with the nucleus results in growth. Figure 12.5 shows the results of a two-dimensional simulation grown on a

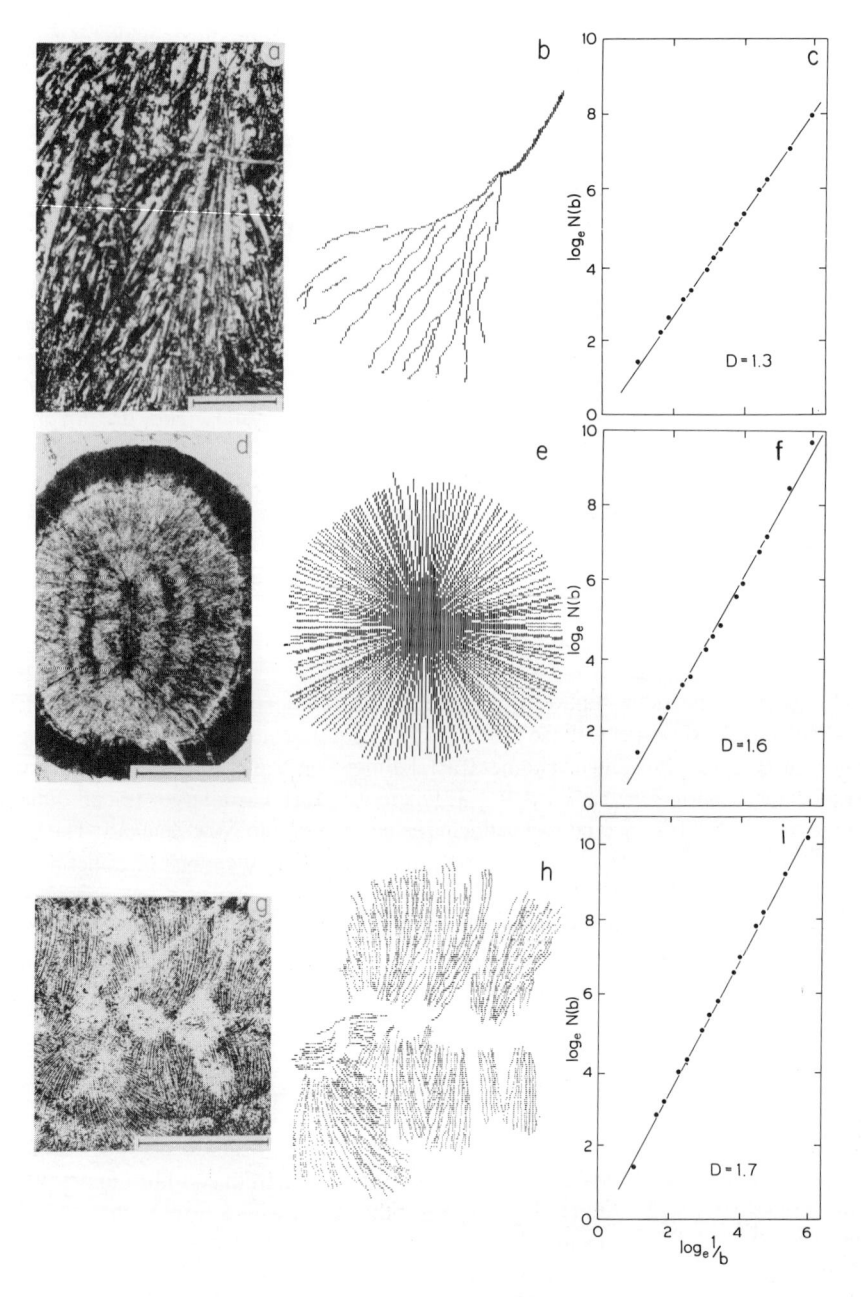

FIGURE 12.7. Images of plagioclase crystals grown far from equilibrium in a pillow metabasalt of the Abitibi greenstone belt Holloway Township, Ontario, Canada. At high undercoolings, crystals have the characteristic morphology (spherulites) of (d) or (g), which are composed of radiating arrays of very fine fibrous crystals, each one having a distinct crystallographic orientation from its neighbor. Note that the image in (a) shows coarser branching more typical of lower degrees of undercooling than the spherulites. The digital drawings (b), (e), and (h) are not so detailed as those used at a larger scale for the digitization. The fractal dimensions follow: (a) $d = 1.3$ over the range of approximately 0.005–0.7 mm; (d) $d = 1.6$ over the range of approximately 0.005–0.6 mm; (g) $d = 1.7$ over the range of approximately 0.005–1.0 mm.

12. MINERAL CRYSTALLINE IN IGNEOUS ROCKS

square lattice. The fractal dimension is 1.6, and the pattern is composed of branches separated by voids that become progressively larger with distance from the nucleus of growth.

Although this algorithm yields fractal patterns and provides an analogue of far from equilibrium mineral growth, it can be modified to produce more meaningful results. The growth algorithm of Witten and Meakin (1983) was used by Fowler, and others to understand disequilibrium mineral growth better. Unlike pure DLA, this model is realistic because it includes a finite number of nuclei and walkers and an anisotropy parameter. Note that the natural image (compare Figs. 12.8a, b, and d) is visually and quantitatively very similar to the simulated image $d = 1.72$ versus $d = 1.73$ (compare Figs. 12.8c and e).

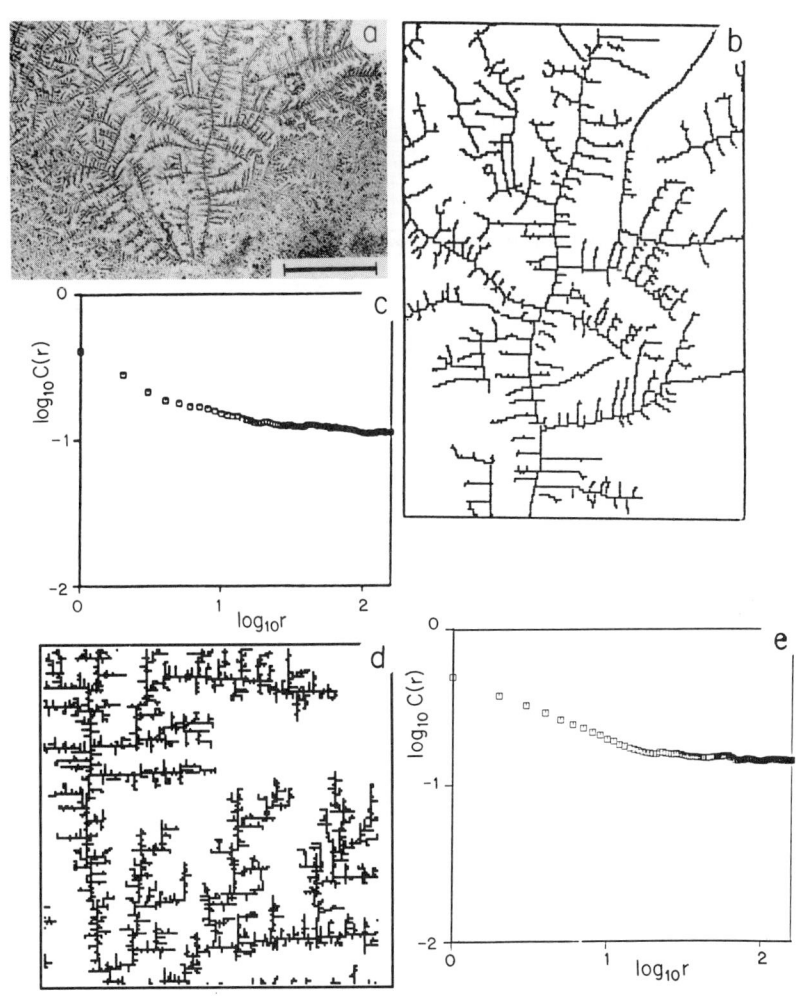

FIGURE 12.8. (a) An image of branching olivine from an altered pillow basalt, scale bar = 0.1 mm. (b) Its digitization; (c) the correlation function plot, slope = 0.28 and $d = 1.72$. (d) and (e) A DLA simulation of the olivine and its correlation function plot, respectively, $d = 1.73$ (Fowler and others, 1989). [Reprinted by permission from Nature **341**, 135 (1989). Copyright © 1989 Macmillan Magazines Ltd.]

Both correlation function plots (see Figs. 12.8c and e) have a cross-over from fractal to constant density behavior; that is, the slope of the correlation function plot tends to zero at a large cluster radius. This occurs because at low radii, the fractal morphology is efficient at trapping pixels (or species from the melt), since the ratio of the crystal surface area to liquid volume is large. At large radii, the distance between branch tips becomes large, so that this ratio is smaller, and hence interbranch diffusion and in-filling growth are possible.

The interpretation of the cross-over has led to the development of a kinetic model based on the fact that at the cross-over from fractal to constant density distribution in the simulations, the average displacement of a walker during the mean time between two growth events (dt) is on the order of the size of the cluster (Witten and Meakin, 1983). This can be expressed as $r_c = \sqrt{D\,dt}$, where r_c is the cluster radius at the cross-over and D is the species diffusivity (cm^2/s). The DLA can be scaled to the actual mineral growth, and using a reasonable value of D, instantaneous growth rates at the cross-over can be calculated. The concentration of species at the time of cross-over is calculated (Fowler and others, 1989) using the estimated dt and Smoluchowski kinetics, which account for diffusion-limited conditions in binary reactions based on the assumption that the concentration of one reactant is infinitesimal. Witten and Meakin (1983) showed that for the simulations, fractal growth proceeds until the pixel concentration in the periphery of the growing cluster is equal to their initial pregrowth matrix concentration. Thus estimates of the initial concentration of species in the system are obtained by interpolating the value of C_r on the ordinate of the correlation function plot, corresponding to the concentration at the cross-over C_{sat}.

12.6. SUMMARY

It has been shown that the fractal dimension provides a quantitative measure of crystallinity and simulations of fractal growth mechanisms model far from equilibrium mineral growth. The fractal method needs to be extended to include self-affine and multifractals in order to document and understand other textural parameters, such as mineral morphology, zoning patterns, and mineral distributions.

ACKNOWLEDGMENTS

Thanks are due to Dan Roach for writing the computer programs, to Bill Fyson for use of the computer, and to Edward Hearn for preparing some of the figures. Thanks are due to F. P. Agterberg, C. C. Barton, and C. E. Krohn for thoughtful reviews. This work was partially supported by NSERC funding and the University of Ottawa, for which the author is grateful.

REFERENCES

Avnir, D., ed. *The Fractal Approach to Heterogeneous Chemistry, Surfaces, Colloids, Polymers*. (John Wiley & Sons, Chichester, 1989).
Barnsley, M., *Fractals Everywhere* (Academic Press, San Diego, 1988).

Fowler, A. D., Jensen, L. S., and Peloquin, A S., *Can. Mineral.* **25**, 275 (1987).
Fowler, A. D., Stanley, H. E., and Daccord, G., *Nature* **341**, 134 (1989).
Kaye, B. H., *Powder Technol.* **21**, 1 (1978).
Lofgren, G., *Am. J. Sci.* **274**, 243 (1974).
Mandelbrot, B. B., *The Fractal Geometry of Nature* (W. H. Freeman, New York, 1982).
Nittman, J., and Stanley, H. E., *Nature* **321**, 663 (1986).
Orford, J. D., and Whalley, W. B., *Sedimentol.* **30**, 655 (1983).
Vicsek, T., *Fractal Growth Phenomena* (World Scientific Publishing Company, Singapore, 1989).
Witten, T. A., and Meakin, P., *Phys. Rev.* **B28**, 5632 (1983).
Witten, T. A., and Sander, L. M., *Phys. Rev. Lett.* **47**, 1400 (1981).

13

Fractal Structure of Electrum Dendrites in Bonanza Epithermal Au-Ag Deposits

J. A. Saunders and P. A. Schoenly

13.1. INTRODUCTION

Epithermal vein and disseminated Au-Ag deposits hosted by Tertiary volcanic rocks account for a significant amount of the precious metals extracted in the Great Basin region of the western United States. Locally extraordinarily rich (bonanza) shallow epithermal vein deposits consist of crustiform bands containing varying proportions of fine-grained silica (SiO_2), adularia ($KAlSi_3O_8$), and electrum (Au-Ag alloy) that were typically deposited at temperatures <250°C (482°F) (Buchanan, 1981; Saunders, 1994). Within some deposits, the bonanza ore zones containing >1 kg Au/tonne occur at approximately the same elevation. This has been cited as evidence of hydrothermal fluids boiling over a narrow vertical range (Buchanan, 1981; Berger and Eimon, 1983). Highest grade ores commonly contain electrum dendrites in a matrix of opal or very fine-grained silica recrystallized from opal (Saunders, 1994). Surface features of electrum dendrites and their intimate association with former silica colloids (now opal) indicate that dendrites formed by aggregation of colloidal particles (Saunders, 1990, 1994).

Gold colloids have been studied in the laboratory by chemists and alchemists for more than a millennium, and general aspects of nucleation, growth, and aggregation of these spherical particles have been known for some time (Turkevich, 1959; Enustun and Turkevich, 1963). More recently, the kinetics of gold colloid aggregation has been studied in detail (Weitz and Huang, 1984; Weitz and Oliveria, 1984; Lin and others, 1989), indicating that gold colloids aggregate by either rapid diffusion-limited aggregation (DLA) or much slower reaction-limited aggregation (RLA). The former process occurs when repulsive

---

J. A. Saunders and P. A. Schoenly • Department of Geology, Auburn University, Auburn, Alabama.

Fractals in the Earth Sciences, edited by Christopher C. Barton and Paul R. La Pointe. Plenum Press, New York, 1995.

electrostatic forces between particles are negligible, thereby leading to high-particle sticking efficiencies. In RLA repulsive forces retard particle sticking (Lin and others, 1989). Van der Waals forces initially link gold particles, but these bonds quickly evolve to stronger and probably irreversible metallic bonds (Weitz and Huang, 1984; Weitz and Oliveria, 1984).

Aggregation of colloidal gold particles of relatively uniform diameter in the laboratory produces structures exhibiting a fractal nature: DLA produces tenuous dendritic structures whose two-dimensional projection have a fractal dimension (D_f) of <2, whereas RLA produces more compact aggregates with $D_f > 2$ (Lin and others, 1989). The fractal dimension obtained for the diffusion-limited case is significantly less than the theoretical value for three-dimensional DLA aggregates of individual particles ($D_f = 2.5$) (Meakin, 1983; Vicsek, 1992), which is interpreted to be the result of colloid clustering prior to diffusion-limited cluster aggregation (Weitz and Oliveria, 1984).

In Chapter 13, we describe the nature and origin of electrum dendrites that comprise the bonanza ores in two epithermal vein deposits. Recent advances in understanding colloid aggregation mechanisms through laboratory studies and computer simulations provide a framework for interpreting the origin of natural dendrites in bonanza epithermal gold deposits.

13.2. GEOLOGIC SETTING

Bonanza ores from the Sleeper and National deposits of northern Nevada (see Fig. 13.1) were chosen for study because of the ubiquitous presence of electrum dendrites in the highest grade ores and because of the numerous samples available for study. Both the Sleeper and National deposits are spatially and temporally associated with Miocene rhyolites that comprise part of an extensive bimodal suite of volcanic rocks in the northern Great Basin region emplaced around 17–15 Ma (Noble and others, 1988; Vikre, 1985; Nash and others, 1991). Sleeper and National are representatives of the adularia-sericite class (Heald and others, 1987) of volcanic-hosted epithermal deposits (also called low-sulfidation by White and Hedenquist, 1990). This class of epithermal deposits typically have Ag:Au ratios <1, and it contains relatively minor amounts of base metal sulfide minerals in the bonanza ores. In typical epithermal systems, both the Ag:Au ratio and base metal content increases with depth (Buchanan, 1981). By analogy to present-day geothermal systems

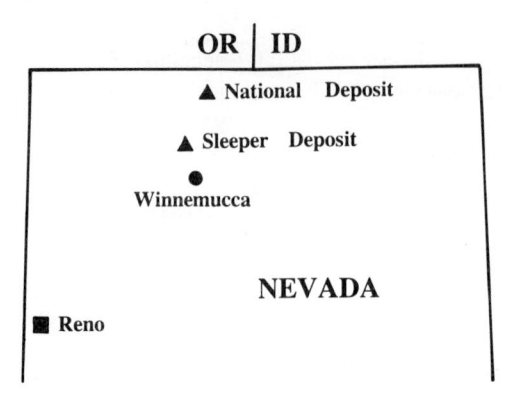

FIGURE 13.1. Map showing the location of the Sleeper and National deposits.

13. ELECTRUM DENDRITES IN BONANZA EPITHERMAL AU-AG DEPOSITS

(Hedenquist, 1991), the presence of adularia (KAlSi$_3$O$_8$) and calcite (CaCO$_3$) in these epithermal veins indicates that the hydrothermal solutions were near neutral to alkaline and probably boiling, although direct evidence of the latter is rarely preserved (Bodnar and others, 1986). Boiling in geothermal or epithermal systems occurs when convectively driven hydrothermal solutions encounter lower pressures as they rise through the shallow curst. Boiling can lead to the precipitation of amorphous silica (Fournier, 1985, 1986), and it is extremely effective in precipitating gold, due to the destabilization of the Au(HS)$_2^-$ complex as H$_2$S is partitioned into the vapor phase (Drummond and Ohmoto, 1985; Brown, 1986; Romberger, 1988; and Spycher and Reed, 1989). In addition, boiling can agitate, cool, and increase the salinity and pH of the remaining liquid fraction, all of which can enhance colloid aggregation (Saunders, 1990). Current models for epithermal systems relate the onset of precious metal deposition to the initiation of boiling (see Fig. 13.2).

13.3. ELECTRUM TEXTURES

Bonanza ores from the Sleeper and National deposits are composed of multiple millimeter- to centimeter-scale bands deposited along the existing vein wall, which consist of fine-grained silica and variable amounts of electrum. In the richest ores, multiple gold rich bands alternate with barren bands, indicating that gold deposition was a cyclic event. Gold rich bands commonly exceed 50 volume percent electrum, and they contain coalesc-

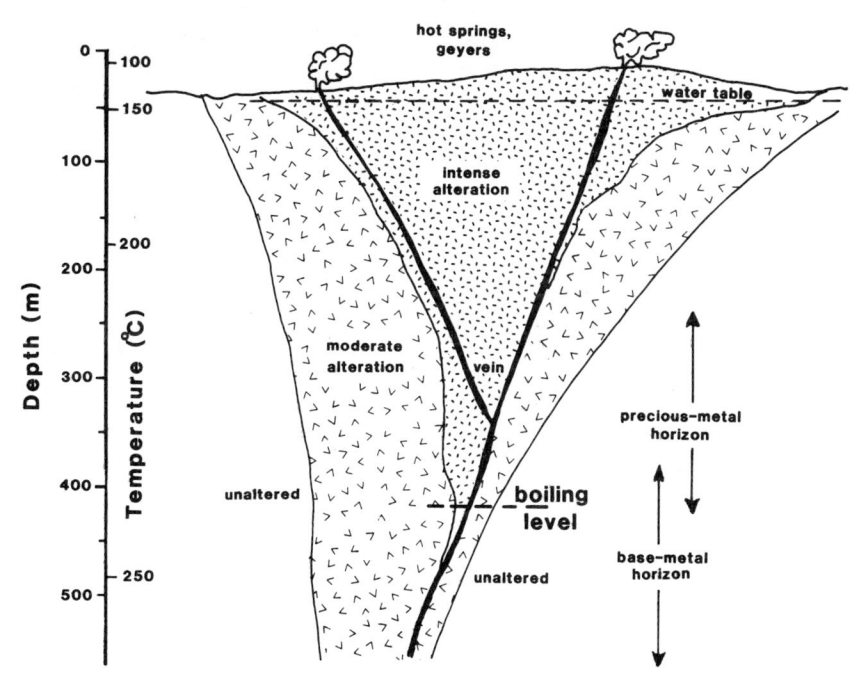

FIGURE 13.2. Diagrammatic cross section through the upper part of an epithermal vein system in volcanic rocks, showing the intensity of hydrothermal alteration and location of precious metal deposition (modified from Buchanan, 1981). Boiling level shown is for a low-salinity solution under hydrostatic pressure conditions.

ing dendrites. Locally isolated dendrites up to 4 cm in height project orthogonally out from the substrate (see Fig. 13.3a), which was the vein wall at the time of dendrite growth. Branches of these ramified structures (e.g., Fig. 13.3a) can be shown to be mutually connected by using an electrical resistance meter. Dendrite branches are in turn branched, illustrating the fractal nature of the structures. Colloidal silica was apparently deposited during dendrite growth, and it in-filled spaces between branches. However cavities locally exist between branches, suggesting silica in-filling was not completely efficient due to branch shading.

Many of the dendrites exhibit a rough radial symmetry, are hemispherical in shape, and their grain size commonly coarsens outward (see Figs. 13.3b, c). Examining dendrite surfaces etched with hydrofluoric acid to remove encrusting silica reveals a botryoidal surface devoid of crystal faces, although a few rare electrum crystals are present (see Fig. 13.3d). We interpret the noncrystalline botryoidal dendrite surfaces to result from the

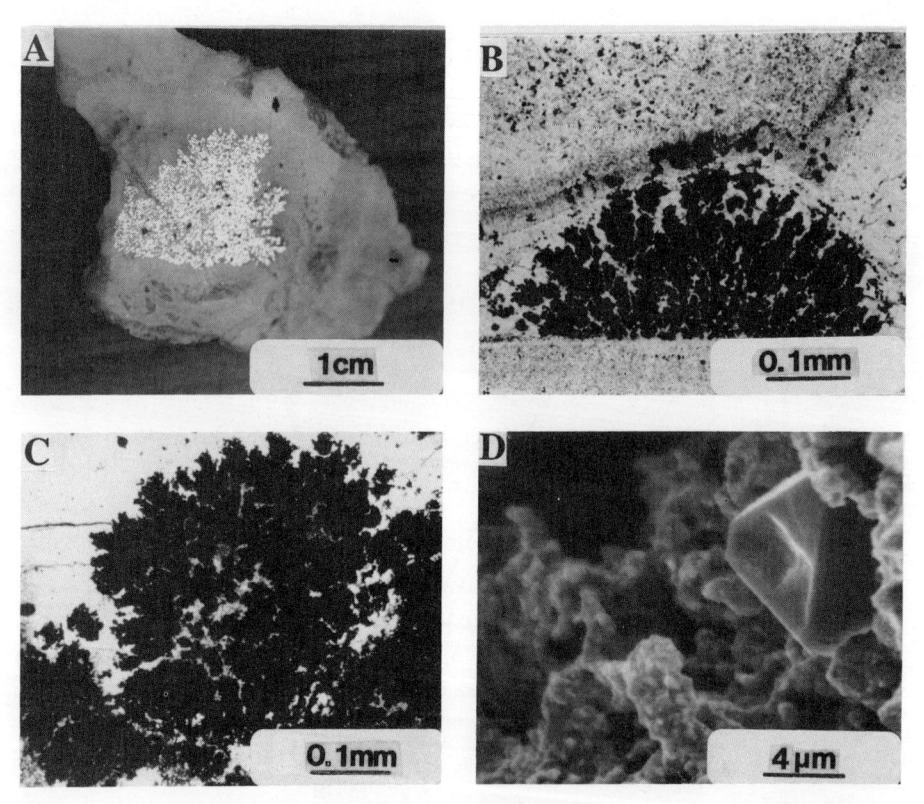

FIGURE 13.3. (a) Photograph of a single electrum dendrite from the National deposit, Nevada. The dendrite apparently grew from a single seed particle embedded in the vein wall. (b) Photomicrograph of a cross-section through a radiating electrum dendrite (*black*) in fine-grained silica from the Sleeper deposit, Nevada, oriented perpendicular to the vein wall (transmitted light). (c) Photomicrograph of a cross-section through an electrum dendrite similar to (b), but oriented parallel to the vein wall (transmitted light). (d) The scanning electron microscope image of electrum dendrites exhibiting botryoidal surfaces, which suggests dendrites formed by aggregation of spherical particles. Note rare octahedral crystal of electrum for contrast.

aggregation of spherical electrum colloids in the ~10–100 nm size range, where the interiors have since recrystallized, leaving only surface features as evidence of the former existence of the spheroids.

In the early part of this century, Lindgren (1915, 1933) and Boydell (1924) proposed that gold colloids may have played a role in bonanza ore formation in the National (Nevada) and Republic (Washington) districts based on ore textures. This concept was supported by laboratory studies (Frondel, 1938) demonstrating that gold colloids could exist under hydrothermal conditions (100°–350°C, 212–662°F) when colloidal silica was present in the solution. Frondel (1938) showed that an increase in salinity, pH changes, cooling, or mixing with other solutions could cause the aggregation of hydrothermal gold colloids. More recently Saunders (1990) documented the opaline nature of silica codeposited with electrum at the Sleeper deposit. He suggested that both silica and electrum were transported to the deposition site as colloids. Recognizing that dendritic electrum textures documented here are a product of colloid aggregation and their occurrence within the highest grade ore zones indicate that this nonequilibrium process is an important ore-forming mechanism in these systems. Furthermore it provides a plausible explanation of how a hydrothermal fluid can deposit a layer that is ~50% electrum during a specific time interval.

13.4. FRACTAL DIMENSIONS OF NATURAL DENDRITES

To evaluate the fractal nature of the electrum dendrites from the Sleeper and National deposits, photomicrographs taken in reflected light of dendrites shown in Figs. 13.3a–c were analyzed by a completely automated procedure to determine their fractal dimensions. A digital scanner was used to scan photomicrographs into digital graphics images. By changing the sensitivity of the scanning procedure, it was possible to enhance the resolution between electrum and the surrounding mineral grains. The resulting digital images of photomicrographs do not represent a true particle-by-particle duplication of the actual dendrites, but they reproduce their general morphology. Once produced the graphic images were imported into a graphics enhancement program. The enhancer removed all remaining stray background computer pixels that had been improperly registered as belonging to the dendrite. At the end of this process, a separate computer program analyzed the digital images for their fractal dimension. The fractal dimensions were calculated by the sandbox method of Vicsek (1992), which determined the pixel distribution within the entire dendrite structure. In this method, a box of linear size R is placed at the center of the image, and the number of particles (in this case, pixels) within this box are counted. As R is increased in size over the length of the image, $N(R)$ is calculated separately each time. For an object to be fractal, $N(R)$ should scale to R with a power law distribution by the equation

$$N(R) = R^D$$

where D equals the fractal dimension (Mandelbrot, 1983). The $N(R)$ and R are then plotted on a logarithmic graph. To determine the fractal dimension properly, the slope of a best fit line through only those data points displaying power law scaling equals the fractal dimension. Figure 13.4 shows the output from the sandbox calculations and the resulting fractal dimensions for the three dendrites. The fractal dimensions are 1.77 and 1.91 for the cross sections of the Sleeper dendrites perpendicular and parallel to the growth surface,

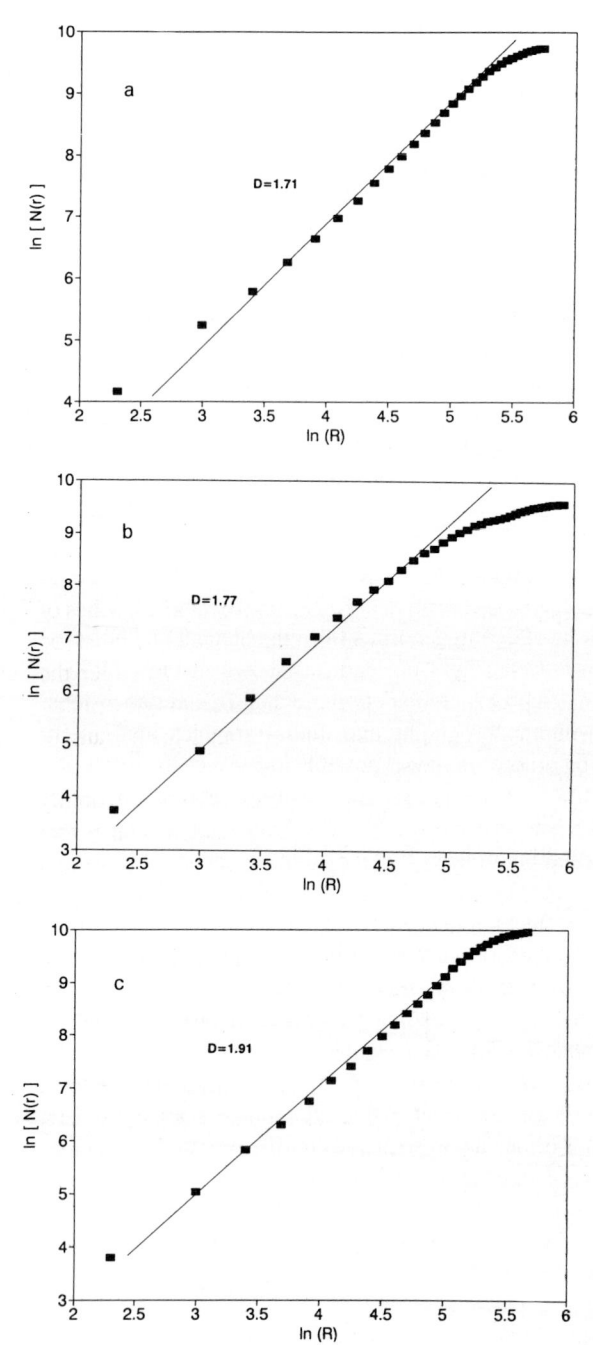

FIGURE 13.4. Log-log plots of sandbox calculations of digitized (reflected light) images of dendrites shown in Figs. 13.3a–c, which correspond to a–c here.

respectively, and 1.71 for the National dendrite. Note that the fractal dimensions calculated are values for two-dimensional slices through three-dimensional objects. An estimate of the three-dimensional fractal dimension can be attained by adding 1 to the D_f value for $d = 2$ (d = Euclidean dimension) provided that $D_f < d$ (Vicsek, 1992). Therefore the electrum dendrites have D_f values 2.77, 2.91, and 2.71, respectively, in $d = 3$.

In the plots of $N(R)$ versus R, all three dendrites scale well to a certain value of R. Past this point, R no longer scales to $N(R)$ as a function of D_f. This feature is not due to R surpassing the linear size of the structure, since the analytical procedure did not use values of R greater than the true linear size. Instead the change is apparently a genuine feature of the natural structures. This behavior is less pronounced in the dendrite from the National deposit than it is in the Sleeper dendrites.

Although the natural dendrites shown in Fig. 13.3 are visually similar to dendrites formed by DLA, their fractal dimensions are significantly higher than the theoretical value of 2.5 for three-dimensional DLA dendrites (Meakin, 1983). The electrum dendrite from the National deposit has the smallest D_f of the three, and this can be verified qualitatively by visually comparing this dendrite to a computer-simulated DLA structure. However the National dendrite's branches are more curved and closely spaced than in normal DLA. The D_f values demonstrate that although branches qualitatively looks similar to DLA, the branch spacing is atypical and subsequently leads to a higher value of D_f. Similarly branches from the two dendrites from Sleeper look much thicker, which in a three-dimensional view would occupy a much greater area than the loose and tenuous branches of simulated DLA structures. The size of the branches is counterbalanced by the fact that the fjords between the branches are larger and have more linear continuity than either the National dendrite or ideal DLA structures. The net effect of these differences is that a higher D_f value is attained because of branch width, but it is counterbalanced somewhat because of fjord size.

In true DLA, ions or molecules diffuse through a melt or aqueous solution, randomly collide with the growing dendrite, then precipitate on the surface. In contrast it appears that the natural electrum dendrites form from the aggregation of colloidal particles of finite and variable size (e.g., ~10–100 nm). We hypothesize that the gross dendrite morphology is a product of random-particle trajectories, resulting in a DLA-like dendrites but D_f values in excess of 2.5 result from departures or modifications from true DLA. We envision that the relatively large and variable particle size, coupled with depositional and postdepositional processes, can account for an increase in the fractal dimension over the DLA value. Given the turbulent nature of a boiling hydrothermal fluid, an important postdepositional process may be fluid shear, which causes more tenuous branch tips to break or bend and results in thicker branches and raises the fractal dimension of colloid aggregates (Lin and others, 1990; Oles, 1992). In the following section, we present results of computer simulations designed to evaluate particle randomness and depositional processes.

13.5. COMPUTER SIMULATIONS

The tailing off of the scaling exponent, the high values of D_f, and differences between natural dendrites and ideal DLA aggregates provide insights into the aggregation mechanism for these dendrites. In normal DLA aggregates, particle size is constant, and structures are tenuous with narrow and highly random branches. Such structures have been simulated by computers using a model originally proposed by Witten and Sander (1981).

In electrum dendrites from the Sleeper deposit, a limited number of principal branches dominate the structure; these have a pronounced radial orientation and typically coarsen outward (see Fig. 13.3b). These features are not completely compatible with the ideal DLA model. Near the nucleus of the structure, branches are more curved and tenuous, which is particularly visible in true cross sections of dendrites. In contrast National dendrite branches do not thicken outward as much; instead dendrite branches are closely spaced, narrow, and the fjords are small and commonly noncontinuous. Thus it appears that National dendrite resembles true DLA aggregates more closely than the Sleeper dendrites, but the spatial distribution of National dendrite branches is atypical of true DLA structures.

A diffusion-limited aggregation algorithm modified to reproduce the distinctive characteristics of the electrum dendrites was used to simulate various factors that could produce these differences. The simulation followed the typical DLA random-walk model; however several variables were introduced in an attempt to reproduce the distinctive characteristics of the dendrites. Two modifications to the DLA model proved effective in simulating the gross morphologic characteristics of the natural electrum dendrites. The first model was designed to simulate natural surficial modifications (e.g., restructuring due to fluid shear) during dendrite growth. This algorithm specified a predetermined number of adjacent particles (pixels) on a dendrite branch that had to be contacted before the incoming particle was allowed to aggregate. This adherence factor was used to grow aggregate when one, two, and three adjacent particles were required to be present (see Fig. 13.5a–c). The fractal dimensions of the resulting aggregates for $d = 2$ were 1.51, 1.71, and 1.77 for an adherence factor of 1, 2, and 3, respectively. The increase in D_f with an increasing adherence factor reflects the denser nature of the aggregates. Computer-generated aggregates with an adherence factor of 3 correspond especially well with the cross section of the Sleeper dendrite in photomicrographs taken in transmitted light (e.g., Fig. 13.3b), whereas aggregates with an adherence factor of 2 have a D_f value similar to the National dendrite for $d = 2$. In the latter case however, morphologies of the computer and natural aggregates are not so similar.

While computer-generated structures approximate relative branch widths, they do not have similar fjord morphologies. To simulate this difference, a second algorithm modification was added in which the randomly walking particles were of random size (see Fig. 13.5d). This modification resulted in aggregates with D_f values of approximately 1.65 for $d = 2$. While this D_f value is less than that of the National dendrite, the simulated dendrite shows how variable particle size can result in highly irregular branch orientations. For example, large particle deposition can affect dendrite structures by forming a bridge between adjacent branches, thereby leading to their merger.

Based on analysis of the fractal dimensions of natural electrum dendrites, it appears that they did not grow by a true diffusion-limited aggregation process. However visual similarities between natural dendrites and ideal DLA structures, coupled with random-walk computer simulations of the natural dendrites, indicate that a modified DLA process was responsible for electrum dendrite formation. Similarly Chopard and others (1991) showed that natural two-dimensional MnO_2 dendrites (pseudofossils) have D_f values commonly exceeding the theoretical DLA value, and the researchers also concluded that these dendrites were formed by a modified diffusion-limited aggregation process.

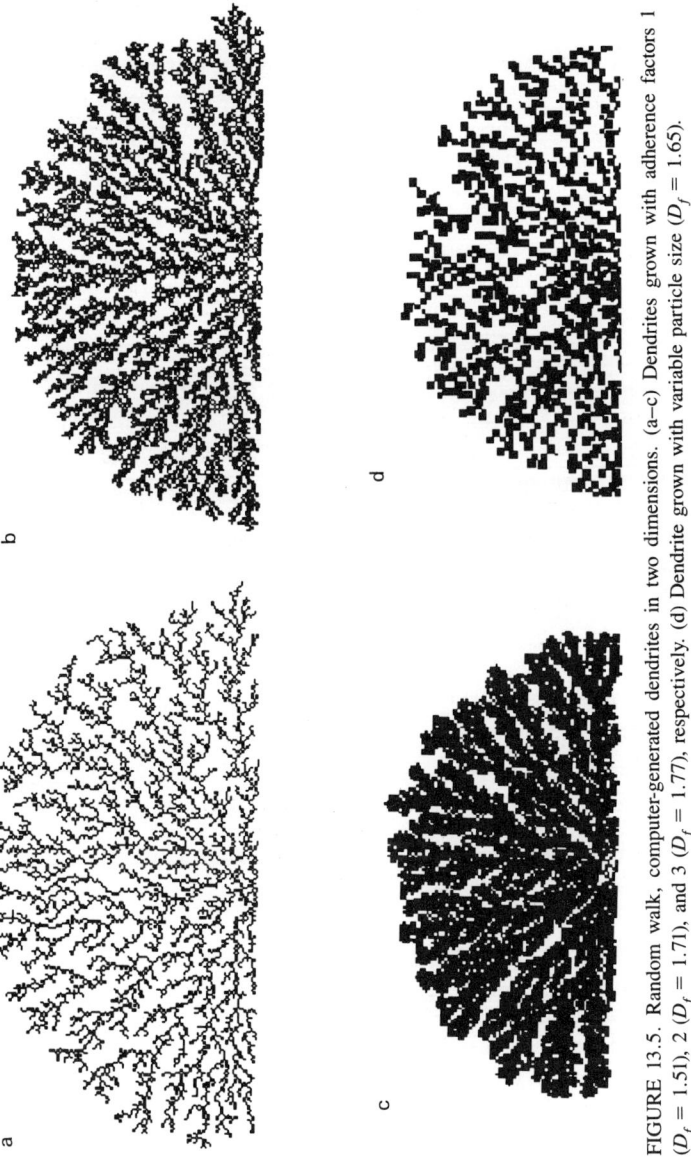

FIGURE 13.5. Random walk, computer-generated dendrites in two dimensions. (a–c) Dendrites grown with adherence factors 1 ($D_f = 1.51$), 2 ($D_f = 1.71$), and 3 ($D_f = 1.77$), respectively. (d) Dendrite grown with variable particle size ($D_f = 1.65$).

13.6. CONCLUSIONS

Natural three-dimensional electrum dendrites from the Sleeper and National deposits apparently grew from the aggregation of colloidal particles in the 10–100 nm range; these dendrites have fractal structures. Their fractal dimensions are significantly higher than the theoretical value for diffusion-limited aggregation. However based on their basic similarity to DLA aggregates, it appears that colloidal particles had random trajectories as they collided with growing dendrites. Thus we conclude that growth of natural electrum dendrites is best described by a modified DLA process, where aggregation of relatively large and variably sized particles led to an increase in D_f values above that for pure DLA. Furthermore it is likely that surficial and postdepositional processes, such as fluid shear, may also have increased the fractal dimension of natural dendrites.

ACKNOWLEDGMENTS

We thank Robert Cook, Steve Schoenly, Ed Coogan, and William Utterback for technical assistance and Sam Romberger, Tamas Vicsek, and David Weitz for comments on earlier versions of this chapter. This research was supported by the NSF (EAR-91171713) and the AMAX foundation.

REFERENCES

Bergen, B. R., and Eimon, P., in *Cameron Volume on Unconventional Mineral Deposits* (W. C. Shanks III, ed.), SME-AJME (1983), pp. 191–205.
Bodnar, R. J., Reynolds, T. J., and Kuehn, C. A., *Rev. in Econ. Geol.* **2**, 73 (1986).
Boydell, H. C., *Inst. Min. Met. Trans.* Part 1, **34**, 145 (1924).
Brown, K. L., *Econ. Geol.* **81**, 979 (1986).
Buchanan, L. J., *Ariz. Geol. Digest* **14**, 237 (1981).
Chopard, B., Herrmann, H. J., and Vicsek, T., *Nature* **353**, 409 (1991).
Drummond, S. E., and Ohmoto, H., *Econ. Geol.* **80**, 126 (1985).
Enustun, B. V., and Turkevich, J., *J. Amer. Chem. Soc.* **85**, 3317 (1963).
Fournier, R. O., *US Geol. S. Bull.* **1646**, 15 (1985).
———, *Rev. In Econ. Geol.* **2**, 45 (1986).
Frondel, C., *Econ. Geol.* **33**, 1 (1938).
Heald, P., Foley, N. K., and Hayba, D. O., *Econ. Geol.* **82**, 1 (1987).
Hedenquist, J. W., *Geochim. Cosmochim. Acta* **55**, 2753 (1991).
Lin, M. Y., Lindsay, H. M., Weitz, D. A., Ball, R. C., Klein, R., and Meakin, P., *Nature* **339**, 360 (1989).
Lin, M. Y., Kelin, R., Lindsay, H. M., Weitz, D. A., Ball, R. C., and Meakin, P., *J. Coll. I. Sc.* **137**, 263 (1990).
Lindgren, W., *US Geol. S. Bull.* **601**, (1915).
———, *Mineral Deposits*, 4th ed. (McGraw-Hill, New York, 1933).
Mandelbrot, B. B., *The Fractal Geometry of Nature* (W. H. Freeman, New York, 1983).
Meakin, P., *Phys. Rev. A* **27**, 1495 (1983).
Meakin, P., Kertesz, J., Vicsek, T., *J. Phys. A* **21**, 1271 (1988).
Nash, J. T., Utterback, W. C., and Saunders, J. A., in *Geology and Ore Deposits of the Great Basin* (G. L. Raines, R. E. Lisle, R. W. Schafer, and W. H. Wilkinson, eds.) (Geological Society of Nevada, 1991), pp. 1063–84.
Noble, D. C., McCormack, J. K., McKee, E. H., Silberman, M. L., and Wallace, A. B., *Econ. Geol.* **83**, 859 (1988).
Oles, V., *J. Coll. I. Sc.* **154**, 351 (1992).
Romberger, S. B., *US Geol. S. Bull.* **1857-A**, A9 (1988).

Saunders, J. A., *Geology* **18**, 757 (1990).
———, *Econ. Geol.*, **89**, 628 (1994).
Spycher, N. F., and Reed, M. H., *Econ. Geol.* **84**, 328 (1989).
Turkevich, J., *Am. Sci.* **47**, 97 (1959).
Vicsek, T., *Fractal Growth Phenomena*. World Scientific, Singapore (1992).
Vikre, P. G., *Econ. Geol.* **80**, 360 (1985).
Weitz, D. A., and Huang, J. S., in *Kinetics of Aggregation and Gelation* (F. Family and D. P. Landau, eds.) (Elsevier, New York, 1984), pp. 19–27.
Weitz, D. A., and Oliveria, M., *Phys. Rev. Lett.* **52**, 1433 (1984).
White, N. C., and Hedenquist, J. W., *J. Geochem. E.* **36**, 445 (1990).
Witten, T. A., and Sander, L. M., *Phys. Rev. Lett.* **47**, 1400 (1981).

Index

Abyssal hills, 107–108, 115–117, 119, 120, 123, 125–127
Anistropy, 94, 118, 127, 133, 136, 171, 247
  parameters, 118
Antipersistence, 145

Box-counting, 65, 68
  algorithms, 14, 240, 241
  method, 69, 78, 81, 82, 84–86, 136, 145, 146–149, 158–160, 169
Brownian motion, 21, 22, 79, 111–113, 115
Brownian fractional, 132
Brownian trails, 153

Cantor dust, 18, 19, 71, 153, 154, 167, 176
Cellular automata, 2
Chaotic behavior, 32, 34
Chemical variation within crystals (zoning), 237
Climatology, 55, 56, 58
Crustal fractures, 180
Crustal shear zones, 182, 202

Dielectic breakdown, 180
Diffusion-limited aggregation (DLA), 180, 242, 247, 248, 257–258, 260
Disequilibrium textures, 245
Divider method, 78, 82–83, 84, 86, 92
Dynamical systems, 228

Euclidean dimension, 110
Euclidean geometries, 96

Euclidean models, 89
Euclidean shapes, 90
Euclidean surfaces, 91, 98, 99, 101–103, 132, 133

Fault
  gouge, 5
  offsets, 214–217, 223
  patterns, 205
Fluid flow, 141, 143, 144, 166, 179
Fourier coefficients, 22
Fourier series transformations, 237
Fourier spectral approach, 25
Fourier transform, 22, 25, 26, 80
Fracture networks, 141, 143, 144, 150, 162–166, 171, 176, 179, 180
Frequency-magnitude, 12, 13
Frequency-size distribution, 1, 10, 14, 38

Gausian distribution, 21, 22, 26, 85
Gaussian noise, 22, 42, 63
Gaussian model, 119
Gaussian topography, 125
Gold colloid aggregation, 251, 252
Grinding limit, 200
Gutenberg–Richter
  frequency magnitude relation, 138
  power law, 227–229, 231, 232

Hausdorff–Besicovitch dimension, 66
Hausdorff dimension, 66, 67
Hurst's law, 42, 47–50

Isostasy, 134, 135

Koch curves, 147, 238
Koch island, 1, 17
Korcak empirical relation, 3

La Place's equation, 2
Laplacian growth, 180
Log–log diagrams, 218
Log–log plots, 28, 81, 82, 103, 132, 149, 150, 232
Log–log space, 146, 151
Log normal, 3, 7, 11
Lopez Fault gouge, 188, 192, 193
Lorenz equations, 36, 37
Lorraine coal basin, 208–210, 217, 218, 220, 221, 223
Lorraine coal mines, 207
Lyapunov exponent, 233

Macrometerology, 56–58
Mechanical deformation, 141, 143
Menger sponge, 6, 7, 19, 170
Monte Carlo approach, 69

Network parameters, 205

Omori power law, 227

Pavements, 155
Percolation models, 166
Percolation threshold, 166
Persistence, 145
Pillow basalt, 241–243, 245
Pixels, 239–241, 245, 248
Poisson distribution, 194
Poisson manner, 157, 158
Poisson process, 167, 169
Power law, 3, 5, 7, 10, 11, 17, 79, 113, 116, 120, 121, 136, 138, 143, 157, 183, 180, 216, 223, 228, 229, 232, 233
Power law distribution, 230
Power spectral density, 78, 84, 93, 115
Power spectrum, 78–80, 113, 114, 117, 124
Pox diagram, 47–50, 55

Random behavior, 37
Random fractals, 153
Random distribution, 189, 231, 233
Random noise, 138, 239
Random reversals, 37
Random simulation, 20; see also Poisson distribution
Random topography, 108, 109
Random walk, 111, 245
  model, 258
Rosin–Rammler distribution, 193, 194
R/S analysis, 41–53, 55

Scale invariance, 1, 2, 10, 11
Scaling properties, 78, 79
Self-affine fractals, 77, 79, 80, 82–86, 94, 111, 112
  definition of, 28
  properties, 241
Self-affine scaling, 90, 91, 96, 103
Self-affine surfaces, 90, 91, 93, 95, 102
Self-organized critical phenomena, definition of, 36
Self-organized criticality, 228, 229, 233, 235
Self-similar distributions, 183
Self-similar fractals, 79, 94, 170
Self-similarity topography, 109
Self-similar operation hydrology, 63
Self-similar scaling, 90, 96, 103
Self-similar structures, 227
Self-similar surfaces, 90, 93, 95, 99, 102
Semivariograms, 150
Sierpinski carpet, 19, 170
Sierpinski gasket, 181, 195
Size-frequency distribution, 3, 4
Sleeper and National deposits, 252, 253, 255, 260
Spatial series, 77
Spectral method, 78, 80, 83, 92, 132
Stochastic fractals, 153
Subduction zones, 100

Time series, 77, 78
  analysis, 92
Transitions
  abrupt, 131–134, 136, 139
  gradual fractal, 131

Vacuum-impregnated sample assemblies, 198
Variogram, 131
von Koch curve, 109–111, 122, 123; *see also* Koch curve

Wallowa Mountains, 28, 30
Water Resources Research, 41
Willamette lowland, 28, 30

DUE DATE